Introductory
Biomaterials

Introductory
Biomaterials
An Overview of Key Concepts

Lia Stanciu

Susana Diaz-Amaya

ACADEMIC PRESS

An imprint of Elsevier

Academic Press is an imprint of Elsevier
125 London Wall, London EC2Y 5AS, United Kingdom
525 B Street, Suite 1800, San Diego, CA 92101-4495, United States
50 Hampshire Street, 5th Floor, Cambridge, MA 02139, United States
The Boulevard, Langford Lane, Kidlington, Oxford OX5 1GB, United Kingdom

Notices
Knowledge and best practice in this field are constantly changing. As new research and experience broaden our understanding, changes in research methods, professional practices, or medical treatment may become necessary.

Practitioners and researchers must always rely on their own experience and knowledge in evaluating and using any information, methods, compounds, or experiments described herein. In using such information or methods they should be mindful of their own safety and the safety of others, including parties for whom they have a professional responsibility.

To the fullest extent of the law, neither nor the Publisher, nor the authors, contributors, or editors, assume any liability for any injury and/or damage to persons or property as a matter of products liability, negligence or otherwise, or from any use or operation of any methods, products, instructions, or ideas contained in the material herein.

Library of Congress Cataloging-in-Publication Data
A catalog record for this book is available from the Library of Congress

British Library Cataloguing-in-Publication Data
A catalogue record for this book is available from the British Library

ISBN: 978-0-12-809263-7

For information on all Academic Press publications visit our website at
https://www.elsevier.com/books-and-journals

Publisher: Katey Birtcher
Acquisitions Editor: Steve Merken
Editorial Project Manager: Chris Hockaday/Naomi Robertson
Project Manager: Rukmani Krishnan/Manikandan Chandrasekaran
Cover Designer: Matthew Limbert

Printed in the United States of America

Last digit is the print number: 9 8 7 6 5 4 3 2 1

Working together
to grow libraries in
developing countries

www.elsevier.com • www.bookaid.org

Contents

Preface

In 2006, when I decided to develop a Biomaterials course for Materials Science & Engineering students at Purdue University, I started to search for a textbook that would tell a comprehensive story about how materials play their well-known major role in the development of virtually all biomedical devices. It has been difficult to find a textbook that covered the structure and properties of materials used for implantation, while also teaching about other ways they are being used, such as in tissue engineering, drug delivery, or point-of-care diagnostics. At the same time, another challenge arose from difficulties of explaining biology concepts to engineering students, who were the ones taking the course. These students did not have significant background in biology and also seemed to be unsure of whether they could understand such concepts within the time frame of a semester course, which made some of them avoid taking a Biomaterials class.

Biomaterials textbooks I consulted had strengths in different areas. Some had a strong focus on host and tissue response and interactions with biomaterials, but were lacking information on artificial materials structure and properties, or on biomedical device design requirements. There are other textbooks where concepts of tissue engineering are emphasized but where the depth of the concepts narrowed the scope too much for the book to be used as an introductory biomaterials textbook. Yet a third type of textbook did cover artificial biomaterials structure and properties, but deemphasized the tissue interactions, tissue and protein structure and properties and the role of the biological environment on performance. My search for a textbook that could serve as a comprehensive, albeit survey-ish, story of what biomaterials are, what their properties and various applications are, and the effect of the environment on their performance led to the design of my course by using a large array of information from different sources, rather than of one textbook. It was also at that time that the decision to write a textbook with the content I wish I had for my course took shape. Years later, this textbook's co-author, Dr. Diaz-Amaya, brought a fresh, updated perspective to the content, with her strong expertise in both Materials Science and Biology, and became critical for the successful completion of this project.

This introductory textbook provides an overview of the main concepts that the authors believe will form a solid knowledge basis for science and engineering students from which they could explore their interest in a specific Biomaterials area in more depth in focused courses. Although initially designed for and populated by Materials Engineering students, the course that this textbook is based on became very popular with students from all engineering disciplines. It is intended to offer fundamental understanding of what makes a biomaterial successful for both implantation and other biomedical applications, such as drug delivery, tissue engineering, or bioanalytical devices.

As the title suggests, this is very much an introductory textbook, designed for science and engineering students from across disciplines, who do not come with any prior specific knowledge on Biomaterials but have a general understanding of materials processing and properties and general chemistry knowledge. The reader is invited to be part of a story, starting with a short history of Biomaterials and moving through fundamental concepts of Materials Science & Engineering, starting from structure and bonding and going through all of the main classes of materials: Metals, Ceramics, Polymers, and Composite Materials. While students are used to learning about how these materials are being applied in fields such as automotive, aerospace, or energy, the text will bring a new point of view, toward understanding of how their properties could be controlled in such a way that would make them applicable in the larger healthcare field. The journey then guides the reader toward gaining basic knowledge on how these biomaterials are affected by, and in turn will affect, the tissues they are often interfacing with or repairing. Here, the reader will learn about the molecular to macro level structure of tissues, from protein structure and its significance for tissues properties and biomaterials interactions to the body's immune response and defense mechanisms. Finally, the fundamentals learned on both artificial biomaterials and tissues or otherwise natural materials come together in an overview of the most common applications in soft and hard tissue replacement and repair, temporary implants, and biomaterials for bioanalytical devices.

Each chapter includes examples of solved relevant questions, along with a list of problems that the reader can use to practice their understanding of the concepts provided in that particular chapter. Lecture slides are also included. It is our hope that this will become a textbook that truly serves as a comprehensive starter's tool, which will allow students to have a solid understanding of the main concepts, issues, and complexities involved in Biomaterials Science.

Teaching Ancillaries for Instructors

To access the solutions manual, PowerPoint lecture slides, and image bank containing all figures from the book, please visit the book's web page on Elsevier Educate, a dedicated website for instructors examining or adopting Elsevier textbooks: https://inspectioncopy.elsevier.com/book/details/9780128092637

Acknowledgments

I am thankful to Purdue University for granting a one-semester sabbatical leave, which helped me make significant progress in writing this textbook. I am also grateful for my husband's support while all evening family time was devoted to writing.

Lia Stanciu

I had no idea about the scope of work and amount of dedication that writing this book would entail. Today, I can only be grateful for all the amazing people surrounding my life, who patiently supported this effort and created the possibilities to take to fruition this ambitious project.

I would like to express my sincere gratitude to Lia Stanciu, for her trust, motivation, continuous support, and generosity through this journey. I could not have imagined having a better mentor and role model. To my loving family, for their endless support and inspiration. A big shout-out to my husband and my son—Pablo and Martin—who eased the way through long nights and full weekends of writing, all with warm hugs and hot coffee. To all the students who took the introductory biomaterials class: you played a critical role here, by helping us to refine the materials developed for this textbook and prioritize the topics included herein.

Last but not least, I want to give my gratitude to the editorial team and reviewers, who gave their best to support our vision. Thank you all for your effort and hard work.

Susana Diaz-Amaya

Introduction

Learning objectives

This introductory chapter of the textbook will enable students to:

- Become familiarized with the main terms encountered in the study of biomaterials.
- Acquire information on routes from past to the present of how biomaterials were developed and used and how they are currently integrated into biomedical devices.
- Gain basic understanding on the needs of the biomedical device market.
- Gain an introductory-level understanding on how biomaterials properties can influence the performance of the biomedical devices for which they are used.

Historical background on biomaterials

Any material incorporated into devices that either replace a part or a function of the body successfully or in any way aid healing of injury or disease is defined as a biomaterial. While the term "biomaterial" was coined only relatively late in the 20th century (more precisely in the 1960s), when researchers both in Europe and the USA founded specific scientific societies dedicated to this field, biomaterials have a much longer history.

It was from the beginning of humanity that people started to make use of various natural materials available in their environment to aid in survival. However, it is difficult to pinpoint the exact moment in history when humans transitioned to using materials to replace or restore the function of various organs. Archeological evidence is available more for dental applications and for orthopedics, and less for other possible biomaterials applications (e.g., ophthalmological, vascular etc.).

Archeological literature speaks of a spear point found embedded into the hip of a man from the Washington area of the United States (the "Kennewick Man"), whose remains were estimated to be approximately 9000 years old (Fig. 1.1). This served as evidence that humans, albeit not always on purpose, discovered the ability of the body to tolerate foreign objects, from the earliest time in history.

While prehistorical people likely did not think of using foreign materials to restore tissue or organ function, the Kennewick Man's embedded spear point is often cited in literature as the unintentional implantation of a foreign material in the human body, ultimately with good tolerance.[2,3]

Introductory Biomaterials. https://doi.org/10.1016/B978-0-12-809263-7.00001-9

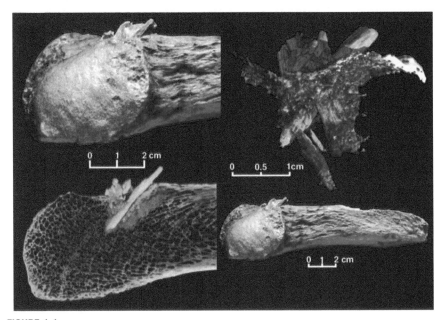

FIGURE 1.1

A stone spearpoint found embedded in the hip of the Kennewick man's skeleton.[1]

Dental applications seemed to be the first and easiest examples of making intentional use of materials to improve function. These include wires to hold teeth together, using materials for dental implants, or filling materials. More advanced uses, such as in the cardiovascular area or orthopedic surgery, were not available until much later, when advances in medicine made such surgeries possible. Osseointegration, at least for dental applications, however, seems to have been achieved by the Mayans with nacre teeth made out of sea shells, which have a hierarchical structure with excellent mechanical properties[4] (Fig. 1.2). Osseointegration is a term used to describe the formation of a direct bond between living bone tissue and the surface of an implant used in orthopedics or dentistry. For example, osseointegration is achieved when a device used to repair bone fracture, or for dental implants, such as a screw or a nail, is used as part of restoring anatomical function. The same result was reported to have been reached with a wrought iron dental implant in France about 200 CE.

According to historical literature, it is possible that the first biomaterial used in dental applications was gold (Au). A so-called "dental prosthesis" including a combination of gold wire and animal teeth to replace human teeth was found dated around 2600 years ago and ascribed to the Etruscan civilization. The chemical inertness of gold makes it an ideal material to be used in dentistry, and this was confirmed by a long history of it being used for this purpose. Besides its chemical inertness, gold is also malleable and ductile and thus can be easily processed in such applications.

Surgical sutures are another type of biomedical "device" used even possibly as early as the Neolithic era to close wounds. Catgut and silk were materials of choice dating from 1600 BCE and have been used into the 20th century. Sutures were used not only to close flesh wounds but also to fix tendons and shoulders.

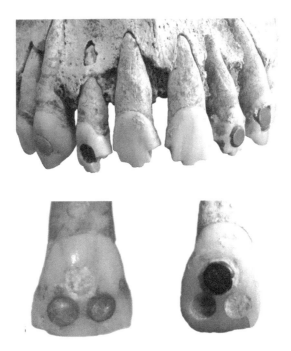

FIGURE 1.2

Depiction of nacre teeth made out of sea shells. The nacre was used for decorative rather than repair purposes.

Reproduced from H. Yi, F. Ur Rehman, C. Zhao, B. Liu, N. He, Bone Research 2016, 4.

https://doi.org/10.1038/boneres.2016.50.

In modern times, prior to any biomaterial being implanted in or in contact with the human body as part of a biomedical device, it needs to undergo a process called sterilization. This term encompasses a set of treatments that result in the biomaterial becoming free of any infection-promoting agents, such as bacteria. However, all the early historical attempts to use biomaterials to correct a health issue in any part of the body were done with no such sterilization taking place. It was only in 1876 that Dr. Sir Joseph Lister introduced the idea of sterilization to prevent infection when he pioneered antiseptic surgery. Because of lack of sterilization, many attempts to implant artificial materials before this historical point failed due to infection. The first sterile sutures, still made of silk or catgut, were introduced in 1887. Polymeric-based sutures made of polypropylene were only introduced in 1969, followed by Vicryl, a suture that can be absorbed by the body, in 1974. Polydioxanone sutures came into use in 1982, and antibacterial sutures followed these in the 2000s, as well as, more recently, more sophisticated suturing devices.

Other early use of biomaterials includes eye prostheses made out of gold or silver, followed by half-cut shells in the 18th century, in France. These prostheses were not truly functional, generally heavy, and difficult to use. Glass eye prostheses followed in the 18th century in Germany, having the disadvantage of degradation and producing irritations to the patients due to the lead content of glass.

Wood, stainless steel, and leather were materials of choice to fabricate artificial limbs starting from the antiquity, with the first ones originating in Egypt. An example of archeological evidence is given by a mummified body of an Egyptian woman whose big toe was amputated and replaced by a synthetic toe made out of wood and leather. Evidence showed that indeed, this prosthesis allowed her to walk. Stainless steel and other metals were used to replace amputated limbs way into the 20th century, with serious problems encountered due to infection, inflammation, or allergic reactions, in the absence of serious attention being given to the host response before the more recent times.

Biomaterials properties and device performance

The study of biomaterials is the avenue toward the engineering design of safe and effective medical devices. While the processing-structure-properties triad is very familiar to the materials scientist, the specific properties relevant for the study of biomaterials include additional issues that are described by terms such as host-tissue interactions, surface properties, corrosion and polymer leaching, and biomechanics. All of these are terms that are prevalent in the study of biomaterials and will be defined and discussed in more depth in future chapters of this textbook. Understanding the underlying principles guiding the properties of biomaterials is key toward their successful implementation into biomedical devices. Thus this textbook includes material that aims to help the reader understand the relationship between the properties, function, and structure of biomaterials. Three major areas of study on the subject of biomaterials can be addressed: biological materials (e.g., tissues and their components), biomaterials that can be implanted in the body or come in close contact with the body (biomaterials for surgery, drug delivery, or tissue engineering), and the interaction between the two categories in vivo.

Table 1.1 lists the main classes of biomaterials that can be used in implantable devices. Generally, for implants that aim to replace a tissue or a function in a part of the body, all three main classes of materials (metals, ceramics, polymers), as well as their composites, are applicable, with tissue compatibility and mechanical resistance being at the forefront of the selection and design process. All of these materials properties and their use as biomaterials are discussed in detail in future chapters (Chapters 3 to 5).

Generally, biomaterials used in surgery or implantation can be either permanent or temporary—also sometimes called transient. Permanent implants are designed with the goal of, as the name suggests, remaining permanently at their place of implantation and serve the function they were designed to restore. However, in some cases, such as, for example, when a child who did not finish the growth process needs a cardiovascular stent to be implanted to correct a congenital defect, it is desirable that the biomedical device serve a function only temporarily and should be designed to slowly dissolve in the body without leaving behind any foreign material or eliciting any negative tissue effects. There are also situations when an implant that was meant to be temporary and intended to be removed after healing turns into a permanent one if for any reason the removal did not happen, such as, for example,

Table 1.1 Materials used in in vivo biomedical devices.

Materials	Advantages	Disadvantages	Examples
Metals (Ti and its alloys, Co-Cr alloys, Au, Ag, stainless steels, etc.)	High mechanical strength, tough, ductile	Corrosion, high density	Joint replacements, dental root implants, pacer and suture wires, bone plates and screws
Ceramics (alumina, zirconia, calcium phosphates, carbon)	Chemically inert, tissue compatible	Brittle, weak in tension	Dental and orthopedic implants
Polymers (nylon, silicone rubber, polyester, polytetrafluoroethylene, etc.)	Resilient, easy to fabricate, some highly tissue compatible	Low strength, prone to deformation, degradation concerns	Sutures, blood vessels, other soft tissues, hip socket, ear, nose, etc.
Composites (carbon-carbon, wire- or fiber-reinforced bone cement)	High strength, tailor made	Difficulties in fabrication	Bone cement, dental resin

a bone rod repairing a fracture remaining in place after complete healing of the fracture occurred.

Permanent implant examples include those used in the musculoskeletal system, such as joints or permanently attached artificial limbs. In the cardiovascular system, prosthetic heart valves, pacemakers, and artificial arteries and veins are all also designed using biomaterials. Other body systems where biomaterials are used as part of biomedical implants are the respiratory system, digestive system, genitourinary system, nervous systems, and the eye. Soft tissues are repaired with sutures, and implants are used for cosmetic reasons (maxillofacial, breast, eye, etc.). All of these devices will be discussed in more detail in later chapters.

Temporary implants, on the other hand, are implants that first support tissue healing and later are either removed or dissolve in the body, leaving behind healthy, prosthetic-free tissues. These include extracorporeal devices that can replace organ function (e.g., kidney, liver, catheterization), temporary artificial skin, temporary bone fixation devices (e.g., screws, pins, staples, sutures), or soft tissue sutures. The implants that are physically removed after tissue healing are generally manufactured from either metals and metallic alloys, ceramics, or polymers, which are not designed to dissolve in the body. The temporary implants that dissolve in the body are made out of what we call bioresorbable materials, which are further defined in the Second-Generation Biomaterials section and presented in detail in Chapter 12.

The success of a biomedical device that can also be implanted in the human body is predicted by three major factors: (i) the properties (chemical, biological, surface, microstructural, mechanical) of the implant, (ii) the health condition of the recipient, and (iii) the competency of the surgeon.

Biocompatibility is another term that has to be understood in the context of the usage of a biomaterial in a biomedical device and not as an intrinsic material property. More specifically, biocompatibility is not only related to the lack of systemic or local toxicity of a given biomaterial, but it ultimately defines the successful integration and application of said biomaterial in the clinical environment. Biocompatibility is therefore a system-related and not a material-related term and involves understanding the biomaterials interaction with biological materials (cells, muscles, ligaments, fat, bones, organs, etc.) as well as the relationship among the biomaterial properties (chemical, mechanical, pharmacological, surface) and their behavior in a specific physiological environment. These behaviors can include cell lysis, systemic reaction, corrosion, degradation, toxic leaching, protein deposition, thrombus formation, encapsulation, calcification, tissue adhesion, etc. All of these terms, as well as their relevance for the study of biomaterials and their applications, will be defined and discussed throughout this textbook. In short, a biomaterial cannot be called biocompatible unless this term is placed in the context of its performance in a biomedical device when this device ultimately functions in vivo in the manner in which it was intended.

All the topics that are being investigated in biomaterials science for the purpose of including a material into a surgically implantable device that helps restore or improve a physiological function ultimately lead to the goal of obtaining a maximum performance during its use as intended. Thus the reliability of an implantable biomedical device is critical. Implants can fail in several ways. Some of the most common implants are joint replacements. For these implants, there are several factors that can lead to their failure: (i) fracture; (ii) infection; (iii) wear; and (iv) implant loosening. All of these factors need to be given the attention necessary at every step of the way to ensure that their effect on lowering performance is minimized. The success of a biomaterial when integrated into an implant inside the body will depend on each of these factors, as well as variable and less controllable factors such as, for example, human error. The reliability of a biomedical implant composed of a certain biomaterial can be expressed as:

$$r = 1 - f \tag{1.1}$$

where r is the reliability and f is the probability of implant failure.

In the case of implant failure, either one or more factors play a role. If even one factor involved in the overall implant functioning, f, has a value of 1, translating into surefire failure, then the total reliability of the implant, calculated as the product of the individual reliabilities, $r_i = 1 - f_i$, will be zero. It is thus critical that each potential factor for failure is[5]

$$r_t = r_1.r_2.r_3 \ldots r_n \tag{1.2}$$

In summary, each of the main materials classes—metals, ceramics, polymers—can be used as part of an implantable biomedical device. In this book, we will investigate each of these classes of materials in the context of their use in surgery. Because mechanical failure has a large effect on reliability, we will also be presenting the basic mechanical properties of each class of materials in the context of their use in medicine. Fatigue properties, for example, are significant as far as implants withstanding cyclic loading are concerned. Examples include but are not limited to vascular stents, pacemakers, or joint replacements. Wear is another concern that this book will delve into. Friction is present in many implanted devices, the most prominent example being that of hip replacement implants.

Biomaterials classification

There are countless definitions of what constitutes a biomaterial that evolved throughout history of the field. The term *biomaterials* can be thus very broadly interpreted, as it can be seen not only in the context of surgical use of materials, but also encompass the vast world of biological materials. Biological materials can include anything from peptides and enzymes, to proteins, nucleic acids, cellular components, and extracellular matrix, to tissues and whole organs. What is, after all, our goal when studying, developing, and eventually presenting to the world new emerging biomaterials?

For providing an overall picture of what this book is intending to achieve, we will provide a limited definition of the term *biomaterials*. Thus for the purpose of this textbook only, "biomaterials are any materials that are used to fabricate devices that replace and restore the function of natural living tissues and organs in the human body."

First-generation biomaterials

Once the sterilization was invented, a new era began, with the development of the so-called first-generation biomaterials. These biomaterials were generally materials that were borrowed from other fields based exclusively on their mechanical properties and ability to serve a purpose and replace an organ or function. However, in the field of biomaterials today, the term *biocompatibility* is one that is ubiquitous, and this concept was not addressed at all with the first generation of biomaterials. One can find a large variety in the definition of biocompatibility in published materials. One easy way to understand this term's meaning is that a biomaterial can be called biocompatible only when it has the ability to perform in vivo (in the human body), when included in a biomedical device, in the manner that was intended and without inducing any negative effects on the tissue, either locally or systemically. Further, this means that all of the performance parameters need to be met for a biomaterial to be considered truly biocompatible, not only the lack of detrimental effects on tissues locally or systemically. For example, if a biomaterial does not produce any negative tissue reactions but fails to work under the normal expected use of a device of which it is a part, the compatibility criterion is not met.

For first-generation biomaterials, however, the only criteria for material selection that touched the idea of biocompatibility at the time was that these materials should be nontoxic and as inert as possible to avoid what we call tissue host response—any harsh and negative effects of the biomaterial on the surrounding tissue, such as inflammation. Nevertheless, many first-generation biomaterials are still in use today in biomedical devices, especially in orthopedics, for example in joint replacements.

Hip implants—artificial implants that are replacing the function of the patient's hip—are necessary when a patient has either suffered an injury or the cartilage in the hip area has deteriorated severely because of age or disease. It was Philip Wiles who performed the first total hip replacement surgery in 1938, which replaced both the socket and the femoral component with stainless steel components. However, Charnley originated the first total hip replacement system adopted by the community. This included a stainless steel stem and a PTFE polymeric cup. One version of these systems included an alumina ceramic head attached to a stainless steel stem. Stainless steel, titanium alloys, and cobalt-chromium alloys are still in use today for total hip replacement; however, these implants often also include parts made of ceramics or polymers. Sometimes, failure with such devices has been registered due to various reasons, such as mechanical failure of one of the components under regular functional stresses. An example of such an implant failure is shown in Fig. 1.3. In this figure, a hip implant is shown, which is composed of a metallic stem and a ceramic head. Although after surgery (left image) the implant is in excellent condition, a radiograph taken after 6.25 years of being in use shows that the ceramic head

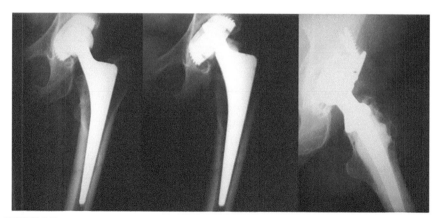

FIGURE 1.3

Radiographs of the hip with a fractured ceramic head. *Left:* directly postop. *Middle:* Fracture of ceramic head at 6.25 years postop. *Right:* After revision arthroplasty.

Reproduced with permission from Beckmann et al.[6]

had fractured. We will learn in future chapters that one particular problem with ceramics is their brittleness, which makes them prone to cracking. In general, the study of biomaterial takes into account all of the materials properties criteria, for each class of materials, which need to be considered when thinking in the context of long-term implant function.

Second-generation biomaterials

With the advances of thought in the biomaterials field, the second generation of biomaterials included not just materials that are nontoxic but materials with bioactivity or bioresorbability characteristics. The term *bioactivity* means that these materials can interact with the tissues in such a way that a positive biological reaction is induced. This typically refers to the bonding between tissues and the biomaterial interface. An example would be osseointegration, in which a biomaterial is made bioactive by inducing porosity or including a porous coating (e.g., coating with hydroxyapatite—a form of calcium phosphate present as one of the bone components), thus encouraging bonding with the bone tissue. Another example of such a positive effect being induced through rational design is the fabrication of breast implants that are coated with a thin, porous layer that encourages integration within the surrounding tissue. The term *bioresorbable* refers to the ability of the biomaterial to slowly degrade within the physiological environment after implantation, while new healthy tissue forms.

Polymeric materials and metallic materials that are used to fabricate cardiovascular devices can be also made bioactive by using coatings that help their integration and adequate functioning. Bioactive glasses, which are materials made of mixtures of amorphous oxides and that have the ability to form a direct bond with the bone tissue, are yet another good example of what a second-generation biomaterial is. In addition, there are other bioactive ceramic materials, as well as glass-ceramics and composites that fit this definition.

Besides bioactivity, which implies the formation of strong bonds between the biomaterial and the adjacent tissue, second-generation materials can be bioresorbable. The term *bioresorbability* is sometimes replaced with *biodegradability* or *bioabsorbability* and refers to the ability of a material to provide structural support in the body for a certain necessary amount of time, followed by dissolving fully in vivo (in the body) without leaving behind any by-products that would produce either local or systemic adverse effects (e.g., toxicity, inflammation, carcinogenic potential, etc.). As mentioned in the Historical Background on Biomaterials section, bioresorbable biomaterials can be of interest when a biomedical device is needed temporarily, but it is desirable for it to disappear from the body after the tissue healing is completed. For example, bone defects could be repaired by using a bioresorbable artificial bone material, based on calcium phosphate materials. Bone, as we will learn in Chapter 6, has the ability to heal completely after injury. Unless there is a bone disorder present, a bone defect can thus be repaired by using artificial bone materials that can serve as support while the existing bone tissue surrounding the injury site is healing and producing new bone, which is gradually replacing the artificial bone mineral. Bioactivity and bioresorbability will be described in more detail in Chapters 4 and 12. Some examples of what we call second-generation biomaterials, including bioactive materials, include but are not limited to: (i) heparinized cardiovascular devices (e.g., stents) for improving hemocompatibility (compatibility with blood) and decreasing the chance of blood clots (thrombus) being formed; (ii) bioactive glasses, which can form direct bonds with the bone tissue; (iii) biodegradable polymers such as poly-lactic acid (PLA) and polyglycolic acid (PGA); (iv) bioresorbable Mg alloy stents that leave behind prosthetic-free tissue after resorption; and (v) synthetic hydroxyapatite (HA), used either as porous resorbable coatings on implants or in resorbable artificial bone formulations.

Each of these second-generation biomaterials and their applications are discussed in detail in future chapters.

Third-generation biomaterials

Third-generation biomaterials go beyond both nontoxicity and ability to replace an organ or tissue (first-generation biomaterials) and bioactivity (second-generation biomaterials). In addition to biocompatibility, bioactivity, or bioresorbability, third-generation biomaterials are engineered to induce healing. This advance came in the wake of advances in the study of immunology, which allowed the discovery of designs that stimulate cellular activity, thus aiding tissue healing. Most third-generation biomaterials are polymeric.

Polymers, having the ability to be relatively easily modified through chemistry, can change both structure and function in ways that program cell proliferation, which in turn leads to some control over the production of extracellular matrix. Fig. 1.4 shows an example of the biomimetic design and integration with bone tissue of biomimetic materials made of chitosan, cellulose, and HA. Other examples include bioactive glasses and porous foams that can act as scaffolds with the ability to activate genes, which in turn can stimulate tissue regeneration. Such materials make the object of fields such as tissue engineering and in situ tissue regeneration.

Who studies biomaterials and how do they use the knowledge?

Although the study of biomaterials is far from new, the topic received most attention in the past 30 to 40 years, as various advances in medicine and medical technologies opened countless exciting opportunities in the biomedical industry, where biomaterials find their uses. Historically, universities did not have biomedical engineering departments, but new departments are now being created everywhere. Moreover, other academic units such as materials science and engineering, chemical

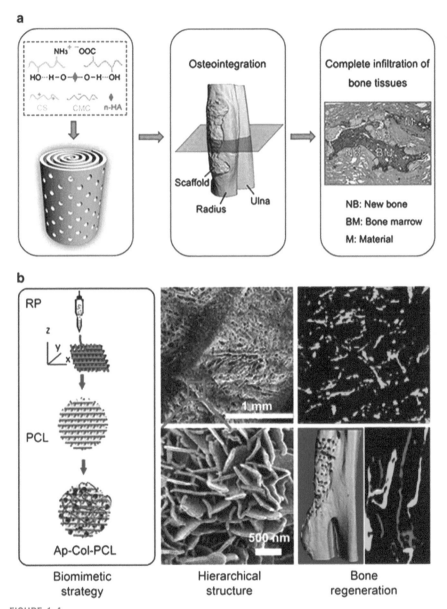

FIGURE 1.4

(A) Biomimetic spiral-cylindrical scaffold based on hybrid chitosan/cellulose/nano-hydroxyapatite membrane. (B) Biomimetically ornamented rapid prototyping fabrication of an apatite—collagen—polycaprolactone composite construct with nano—micro—macro hierarchical structure.

Reproduced with permission from Yi et al.[7]

engineering, or even computer engineering include in their undergraduate curriculum and graduate studies topics that are relevant to the biomaterials field.

Biomaterials are used mainly in the biomedical industry. In addition to advances in health care technologies, an increase in the life span of the population leads to more opportunities for enhancing quality of life of the older generation through careful medical attention.

Biomedical engineers work on using biomaterials to bring to life new devices or medical equipment that can replace tissues and organs, improve their function, and repair damage created by injury or disease. On another level, they work on the creation of new detection tests that can aid in the earlier diagnosis of various diseases, from viral or bacterial infections to cancer. Early detection and intervention is critical for patient prognosis, and thus advances in the diagnostics area are critical for improving or even saving peoples' lives.

Not only biomedical engineers are involved in the study of biomaterials. Since the principles of biology and biochemistry are highly relevant when the goal is to alleviate or solve a medical issue, biomedical engineers have to work in close interaction with chemists, biologists, medical personnel, and, last but not least, the general public. Short- and long-term studies on the success of a certain biomedical device depend on the patient-reported data and cooperation. Working in the field of biomaterials is participating in a vast experiment where data is collected at various time points but not centralized in one single location.

Besides designing medical devices and equipment, working in the field of biomaterials involves material testing, device installation and repair, as well as observing the effectiveness and safety of medical devices. These scientists and engineers need to have a background in both engineering and medicine, as well as knowledge of mathematics, computer science, physics, chemistry, and biology. While biomaterials may refer to the materials that are used in biomedical devices, their successful application depends on sub-areas such as biomechanics, bioinstrumentation, systems physiology, or rehabilitation engineering.

The biomedical industry is broader than just the medical device community. It includes the biotechnology and pharmaceutical, as well as medical device and diagnostics segments. Thus the needs for its success are just as varied. It is clear that the highly trained human capital is one of the critical components for advancement, in addition to the hurdles brought by a tremendously lengthy and difficult regulatory process approval, funding models, and risks related to science and ethics.

Among the four main segments of the biomedical sector, the pharmaceutical field is very well developed and has well-established companies that have been successful for a long time. On the other hand, the relatively newer biotech sector is making strides in discoveries in medicine. Although the medical device component, which is what most of this textbook is focused on in the upcoming chapters, is much older than biotech, fewer companies are present on the market, mainly due to the highly invasive and life-altering character of such products, which imposes very strenuous regulatory procedures for bringing new products to the market. The genomics revolution gave a boost to the medical device sector, which requires relatively easier regulatory processes unless implantation in the human body is necessary for a particular product, such as in vivo glucose monitors.

The common denominator for the biomedical industry, in general, is that it is driven by science and strongly regulated. There are constant changes in all sectors of the biomedical industry, which lead to continuous advances. If the current trend continues, personalized medicine will become a reality in the next couple of decades. To achieve this and other successes and improve human life in the most meaningful way, there are several needs that must be addressed.

First, it is imperative that continuous funding is available to allow for uninterrupted work on promising projects. This includes funding for basic research through prototype development and ultimately commercialization. The funding models are generally diverse and include both public and private investors. This allows the products that have market value to become selected for further advancement.

Intellectual property protection is also a key need in the industry. Portfolio problems are created by patent expiration and commercialization of generic brands. This is more relevant for the pharmaceutical and biotech sectors.

There are three stages for the growth of a biomedical company: (i) the trial-and-error stage that is funded by federal agencies and angel investors; (ii) product focused, financially supported by venture capital; and (iii) diversification, where commercialization occurred with the initial products and more products are being developed.

Summary

There is a wealth of surgical applications of biomaterials, and each of the main classes of materials finds their use in the biomedical field as part of biomaterials design. In this textbook, we will investigate each of these classes of materials for their use in medicine and the biomedical industry.

Archeological evidence shows that humans made use of various materials in attempts to improve or restore anatomical function, but such early use was largely limited to the dental realm, and the lack of medical advances in surgery did not allow for progress for much of the history.

Only in the 20th century was the term *biomaterials* first put forward. First-generation biomaterials were used exclusively because of their mechanical properties, which needed to be sufficient to replace an organ or function. Beyond mechanical properties, the biocompatibility concept was not considered or discussed. Biocompatibility refers to a biomaterial's ability to meet a host of criteria, outside of purely mechanical properties, when included in a biomedical device implanted in vivo, including, but not limited to, proper functioning with no local or systemic tissue effects. Moreover, all of the performance parameters for proper usage need to be met for a biomaterial to be considered truly biocompatible.

Second-generation biomaterials were selected not only for their lack of negative tissue effects and adequate mechanical properties but also for appropriate bioactivity or bioresorbability characteristics. Bioactive materials can interact with tissues and elicit a positive reaction, for example direct tissue bonding or tissue integration.

Going one step further, third-generation biomaterials are healing-inducing and make use of immunology concepts in their design and selection.

The following chapters will give a broad overview of all of the concepts that are relevant for the reader to have a basic understanding of what are the main biomaterials properties that enable selection of a certain material with a goal of functional integration into a medical device and what are the avenues that need to be followed for a successful laboratory-to-market trajectory for such materials.

References

1. Taylor RE, Kirner DL, Southon JR, Chatters JC. Radiocarbon dates of Kennewick Man. *Science*. 1998;280(5367):1171.
2. Nicastro N. Riddle of the bones: politics, science, race, and the story of Kennewick Man. *Archaeology*. 2000;53(3):69−71.
3. Ratner BD. *Biomaterials Science: An Introduction to Materials in Medicine*. 3rd ed. Amsterdam; Boston: Elsevier/Academic Press; 2013.
4. Sampath V, Huang P, Wang F, et al. Crystalline organization of nacre and crossed lamellar architecture of seashells and their influences in mechanical properties. *Materialia*. 2019;8.
5. Park JB, Bronzino JD, eds. *Biomaterials: Principles and Applications*. Boca Raton: CRC Press; 2003.
6. Beckmann NA, Gotterbarm T, Innmann MM, et al. Long-term durability of alumina ceramic heads in THA. *BMC Musculoskel Dis*. 2015;16:249.
7. Yi H, Rehman FU, Zhao CQ, Liu B, He NY. Recent advances in nano scaffolds for bone repair. *Bone Res*. 2016;4.

Structure and bonding

Learning objectives

In the biomaterials field, all the structure-relationship properties that are important for other areas of Materials Science and Engineering apply. The relationships between atomic structure and chemical bonding, corrosion, and the ultimate interaction between synthetic materials that comprise the implants and the biological materials in tissues play a vital role in the ultimate success of a biomedical device. The processing and its relationship to microstructure, defects, and mechanical properties also play a critical role in the customization of an implant that will perform as intended in the human body.

After reading this chapter on structure and bonding in materials, students will:

- Become familiarized with the basic concepts of atomic structure and bonding in solids.
- Acquire information on crystallography concepts and be able to relate this to the selection of biomaterials for in vivo applications.
- Gain basic understanding on how the processing routes of materials affect the successful implementation of biomaterials in tissue interfacing biomedical implants.
- Gain an introductory level understanding of the main concepts of mechanical properties that are relevant to the performance of biomaterials in load-bearing applications.

Fundamentals of atomic bonding

Most synthetic biomaterials that we will be discussing in this book are solids. Thus an understanding of what keeps all these different types of solid materials together and ultimately imparts their properties is required for the design of biomaterials.[1-5]

Solids are held together by interactions between their composing atoms. Further, the valence electrons of atoms and their energy levels ultimately determine how the atoms interact with each other, forming interatomic bonds. The primary interatomic bonds can be ionic, covalent, or metallic. In addition, the larger units in some solids, molecules, can also interact with each other via what we call intermolecular bonds. Ionic bonds are strong primary bonds, formed via electrostatic attraction between

Introductory Biomaterials. https://doi.org/10.1016/B978-0-12-809263-7.00002-0

cations and anions in solids, such as NaCl, CaO, and $Ca_3(PO_4)_2$. Covalent bonds, another type of primary bond, hold together molecules such as H_2 or Cl_2, which are gaseous, but also polymers relevant for the biomaterials field, such as polyethylene $(C_2H_4)_n$ (where n is the degree of polymerization). Metallic bonds, the third type of primary bonds, include cations and delocalized electrons and are present in metallic materials. As we will see in following chapters, metals and their alloys play a significant role in biomedical devices used in orthopedics or cardiovascular surgery. Primary bonds are strong bonds, and in turn determine the corresponding properties of materials. Secondary bonds, less strong than primary bonds, include hydrogen bonds—very significant in biochemistry—as well as Van der Waals bonds, which are in fact dipole-dipole interactions and are less strong than hydrogen bonds. Table 2.1 shows the range of bond energies for each of the bonding types described here. Bonding energy is the minimum energy required to break apart a certain type of chemical or physical bond between atoms or molecules.

One of the classes of materials widely used as part of biomedical devices are metals and their alloys. They are composed of either one single element or different metallic elements combined together in alloys. The main bond in metals is the metallic bond, which is characterized by a significant number of delocalized electrons that form "electron clouds" (Fig. 2.1). The nonvalence electrons and atomic nuclei form the ionic cores, which are positively charged. The valence electrons in the electron clouds being delocalized are not strongly bound to individual atoms, which in turn imparts metallic properties such as very high electrical and thermal conductivity, as well as their mechanical properties (high strength, coupled with ductility). These mechanical properties make metals good choices to be used in biomedical implants that are load bearing and are thus used extensively in orthopedic applications.

The strength of the metallic bond varies widely, depending on the electronic configuration of the individual metal, and bonding energies range from around 70 to 850 kJ/mol (Table 2.1).

The ionic bond, another strong primary bond, is found mainly in chemical compounds that contain metals and nonmetals together. Metallic atoms have the tendency to donate their outer shell valence electrons to nonmetals and thus acquire a stable electronic configuration and become positively charged. In turn, the

Table 2.1 Bond energies.

Type of bond/attraction	Range of bonding energies (kJ/mol)
Ionic bonds	600–1500
Covalent bonds	200–1100
Metallic bonds	70–850
Dipole attractions	40–400
Hydrogen bonds	10–40

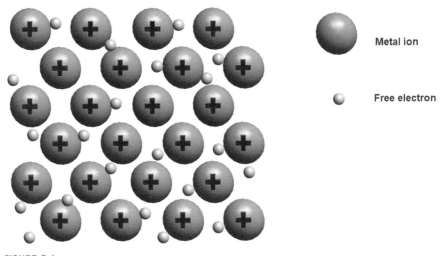

FIGURE 2.1

Schematic representation of metallic bonding.

nonmetallic element receiving the electrons from the metallic element becomes negatively charged. The coulombic interactions between the positive metallic cations and the negative nonmetallic anions are what keep ionic solids together.

The most widely shared example of an ionic solid is table salt, NaCl. Sodium has one valence electron on the 3s level, which it transfers to the chlorine atom. The Na^+ cation now has the stable electronic configuration of neon (Ne), while the chlorine atom becomes negatively charged and has the electronic configuration of argon (Ar). The electrostatic (coulombic) attraction between the cations and anions leads to the formation of a stable ionic solid.

From the biomaterials perspective, materials such as alumina, Al_2O_3, or calcium phosphates, $Ca_3(PO_4)_2$, are the most widely used ionic compounds for many applications. The most well-known calcium phosphate used in biomedical devices is hydroxyapatite, which has the chemical formula $Ca_5(PO_4)_3(OH)$ (Fig. 2.2).

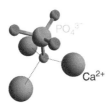

FIGURE 2.2

Schematic representation of ionic bonding in hydroxyapatite.

In alumina, for example, aluminum transfers three electrons to the oxygen atoms and becomes positively charged, while the oxygen atom becomes negatively charged (Fig. 2.3).

In general, an anion and a cation exert an electrostatic force of attraction toward each other that depends on their interionic distance:

$$F_a = -\frac{c}{r^n} \tag{2.1}$$

where r is the distance between ions, and c and n are constants. For electrostatic interactions, when the coulombic force is involved, the value of the n constant is 2.

In this formula, the constant c depends on the valence of the ions and the permittivity of vacuum, as follows:

$$c = \frac{(z_1 e)(z_2 e)}{4\varepsilon_0 \pi} \tag{2.2}$$

where z_1 and z_2 are ion valences, e is the charge of the electron in coulombs, ε_0 is the vacuum permittivity with a value of 8.85×10^{-12} F/m, and r is the distance between the ions.

Ceramic materials, such as those mentioned previously, are ionic solids. The bonding energy in ionic solids ranges between 600 and 1500 kJ/mol. These are large bonding energies, which impart high strength and high melting temperatures to ceramics. On the other hand, ionic materials are brittle and generally electric and thermal insulators. Their high strength makes some ceramics such as alumina or hydroxyapatite useful in some biomaterials-related applications that are exposed to significant mechanical forces (e.g., dental applications, some orthopedic applications). However, the brittleness of ceramics and their high stiffness do not make them suitable for use in implants that are subject to high torsion loads.

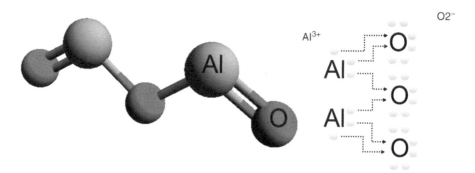

FIGURE 2.3

Schematic representation of ionic bonding in alumina.

Example 2.1

What is the attraction force between Li^+ and F^- if the interionic distance is 2 nm?

Answer The attraction force between two ions is: $F_a = -\frac{c}{r^n}$, where c has the formula given in Eq. (2.2).

The valences of both Li^+ and F^- are 1.

$$F_a = \frac{(z_1 e)(z_2 e)}{4\pi\varepsilon_0 r^2} = \frac{(1)(1)(1.6 \times 10^{-19} C)^2}{4\pi\left(8.85 \times \frac{10^{-12} F}{m}\right)(2 \times 10^{-9} m)^2} = 5.75 \times 10^{-7} N$$

Covalent bonds form through sharing of valence electrons between atoms in proximity to one another. Instead of electron transfer between metals and nonmetals, as in the ionic bond, the covalent bonds form, with some exceptions, between nonmetals, where each atom contributes electrons to the bond. While ionic and metallic bonds are nondirectional, covalent bonding is directional. In the formation of a covalent bond, the participating atoms share their electrons from the s and p orbitals with the goal of attaining a noble-gas electron configuration of eight or two electrons.

The strength of covalent bonds can also vary widely, leading to a wide variety of properties in covalently bound materials, and is in general in the range between 75 and 300 kcal/mol. The covalent bond is directional, which means that in a covalent compound, the bond strength is not the same in all directions. Covalent compounds have low electrical conductivity and can be gases, such as methane, or solids with very strong bonds, such as diamond. Covalent compounds with very high bonding energies have very high melting points and hardness. These are generally inorganic compounds such as GaAs, SiC, or diamond. Diamond is composed of covalently bound carbon atoms and has a melting temperature of around 3550°C.

Most relevant covalent materials from a biomaterials perspective, however, are polymers, which typically do not have the same strength and hardness as the covalent ceramics mentioned previously. Generally, the molecular structure of a polymer is a long chain of covalently bound C atoms, with other covalent bonds between the core C atoms and other nonmetallic atoms or functional groups. Polyethylene is an example of a polymer relevant to biomedical device applications (Fig. 2.4). Bonding in polymers is reviewed in more detail in Chapter 5.

Metallic, ionic, and covalent bonds are primary bonds. Secondary bonds are much weaker (bonding energy of approximately 10 kJ/mol) compared with the primary bonds and include Van der Waals bonds, physical bonds, and hydrogen bonds. These occur not only between atoms but also between molecules, and are present at all times along with the much stronger primary bonds.

Hydrogen bonds form between molecules or functional groups that contain hydrogen. In general terms, a hydrogen bond forms when a hydrogen atom is

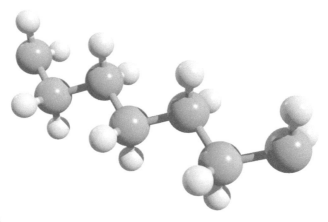

FIGURE 2.4

Structure of polyethylene. Larger darker spheres represent carbon atoms, while smaller lighter spheres represent hydrogen atoms.

covalently bonded to an atom with high electronegativity and when there is another electronegative atom with a lone pair of electrons in the vicinity. These bonds are weaker than the covalent bonds but stronger than, for example, dipole-dipole or dispersion forces. From the biomaterials perspective, the numerous proteins that compose the natural world, as well as a large number of polymers, have structural, mechanical, and chemical properties that are mainly imparted by H bonding. For example, the more H bonds a polymer is capable of forming, the higher its viscosity. A similar trend is followed by the boiling point, in which a compound that forms more H bonds has a higher boiling point than one that cannot form H bonds or has fewer H bonds.

Example 2.2

HF has a smaller mass than HCl. Which compound is likely to have a higher boiling point and why?

Answer Although the compound with a smaller molar mass is expected to have a lower boiling point, HF has a higher boiling point than HCl because it can form more H bonds.

Van der Waals bonds are created due to the existence of atomic or molecular dipoles. In some cases, there is a separation between the positive and negative fragments of a molecule. When this happens, bonds form via the electrostatic interactions between oppositely charged portions of atoms and molecules. These forces are especially relevant to protein-surface interactions in biomaterials.

Crystallography concepts

In the previous section, we described the primary interactions between atoms that make up all materials. Crystallography takes this exercise a step further and helps us understand how exactly these atoms and ions are organized within a certain material. If atoms or ions are arranged in a repetitive fashion in a three-dimensional structure, and over a long-range atomic distance, the material is said to be crystalline. Metals, some ceramics, and some polymers are crystalline. The crystal structure is the way in which atoms or ions are arranged spatially, and the type of crystal structure has a significant impact on the properties of materials. Amorphous materials, on the other hand, lack organization in a long-range atomic structure.

Crystalline materials contain unit cells, which are groups of atoms or ions that form the smallest unit of a pattern that repeats over a long atomic range. To represent crystal structures, atoms or ions are usually represented as spheres, to aid understanding of crystallography concepts. The symmetry of a structure is described by its unit cell, which is the basic building block of a crystalline material. Unit cells are represented mainly by cubes or parallelepipeds. There are three common types of crystal structures encountered in materials: face-centered cubic (FCC), body-centered cubic (BCC), and hexagonal close-packed (HCP). To simplify and streamline an introductory understanding of crystallography concepts, in this chapter we will only present those related to metals. However, as mentioned previously, ceramics and some polymers also have crystalline structures. Their crystallography concepts are, at the fundamental level, very similar to those described in this chapter for metals. However, there are some additional requirements that are involved in understanding the crystal structures of ceramics and polymers. One example of such additional requirement is maintaining the charge neutrality in ionic crystals, which contain both anions and cations, such as is the case of many ceramics, which is not a concern for metals. We will address the crystallography concepts that are specific for ceramics and polymers in Chapters 4 and 5, respectively.

Before we delve into the specifics of crystal structure types, we have to clarify the generally accepted ways that we can depict and visualize these structures. The hard sphere model, which we use here, assumes that atoms or molecules are very small and hard spheres, which have a mass and well-defined radii and which touch each other. In this representation, one diameter of a hard sphere corresponds to the shortest distance between two like atoms. We can, however, for simplicity, represent a crystal structure as a lattice of points corresponding only to the centers of these spheres (Fig. 2.5, left), instead of representing it as a lattice of the entire spheres touching (Fig. 2.5, right). That does not mean that the atoms do not touch, but just that we chose to only represent their centers.

The FCC crystal structure has as its unit cell a cube with all corners occupied by atoms and more atoms located at the center of all cubic faces.

One parameter that is often used to describe the organization of atoms in crystal structures is the coordination number (CN). This is used to represent the number of closest atomic neighbors of an atom in that crystal structure, or the number of

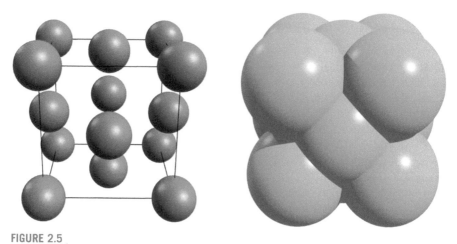

FIGURE 2.5

Schematic of an FCC unit cell.

touching atoms. Another parameter that we need to define is the atomic packing factor (APF). This is the fraction of the volume occupied by the hard spheres in the unit cell. It can be calculated as the sum of the volume of the atoms divided by the volume of the unit cell, and it indicates the maximum possible packing of hard spheres. The APF is relevant for the properties of biomaterials because it influences properties. For example, a biomaterial with a lower APF will be more ductile and less brittle than one with a higher APF, which will make a difference when selecting such materials for use in implants that are load bearing.

Many metals have an FCC structure (Table 2.2). For this structure, the CN is 12. This can be derived from observing that the atom at the center of each of the cube's faces has four nearest neighbors in the corner. Also, there are four other face atoms behind and four others on the faces of the unit cell in the front.

If we consider the atoms modeled like spheres touching one another along the diagonal of a face, the relationship between the cube edge length, a, and the radius of the spheres, R (atomic radius), is:

$$a = 2R\sqrt{2} \qquad (2.3)$$

Knowing that the APF can be calculated by the sum of the volume of the atoms divided by the volume of the unit cell, we now can use Eq. (2.3) to calculate the volume of the unit cell, and then find that the APF for an FCC crystal is 0.74.

Table 2.2 Atomic and mass percent compositions for TiAl6V4 alloy.

Element	Atomic weight	Atoms	Atomic percent	Mass percent
Titanium (Ti)	47.867	1	9.1	11.5754
Aluminum (Al)	26.9815386	6	54.5	39.1489
Vanadium (V)	50.9415	4	36.4	49.2757

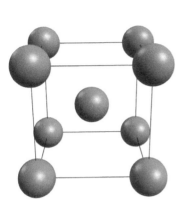 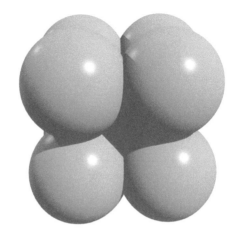

FIGURE 2.6

Schematic of a BCC unit cell.

The BCC crystal structure is also part of the cubic family, with atoms located at the corners of the cube and one additional atom in the center of the cube (Fig. 2.6).

In this case, the relationship between the cube edge length and the atomic radius is given by:

$$a = \frac{4R}{\sqrt{3}} \tag{2.4}$$

Chromium, iron, and tantalum are examples of biomaterials-relevant metals with the BCC crystal structure (Table 2.2). In BCC structures, the CN is 8, while the APF is 0.68; hence, the materials with this structure are less closely packed than those with an FCC structure.

Example 2.3

Nickel has an FCC lattice geometry. The density of Ni is 8.908 g/cm^3. What is the cube edge length?

Answer The average mass of one atom of Ni is:

58.6934 g/mol $\div$ 6.022 $\times$ 10^{23} atoms/mol $= 9.746496 \times 10^{-23}$ g/atom

There are four Ni atoms in the FCC unit cell, with the mass:

9.746496 $\times$ 10^{-23} g/atom $\times$ 4 atoms/unit cell $= 3.898598 \times 10^{-22}$ g/unit cell

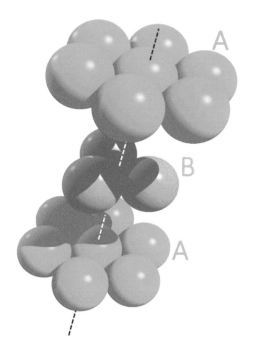

FIGURE 2.7

Schematic of an HCP unit cell.

Given that density is mass divided by volume, we can find the volume of the unit cell:

$$3.898598 \times 10^{-22} \text{ g}/8.908\left(\text{g/cm}^3\right) = 4.376514 \times 10^{-23} \text{ cm}^3$$

Thus the length of the unit cell edge is:

$$4.376514 \times 10^{-23} \text{ cm}^3 = 3.524 \times 10^{-8} \text{ cm } = 3.524 \times 10^{-10} \text{ m}$$

Some metals have a hexagonal symmetry, including titanium, which is widely used in biomedical devices. The HCP structure (Fig. 2.7) has six atoms at the corners of each of the top and bottom faces and an atom in the center. In addition, there is an extra three-atom plane located in between the top and bottom hexagonal planes. The ideal height (c) to edge (a) value (c/a), for the HCP structure is 1.633, but this number is not maintained for all metals. The CN and APF for HCP crystals are 12 and 0.74, respectively, which is the same as for the FCC crystals. Other metals used in biomedical devices with an HCP structure, besides titanium, are zinc and magnesium (Table 2.2).

Example 2.4

Rhenium has an HCP crystal structure, an atomic radius of 0.137 nm, and a c/a ratio of 1.615. Calculate the unit cell volume for this metal.

Answer Unit cell volume, $V_c = c \times 6R^2\sqrt{3} \times R^3$

$$Vc = 1.615 \times 2 \times 6\sqrt{3} \times (1.37 \times 10^{-8} \text{ cm})^3 = 8.63 \times 10^{-23} \text{ cm}^3$$
$$= 8.63 \times 10^{-2} \text{ nm}^3$$

Defects in solids
Point defects

Although we presented the theory behind organization of atoms in the crystal structure, real materials, like anything in life, are not perfect. Crystal imperfections, called point defects, are present in most cases. These point defects include atomic vacancies, self-interstitials, and impurities.

Vacancies are the simplest of these imperfections. The term denotes, as the name suggests, simply a vacant lattice site. More specifically, a location is vacant where an atom should be present. Atomic vibrations, as well as the solidification process, could be causes for vacancy formation. Vibrations can "shake off" and displace atoms from their normal position in a lattice. The number of vacancies at equilibrium can actually be calculated as:

$$N_v = N exp\left(-\frac{Q_v}{kT}\right) \tag{2.5}$$

where N_v is the number of vacancies, N is the number of atomic sites, Q_v is the activation energy for the formation of the vacancy, T is the absolute temperature, and k is the Boltzmann constant (1.38×10^{-23} J/atom-K). It is clear from this expression that higher temperatures will lead to a higher number of vacancies, as the atomic vibrations will increase at higher temperatures. For metals, the concentration of vacancies N_v/N near the melting temperature is around 10^{-4}. That shows that one out of 10,000 atomic sites is likely to be a vacancy.

An interstitial, or self-interstitial, atom is an extra atom in a crystal that is present in the space in between four regularly placed atoms in the structure. Due to this site being normally unoccupied, the placement of an extra atom normally produces significant distortions in the lattice because usually the atomic radius of the interstitial is larger than the dimension of the site where it sits.

In addition to interstitials and vacancies, there are also other impurities present in a material. A level of 100% purity is not attainable in real life. Foreign atoms of other elements than those in the base crystal will be present at trace levels. In the case of metals in general and metals used as biomaterials in particular, most metals are in fact alloys, where impurity atoms are intentionally added during processing to meet certain performance specifications for the intended application. For biomaterials in particular, tuning the mechanical properties and corrosion resistance are usually the main goals when choosing alloy composition.

When impurity atoms are added to a metal, it is sometimes said that a solid solution is formed. In a solid solution, like with liquid solutions, terms such as *solute* (the impurity) and *solvent* (the base metal) are typically used. In other situations, the term *secondary phase* is used to denote the presence of the impurity. When upon the addition of the impurity atom, the crystal structure of the host metal is not changed, a solid solution forms.

Impurity atoms can be present in a host metal as interstitials or substitutional point defects. As the name suggests, substitutional point defects indicate that the impurity atoms replace host atoms. In the case of interstitial impurities, impurity atoms are present within the voids between the atoms.

When discussing alloys, which are relevant for biomaterials, it is important to know their composition. Typically one can express alloy composition as either weight percent (wt%) or atomic percent (at%). To calculate the wt% of an element in an alloy, one needs to calculate the weight of that element relative to the total weight of the alloy. For example, if we consider an alloy that contains only two types of atoms, A and B, the wt% of atom A can be calculated by the following formula:

$$C_A = \frac{m_A}{m_A + m_B} \times 100 \tag{2.6}$$

The at% of an element A in an alloy containing atoms of type A and B can be calculated as the number of moles of that element divided by the total number of moles of all elements in the alloy:

$$N_{m(A)} = \frac{m_A}{M_A} \tag{2.7}$$

where m_A is the mass in grams of the A element and M_A is the atomic weight of the same element.

The concentration of the element A, expressed as at%, is:

$$C'_A = \frac{N_{m(A)}}{N_{m(A)} + N_{m(B)}} \tag{2.8}$$

We can also calculate at% concentrations by using the relative number of atoms of each element to the total number of atoms since one mole of a substance contains the same number of atoms.

A very common alloy used in the biomedical industry is the titanium alloy with the chemical formula TiAl6V4. Table 2.2 shows an example of at% versus wt%, when calculated with the aforementioned formulas for one mole of TiAl6V4.

Line defects

The most notable line defects are the dislocations. These are called linear defects because of their one-dimensional character. There are different types of dislocations, depending on their orientation within the crystal. Understanding the basics of line defects is important for the study of biomaterials due to their importance to plastic deformation and mechanical strength in biomedical alloys, such as TiAl6V4, or 316L stainless steel, which will be discussed in Chapter 3 in more detail.

Edge dislocations are represented by an extra partial plane of atoms that sits between the atomic planes and which does not extend all the way to the end of the crystal but ends within it (Fig. 2.8).

The edge dislocation is centered around the line that can be drawn along the end of the extra plane of atoms. This is called the dislocation line (Fig. 2.9). The dislocation line is perpendicular to the extra half-plane of atoms within the crystal. Because of the presence of the extra plane of atoms in the crystal, local lattice distortions are created. The further away from the location in the lattice, the lowest this distortion, until it eventually disappears.

Screw dislocations, in turn, are defects that are produced by shear stress that pushes part of the crystal out of the center (Fig. 2.9). In the figure, we can see that the upper front part of the crystal is pushed to the right by one atomic distance. There is obviously a crystal distortion induced by this shift, which goes along the dislocation line (see Fig. 2.9). There is a helical path of atoms that form around the dislocation line, which is why these types of defects are called screw

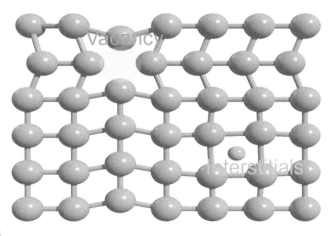

FIGURE 2.8

Representation of an edge dislocation within a crystal, a vacancy, and an interstitial atom; there is an extra half-plane of atoms in between crystal planes.

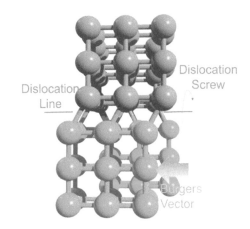

FIGURE 2.9

Screw dislocations (including the dislocation line).

dislocations. In reality, crystals do not contain pure line defects. In most cases, dislocations are mixed dislocations, that is, dislocations that have elements of both screw and edge dislocations.

One measure of the magnitude and direction of the distortion created in the lattice by the dislocations is defined by the Burgers vector, b. For edge dislocations, the dislocation line and the Burgers vector are perpendicular to each other. For a screw dislocation, in turn, they are parallel to each other. For mixed dislocation, the relative orientation of the Burgers vector and dislocation line varies, and it is neither parallel nor perpendicular.

Dislocations are relevant for understanding the properties of some biomedical alloys such as, but not limited to, shape-memory alloys used for orthodontic arch wires or cardiovascular stents. The most common shape-memory alloy is nitinol, an alloy in the NiTi family. Dislocation motion, the ways in which dislocations move, influences the plastic deformation of such alloys. It is outside the scope of this introductory materials to go into the details of different dislocation motion mechanisms, but it is important to remember that these are significant for plastic deformation, which, in turn, for biomedical alloys will influence properties such as Young's modulus and stress-induced phase transformations. Briefly, for a metal or an alloy to deform under stress, the line defects would need to move through the crystals. The less barriers to dislocation motion a biomedical alloy's microstructure has, the easier it can deform. More details on the importance of such parameters are given in Chapter 3.

Another type of crystal defect can be considered to be grain boundaries. Grain boundaries are not typically considered line defects, but rather planar defects separating grains, which have different crystalline orientations. However, because the atoms in the grains are not perfectly aligned, linear defects called grain boundary dislocations will form within the grain boundaries. It is important to observe that while dislocations can move through grains with ease, under mechanical stress and/or thermal treatments, grain boundaries can serve as a barrier to dislocation motion, thus further stopping or delaying deformation of the material. The higher number of grain boundaries is why nanocrystalline alloys tend to have a higher mechanical strength than their regular, microcrystalline counterparts. The dislocation movement is hindered by grain boundaries and there is a larger magnitude of the applied stress that is needed to move a dislocation across a grain boundary. We will learn in Chapter 3 that grain boundaries can have an important effect on corrosion of biomedical alloys as well, and are thus critical to be understood when deciding on fabrication and processing routes of such biomaterials.

Summary

The materials science paradigm shows that structure informs and connects processing and properties. This remains true when materials science is understood and leveraged for the rational design of biomaterials to be used as part of biomedical devices. The structure and type of bonding of materials, as well as crystallography concepts, are all determinant of how a biomaterial can be expected to be processed and further perform, based on its chemical and surface properties, mechanical properties, and its potential for corrosion, degradation, encapsulation, thrombus formation, calcification, toxic leaching, cell lysis, fatigue failure, dislocation, or system reaction. Crystalline solids, such as metals, and some ceramics and polymers are arranged and organized in a long-range atomic structure. Amorphous solids, such as many polymers, lack organization in a long-range atomic structure. The presence of crystalline defects and grain boundaries can play a major role in the final mechanical properties or corrosion resistance of metals and alloys. Therefore gaining an understanding of structure and bonding in materials in general, and biomaterials in particular, is key to understanding the next chapters, which will be focused on specific classes of biomaterials and their applications.

Problems

1. Calculate the attraction force between Ca^{2+} and O^{2-}, which are separated by a distance of 1.25 nm.
2. The force of attraction between the Ca^{2+} and Br^- in $CaBr_2$ is 4.1×10^{-9} N. Calculate the distance between the nuclei of these ions.
3. Why is the boiling point of H_2O higher than that of HF, although water has a smaller molar mass?
4. Graphene is a two-dimensional lattice composed of a continuous arrangement of regular hexagons of carbon atoms. The distance between nearest neighbor atoms is 0.14 nm. What is the graphene's unit cell?
5. Calcium has an FCC structure as a solid. Assuming that it has an atomic radius of 1.97×10^{-8} cm, calculate the density of solid calcium.

6. Krypton crystallizes with an FCC unit cell of edge 5.59×10^{-10} m.
   a) What is the density of solid krypton?
   b) What is the atomic radius of krypton?
   c) What is the volume of one krypton atom?
   d) What percentage of the unit cell is empty space if each atom is treated as a hard sphere?
7. You are given a small bar of an unknown metal. You find the density of the metal to be 4.506 g/cm^3. An X-ray diffraction experiment measures the a edge of the HCP unit cell as 2.95×10^{-10} m and the c edge as 3.58×10^{-10} m. Find the gram-atomic weight of this metal and tentatively identify it.
8. Titanium has an HCP crystal structure, an atomic radius of 0.215 nm, and a c/a ratio of 1.587. Calculate the unit cell volume for this metal.

References

1. Callister WD, Rethwisch DG. *Fundamentals of Materials Science and Engineering: An Integrated Approach*. 4th ed. Hoboken, NJ: Wiley; 2012:910. p xxv.
2. Shackelford JF. *CRC Materials Science and Engineering Handbook*. 4th ed. Boca Raton: CRC Press, Taylor & Francis Group; 2016:634. p x.
3. Shackelford JF. *Introduction to Materials Science for Engineers*. 4th ed. Upper Saddle River, NJ: Prentice Hall; 1996:670. p xiii.
4. Brandon DG, Kaplan WD. *Microstructural Characterization of Materials*. Chichester; New York: J. Wiley; 1999:409. p xiii.
5. Cullity BD. *Elements of X-Ray Diffraction*. 2nd ed. Reading, MA: Addison-Wesley; 1978: 555. p xii.

Metallic biomaterials

Learning objectives

This chapter is dedicated to reviewing the key properties of metallic biomaterials. At the end of the chapter, the students will:

- Acquire a critical perspective of the main applications of metallic biomaterials, potential opportunities, and current challenges.
- Understand metallic biomaterials mechanical properties, including stress-strain relationships, stress shielding, and the fatigue failure.
- Understand how microstructure influences properties in metallic biomaterials.
- Learn the meaning of the term *biocompatibility* as related to metallic biomaterials.
- Learn the main types of metallic biomaterials used for implantation, and what their mechanical, chemical, and physical properties are.
- Understand the main mechanisms of corrosion in metals and its effect on implant properties.

Introduction

Metals and their alloys are widely used as materials of choice in a series of biomedical devices. In the previous chapter, we discussed some of the characteristics of metallic materials. However, intrinsic metal properties are not the sole determinant of implant performance and success. Various metals and alloys have been used in both successful and unsuccessful implant designs.[1] It is estimated that up to 80% of biomedical implants are metallic and the demand continues rising due to the increased life span, overall aging of the population, and long-term effect of nutrition, and as surprising as it sounds, the main driver of the market of metallic implantable materials is the increasing rate of road accidents.[2]

There could be many reasons for a failure occurs, including surgical error, problems related to mechanical design, or error in the use of the implant.[3] Thus once a metal meets the basic property criteria that make it appropriate for use in a certain device, deciding with accuracy which material is superior to another is often close to impossible. In other words, material selection during implant design is only one factor in an otherwise multifaceted challenge.

Introductory Biomaterials. https://doi.org/10.1016/B978-0-12-809263-7.00003-2

As discussed in Chapter 2, in the metallic bond, atoms are packed close together. Thus electrons from one atom are close to the nuclei of other atoms, and electrons are attached to these atoms. Metals are often viewed as positive ion cores with a cloud of valence electrons. These electrons can move freely within the molecular orbitals, and so each electron becomes detached from its parent atom. The electrons are said to be *delocalized*. The metal is held together by the strong forces of attraction between the positive nuclei and the delocalized electrons. Valence electrons are weakly bonded to the positive core and move readily in the metallic lattice.[4] This is the reason why metals can deform significantly without fracture: atoms can slide past one another without completely disrupting the metallically bonded structure. The ability to deform without fracturing is useful in their use as implants in areas of the body where significant loads are applied.[5]

Mechanical properties of metallic biomaterials

Considering the scope of this textbook, we will focus the discussion on the desired mechanical properties of metallic biomaterials when incorporated in medical implantable devices. Metallic materials arrange in crystalline structures formed by tight atomic packing. This characteristic favors the possibility of planes to slip over each other, creating slip systems.[6] The motion of dislocations allows slip-plastic deformation to occur, while the density of these defects along the slip planes will affect the strength and ductility profile of the metallic material.

The mechanical profile of metallic materials can be approached by calculating the deformation/elongation response (strain) of the tested specimen upon the internal distribution of the applied forces on a certain cross-sectional area (stress). Fig. 3.1 presents the typical mechanical response of metallic materials when tension or compression is applied; the plot shows the characteristic linear region (1); within this phase, any deformation occurred is reversed once the stress is removed, fully recovering to its initial condition. The plastic region (2) refers to the phase when the material is subjected to stresses that overcome the yield strength, resulting in permanent distortion or deformation (even when the stress is removed). Finally, the failure region (3) refers to the ultimate fracture of the tested specimen.

According to Hooke's law, within the elastic region (reversible/flexible deformation), the strain experienced by a spring will be proportional to the stress applied. Therefore,

$$F = k * x \tag{3.1}$$

where F = spring force, k = spring constant, and x = spring stretch or compression.

Hooke's proportionality was initially described for springs. However, this concept has been extrapolated to all materials of known surface area and is expressed in terms of the proportionality of stress and stress. The stress (σ) is

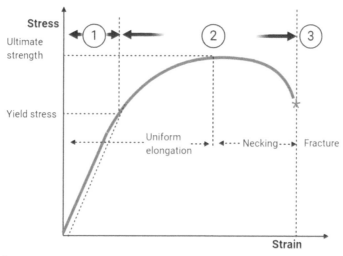

FIGURE 3.1

Illustration of the typical stress-strain curve of metallic materials (from a tension test). Elastic (linear) deformation *(1)*; plastic (permanent) deformation *(2)*; and fracture (failure) *(3)*.

calculated by accounting for the force applied (F) per cross-sectional area (A) of the testing material as follows:

$$\sigma = F/A \tag{3.2}$$

Since stress cannot be measured directly, it needs to be inferred from a measure of strain (ε), and Young's modulus (E). Young's modulus, also known as modulus of elasticity, is the slope of the curve in the linear region (identified as area (1) in Fig. 3.1) and provides information on the stiffness of the tested material. This property is calculated by:

$$E = \sigma/\varepsilon \tag{3.3}$$

where σ refers to the applied stress and is defined as:

$$\sigma = E * \varepsilon \tag{3.4}$$

while ε represents the strain experienced by the specimen, and is defined as:

$$\varepsilon = \Delta L/L \tag{3.5}$$

where $\Delta L/L$ is the material extension or compression per unit length.

Thinking back to the nature of metallic biomaterials, it is to be noticed that their continuous exposure to repetitive cyclic loads might lead to crack formation, corrosion, and further fracture. This process is known as fatigue failure and is the most common mode of mechanical failure experienced by implants. Under fatigue conditions, the number of stress cycles that the material can withstand is inversely proportional to the magnitude of the applied stress. That is, the number of stress cycles the material can withstand increases as the stress intensity is reduced. The stress at which the material can withstand 10 million stress cycles without failure is called the endurance limit.[7]

Example 3.1

A 2.5-cm diameter bar is subjected to a 3000-kg mass. Calculate the engineering stress on the bar in megapascals (MPa).

Answer

$$F = m * a = 3000 \text{ kg}\left(9.81\,\frac{m}{s^2}\right) = 29,430 \text{ N}$$

$$\text{Diameter of the bar} = 0.025 \text{ m}$$

$$\sigma = \frac{F}{A} = \frac{29,430 \text{ N}}{\pi\left(\dfrac{0.025 \text{ m}}{2}\right)^2} = 5.99 \times 10^7 \text{N/m}^2$$

Material properties such as malleability and ductility are also critical to metallic implants. These material properties are tightly related; a malleable material shows the ability to deform under compressive stress by squeezing the specimen together and aligning the shearing forces toward the specimen, as illustrated in Fig. 3.2A, while a ductile material has the ability to deform under tensile stress, undergoing elongation, as the overall forces are pulling away from the specimen, as presented in Fig. 3.2B. The successful performance of implanted metallic devices depends greatly on the ability of the material to undergo plastic deformation under mechanical stress.[8]

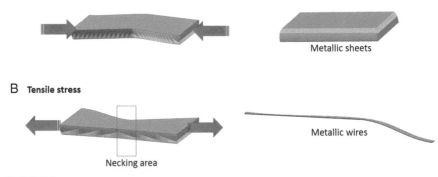

FIGURE 3.2

Typical strain response of metals under compressive (A) and tensile stresses (B). Malleable materials can be hammered into a sheet, while ductile materials can be drawn into wires.

A sense of the material ductility indicates the degree of plastic deformation before reaching fracture; thus the ability to control this property is critical along the implant design process. Ductility might be expressed quantitatively as either percent elongation (%EL) or percent reduction in area (%RA). %EL refers to the stress at fracture in tension, and it is expressed as a percentage:

$$\%EL = \left(\frac{l_f - l_0}{l_0}\right) \times 100 \tag{3.6}$$

where l_0 is the initial length, l_f is the length of the specimen when it breaks (fractures), and $\Delta l = l_f - l_0$ is the change in length after the specimen reaches fracture.

On the other hand, reduction of area (RA) is commonly reported as an additional metric (to %EL) on the deformational profile of the tested material. RA is measured at the minimum diameter of the neck as a better indicator of ductility, accounting for the uneven nature of deformation:

$$RA = \left(\frac{A_0 - A_f}{A_0}\right) \times 100 \tag{3.7}$$

where A_0 is the initial cross-sectional area and A_f is the minimum final area (necking area).

Typically, if the elongation is $\geq 5\%$ the specimen is termed as ductile; otherwise, the material tested is considered as brittle (EL $<5\%$).

Example 3.2
A round sample of a Ti6Al4V (F136) alloy with a 1.3-in radius is pulled to failure in a tensile testing machine. The radius of the sample was 0.732 in at the fracture surface. Calculate the %RA of the specimen tested.

Answer

$$\%RA = \left(\frac{A_0 - A_f}{A_0}\right) \times 100$$

$$A_0 = \pi r^2 = \pi(1.3)^2 = 5.31 \text{ in}^2$$

$$A_f = \pi r^2 = \pi(0.732)^2 = 1.68 \text{ in}^2$$

$$RA = \left(\frac{5.31 - 1.68}{5.31}\right) \times 100\% = 68.4\%$$

As discussed in Chapter 2, metals form crystalline structures with a high degree of organization, which has a critical effect on the properties of the materials. From previous chapters we know that crystal structures can present different arrangements (e.g., face-centered cubic [FCC] and body-centered cubic [BCC] among others). It is important to note here that closely packed planes allow more plastic deformation than those not closely packed. The delocalized nature of the metallic bonds makes it possible for the atoms to slide past each other under stress instead of fracturing. Therefore FCC structures will exhibit increased ductility than BCC structures.

When crystals are organized in the same direction, they are known as grains. Therefore when we talk about alloyed metallic materials, the term *polycrystalline* refers to the interface between two crystals and the existence of a grain boundary. It is no big surprise that almost all engineering alloys are polycrystalline. This composition favors their mechanical strength, since grain boundaries act as barriers to dislocation motion.[9] For example, at 20% strain, the tensile strength of polycrystalline copper is 276 MPa, while for single crystal copper it is 55 MPa. Each grain has its slip planes, which have orientations that are different from neighboring grains, and dislocations cannot propagate to neighboring grains. However, this strengthening mechanism becomes less effective when polycrystalline materials are exposed to high temperatures.

Challenges of metallic materials under physiological conditions

The complexity of biocompatibility

Because the biomaterials are foreign to the in vivo environment, implant biomaterials, metallic and nonmetallic, always initiate a cascade of events when implanted that can be harmful or toxic to living tissues.[10] An inert biomaterial that results in little or no host response is considered biocompatible. Interactive biomaterials, on the other hand, are designed to elicit desirable host responses such as tissue ingrowth, and they are also considered biocompatible. Viable materials are biocompatible as well. They typically incorporate or attract cells that are then resorbed or remodeled (e.g., biodegradable polymeric scaffolds for functional tissue engineering). The last category of biocompatible biomaterials includes replanting materials that consist of native tissue that has been cultured in vitro from cells obtained from a specific patient.[11] As discussed in Chapter 1, the term *biocompatibility* needs to be used carefully and understood in the specific context of both the material and the biological host, accounting for the complexity and uniqueness of the physiological environment where the material is implanted. Therefore biocompatibility is referred to here as a property of a system rather than a material property.

Fatigue failure

Some of the most common examples of biomedical applications of metallic materials are hip replacement, cardiovascular stents, joint replacements, spinal fixation devices, fracture fixation devices, and spinal discs[12]; based on the fact that most of these implanted devices are subjected to repetitive, cyclic loads, careful tuning of the mechanical properties of the metallic materials used is critical to guarantee an adequate long lifetime performance and to avoid failure. As we mentioned

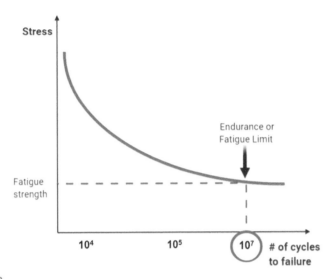

FIGURE 3.3

Illustration of a typical S-N curve, representing the magnitude of alternating stress versus the number of cycles until reaching failure.

previously, the increased life expectancy of the overall population entails the development of implantable materials that endure an adequate long lifetime performance. According to the Centers for Disease Control and Prevention (CDC), the rate of hip replacement more than doubled for those aged 45 to 54 years in 10 years,[13] representing almost 20% of the total hip replacements. Considering the last update on overall life expectancy from the World Health Organization (WHO), in 2016 the average life span was 72 years, which results in lifetime requirements for implants to be above 22 years on average.

Fatigue failure is the most common mode of mechanical failure experienced by implants. Under fatigue conditions, the number of stress cycles that the material can withstand is inversely proportional to the magnitude of the applied stress; that is, the number of stress cycles the material can withstand increases as the stress intensity is reduced. The stress at which the material can withstand 10 million stress cycles without failure is called the *endurance limit*, as illustrated in Fig. 3.3. The steps of fatigue-induced failure are crack initiation, crack propagation, and catastrophic failure.

Stress shielding

Bone tissue is a dynamic tissue that undergoes constant resorption and regeneration as a response to the natural mechanical load; hence the density of the remodeled bone adapts to withstand prevalent daily loads.[14] Stress shielding, also known as stress protection, is the phenomenon occurring when an implanted material used to repair fractures or tissue replacement (e.g., joint replacement) absorbs the mechanical load, preventing a successful remodeling of the surrounding bone tissue, thereby reducing the support of the implant and raising the risk of loosening and micromotion. These conditions are tightly correlated to further complications discussed in detail in the following sections, such as corrosion and fatigue failure. In order to reduce the risk of stress shielding, the material implanted must fulfill an optimal Young's modulus, within the stiffness ranges shown by natural bone tissue. However, achieving a low modulus of elasticity while delivering fatigue strength continues to be a challenging task.[15] In addition to stiffness, material density plays a critical role in the ultimate performance of the implant; the latter is explained by the strong positive correlation between atomic packing factor (APF), density, and stiffness. In other words, increased density confers enhanced strength and stiffness; thus when selecting a candidate material, its physical performance should match as much as possible the profile of the adjacent bone to prevent stress shielding and/or bone loss.[16]

Corrosion in metallic implants

The creation of cracks also leads to corrosion initiation, which further speeds up the failure. Specific for metallic biomaterials, one of the most significant biocompatibility and performance-related challenges is corrosion and degradation in the very aggressive in vivo environment. Corrosion leads to damaged regions of implant surfaces, resulting in decreased strength, as well as to the release of corrosion products, with negative effects on tissue biocompatibility.

There are several types of corrosion, which will be discussed here, including galvanic corrosion (between two metals in a conductive medium), fretting corrosion (at contact sites between materials with relative micromotion), and crevice corrosion.[17]

The physiological (in vivo) environment is highly corrosive; it contains water, oxygen, and multiple ionic species that can easily attack a metallic implant. When corrosion occurs, it results in damaged regions of implant surfaces, leading to decreased material strength and the release of corrosion products into the bloodstream, with negative effects on biocompatibility.

To establish the tendency of each metallic element to oxidize, a standard electrochemical series is employed. The series is obtained by arranging various redox equilibria in order of their redox potentials. For the reaction to proceed, the potential difference needs to be positive. In the electrochemical series, elements with a more positive potential are increasingly inert (e.g., gold), while those with a negative potential are increasingly active (e.g., lithium), as presented in Table 3.1.

As explained previously, most corrosion reactions are electrochemical. The lowest free energy state of many metals in an oxygenated and hydrated environment is that of the *oxide. Corrosion occurs when metal atoms become ionized and go into solution* or combine with oxygen or other species in solution to form a compound that flakes off or dissolves. The physiological environment is very aggressive in

Table 3.1 Electrochemical series of common elements found in biomaterials and physiological environments.

Electrode material		Electrode reaction	E° (V)
Au	Gold	$Au^{3+} + 3e^- \leftrightarrow Au$	+1.43
Ag	Silver	$Ag^+ + e^- \leftrightarrow Ag$	+0.80
Cu	Copper	$Cu^{2+} + 2e^- \leftrightarrow Cu$	0.34
H	Hydrogen	$H^+ + e^- \leftrightarrow H$	0
Pb	Lead	$Pb^{2+} + 2e^- \leftrightarrow Pb$	−0.13
Sn	Tin	$Sn^{2+} + 2e^- \leftrightarrow Sn$	−0.14
Ni	Nickel	$Ni^{2+} + 2e^- \leftrightarrow Ni$	−0.25
Cd	Cadmium	$Cd^{2+} + 2e^- \leftrightarrow Cd$	−0.40
Fe	Iron	$Fe^{2+} + 2e^- \leftrightarrow Fe$	−0.44
Zn	Zinc	$Zn^{2+} + 2e^- \leftrightarrow Zn$	−0.76
Ti	Titanium	$Ti^{2+} + 2e^- \leftrightarrow Ti$	−1.63
Al	Aluminum	$Al^{3+} + 3e^- \leftrightarrow Al$	−1.66
Mg	Magnesium	$Mg^{2+} + 2e^- \leftrightarrow Mg$	−2.37
Na	Sodium	$Na^+ + e^- \leftrightarrow Na$	−2.71
K	Potassium	$K^+ + e^- \leftrightarrow K$	−2.93
Li	Lithium	$Li^+ + e^- \leftrightarrow Li$	−3.05

E°, Standard Electrode Potential.
Data from Vanysek.[44]

terms of corrosion, since it is not only aqueous but also contains chloride ions and proteins,[18] acting as an effective electrolyte that serves to complete the electric circuit. In the human body, the required ions are plentiful in body fluids. Anions are negative ions that migrate toward the anode, and cations are positive ions that migrate toward the cathode. The oxidation reaction by which metals form ions that go into aqueous solution is called the *anodic reaction*. The reduction reaction in which a metal or nonmetal is reduced in valence charge is called the *cathodic reaction*. Electrochemical corrosion reactions involve both oxidation and reduction reactions. Both occur at the same time and rate to prevent charge buildup in metals.

There are several ways in which corrosion can occur. One is galvanic corrosion. This type of corrosion happens between two metals in a conductive medium. This is one of the major types of corrosion for metallic implants since the biological environment is indeed conducive. Thus if two different metallic species are nearby, the danger of galvanic corrosion is high. Galvanic corrosion happens because two different types of metals will have different electrochemical potentials and will thus form an electrical circuit in a conductive liquid.

The potential difference E observed among different metallic materials depends on the concentration of metal ions in solution according to the Nernst equation:

$$E = E_0 - \left(\frac{RT}{nF}\right) \ln(M^{n+}) \tag{3.8}$$

where R is the gas constant, E_0 is the standard electrochemical potential, T is the absolute temperature, F is Faraday's constant, and n is the number of moles of ions.

Therefore the standard electrode potential in an electrochemical cell is calculated as follows:

$$E_0 = E_{reduction} - E_{oxidation} \qquad (3.9)$$

Some metals can be covered with a passivating film that is protective of further attack. The kinetics of reactions have to be considered also, as even if a corrosion reaction is favored thermodynamically, it can proceed slowly. In certain alloys that contain an element that can form a passivating film, such as Cr, grain boundaries can become depleted of this corrosion-resistant metal due to some chemical reactions that occur under certain conditions (e.g., formation of carbides). Fig. 3.4 illustrates the situation when the grains become cathodic and the grain boundaries become anodic, forming a galvanic couple and leading to what we call intergranular or grain boundary corrosion. When a metal or an alloy that is protected from corrosion by a passivating film is exposed to highly corrosive conditions,[19] the highly corrosive fluid medium can penetrate small crevices, such as cracks. In this situation, the passivating film at the crack is destroyed. The crack becomes anodic, while the matrix becomes cathodic. A galvanic cell is thus created and corrosion proceeds.

This principle, the creation of a galvanic couple between two metals or two sites on the same metal that have different electrochemical potentials, is the basis of most corrosion types.[17] Fretting corrosion is the corrosion that is produced at the contact sites between materials with relative micromotion. Thus if there is an implant composed of different metals and shows micromotion between components, fretting corrosion is likely to occur.[20] Crevice corrosion is the type of corrosion that is induced by the presence of scratches or cracks in the surface of metallic implants. In this case, the top of a crack will have a different electrochemical potential compared to the inside of the track, triggering corrosion events.

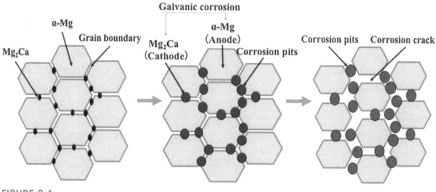

FIGURE 3.4

Micro corrosion cells. Grain boundaries are anodic concerning the grain interior, thereby creating an electrochemical cell.

Reproduced with permission from Wang et al.[41]

Inhomogeneous structure or composition within a metal or an alloy can set up galvanic corrosion even in the absence of fretting or cracks. This type of corrosion can be much more rapid than the corrosion of a single metal. Consequently, the implantation of dissimilar metals is to be avoided. Galvanic action can also result in corrosion within a single metal if there is inhomogeneity in the metal or its environment.[21]

There is also uniform corrosion of a metal, which is not due to the formation of galvanic couples or defects or boundaries in the material, but this is less prevalent in implants. Galvanic corrosion can be much more rapid than the corrosion of a single metal. However, a metal oxide can also be formed in the absence of another metal and in the presence of oxygen. As a rule of thumb, anything that upsets the dynamic equilibrium of the charged double layer might accelerate the oxidation (corrosion degradations) of metallic biomaterials.

Example 3.3

Consider the corrosion of a metal such as zinc. In acid environments (HCl), metallic zinc goes into solution in the ionized form as follows:

$$Zn + 2H^+ + 2Cl^- \rightarrow Zn^{2+} + H_2$$

a) Balance the equation.
b) Write separate equations for the oxidation and reduction half-reactions.
c) Calculate the cell potential.

Answer

a) $Zn + 2HCl = ZnCl_2 + H_2$ or $Zn + 2H^+ = Zn^{2+} + H_2$
b) Chloride ions do not experience a net change. Half-reactions are both written as reduction.

$$Zn^{2+} + 2e^- \; Zn \rightarrow (\text{anode}) \; (E_0 = -0.76)$$

$$2H^+ + 2e^- \rightarrow H_2 (\text{cathode}) \; (E_0 = 0.00)$$

c) $E_0 = E_{reduction} - E_{oxidation}$

$$E_0 = 0 - (-0.76) = +0.76 \text{ V}$$

Pourbaix diagrams

To aid in selecting metals and alloys that could be appropriate to be used in implantation in various areas of the body that have different pH levels, a Pourbaix diagram is of use.[22] This is a plot of regions of corrosion, passivity, and immunity as they depend on electrode potential and pH. Pourbaix diagrams are derived from the Nernst equation, the solubility of the degradation products, and the equilibrium constants of the reaction.

For the sake of definition, the *corrosion region is set arbitrarily at a concentration of* 10^{-6} gram atom per liter (molar) or more of metal in the solution at equilibrium. This corresponds to about 0.06 mg/L for metals such as iron and copper, and 0.03 mg/L for aluminum. *Immunity is defined as the equilibrium between metal and its ions at less than* 10^{-6} *molar.* In the region of immunity, corrosion is energetically impossible. Immunity is also referred to as cathodic protection. In the passivation domain, the stable solid constituent is an oxide, a hydroxide, a hydride, or a salt of the metal. *Passivity is defined as an equilibrium between a metal* and its reaction products (oxides, hydroxides, etc.) at a concentration of 10^{-6} molar or less. This situation is useful if reaction products are adherent. In the biomaterial setting, passivity may or may not be adequate: disruption of a passive layer may cause an increase in corrosion. The equilibrium state may not occur if reaction products are removed by tissue fluid. Materials differ in their propensity to reestablish a passive layer that has been damaged. This layer of material may protect the underlying metal if it is firmly adherent and nonporous; in that case, further corrosion is prevented. Passivation can also result from a concentration polarization due to a buildup of ions near the electrodes. This is not likely to occur in the body, since the ions are continually replenished. Cathodic depolarization reactions can aid in the passivation of a metal by an energy barrier that hinders the kinetics.[23]

In the diagram shown in Fig. 3.5, there are two diagonal lines. The top ("oxygen") line represents the upper limit of the stability of water and is associated

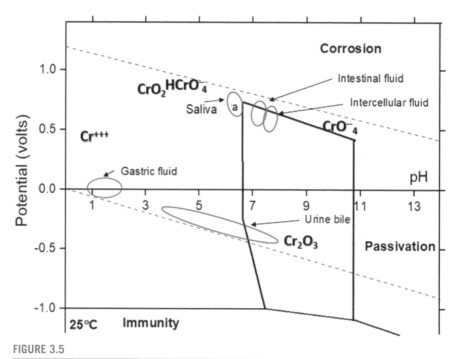

FIGURE 3.5

Pourbaix diagram for chromium, showing regions associated with various body fluids.

with oxygen-rich solutions or electrolytes near oxidizing materials. In the region above this line, oxygen is evolved according to $2H_2O = O_2 + 4H^+ + 4e^-$. In the human body, saliva, intracellular fluid, and interstitial fluid occupy regions near the oxygen line, since they are saturated with oxygen. The lower ("hydrogen") diagonal line represents the lower limit of the stability of water. Hydrogen gas is evolved. Aqueous corrosion occurs in the region between these diagonal lines on the Pourbaix diagram. In the human body, urine, bile, the lower gastrointestinal tract, and secretions of ductless glands occupy a region somewhat above the hydrogen line.

The significance of Pourbaix diagrams is that different parts of the body have different pH values and oxygen concentrations. Thus a metal that performs well in one part of the body may corrode excessively in another part. A pH can change results in tissue that has been injured or infected. In a wound, the pH can be as low as 3.5, in an infected wound as high as 9.0, while in normal tissue the pH is 7.4. The presence of other ions can change the significance of Pourbaix diagrams, and reaction rates are not taken into consideration, which can affect predictions based on Pourbaix diagrams.

To avoid corrosion of metallic implants, care should be taken to use appropriate metals, avoid implantation of different types of metal in the same region, design the implant to minimize pits and crevices, avoid the transfer of metal from surgical tools to the implant or tissue, and recognize that a metal that resists corrosion in one body environment may corrode in another part of the body.

Biomedical alloys

Most orthopedic and oral implants use Ti alloys, stainless steels (SSs), or Co-Cr alloys.[24] Besides, there are other materials used mostly in dentistry, including Au and Au alloys (alloying elements include Cu, Pt, Ag, and Zn). Although Au alloys show outstanding mechanical properties, the minimal composition of pure Au must be at least 75 wt% to avoid corrosion in dental materials that contain Au. On the other hand, Ni-Ti alloys (nitinol) are shape-memory alloys that are used as orthodontic dental archwires. These alloys are also used in arterial blood vessel stents, intracranial aneurysm clips, and even orthopedic implants. Ni-Ti alloys have the property that after the metal is deformed below the transformation temperature, it can snap back to its previous shape following heating.[25] This phenomenon is due to a diffusion of less martensitic phase transformation that is also thermoelastic. The thermoelastic martensitic transformation has the following characteristics: (1) can be initiated by cooling the material below the temperature at which the martensitic transformation begins; (2) both martensitic and austenitic temperatures can be increased by applying stresses before the yield point and the increase is proportional to the applied stress; (3) the material is more resilient than most metals; and (4) the transformation is reversible.

Other metallic materials that have limited uses for implantation are tantalum (biocompatible, poor mechanical properties, high density, tested for bone substitute),[26] Ni-Cu and Co-Pd alloys (used for cancer treatment due to their magnetic properties),[27] Pt, Pd, Rh, and Ir (extremely corrosion-resistant, poor mechanical properties, used as pacemaker tips).[28]

Titanium and its alloys

Attempts to use titanium for implant fabrication date to the late 1930s. Ti was found initially to be well tolerated in cat femurs. Ti is lighter (4.5 g/cm³) than SS (7.9 g/cm³ for 316 SS), cast CoCrMo (8.3 g/cm³), and wrought CoNiCrMo (9.2 g/cm³) alloys. It has good mechanical properties for hard tissue replacement applications. However, instead of pure Ti, its alloy, Ti6Al4V, is widely used to manufacture implants. It also contains small amounts of N, C, H, Fe, and O. The Ti6Al4V alloys are allotropic materials that exist as a hexagonal close-packed (HCP) (α-Ti) structure up to 882°C and BCC (β-Ti) structure above that temperature.[29]

The titanium of commercial purity is also used in some biomedical devices, such as cardiovascular stents or spinal fixation devices, which are less load bearing than orthopedic devices. However, pure Ti does not meet the mechanical properties criteria for use in bone and dental replacement devices.

The addition of Al and V to Ti in the alloy leads to optimized properties. Aluminum stabilizes the α-phase, which promotes excellent strength and oxidation resistance at high temperatures. Vanadium stabilizes the β-phase, which has the advantage that it can be strengthened by heat treatment.[30] The microstructure plays a significant role in the properties of titanium alloys. The α-phase has an equiaxed microstructure, while the β-phase displays a lamellar structure (Fig. 3.6A−B). The

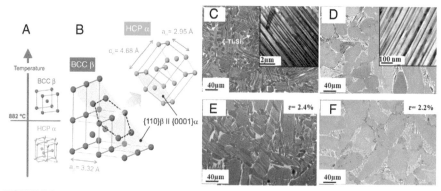

FIGURE 3.6

Typical microstructures of TiAl-based alloys: Illustration on the effect of the β-transus temperature (Tβ) on the alloy crystal structure (A); lattice correspondence between hexagonal close-packed (HCP) α body-centered cubic (BCC) β structures (B). TiAl alloys in cast condition. Scanning electron microscopy and inset transmission electron microscopy images of lamellae microstructure (C); β-microstructure (D); microstructure after creep test at 850°C/150 MPa at a maximal creep strength of ε = 2.4% (E) and ε = 2.2% (F).

Reproduced with permission from Kastenhuber et al.[30]

α-phase, due to its microstructure, has high strength and ductility and good fracture toughness. The lamellar structure of the β-phase brings some advantages in terms of fracture toughness but generally offers lower strength and ductility (as characterized in Figure 3.6 C-F).[31] Finer microstructures are advantageous for increasing the fatigue resistance. Table 3.2 lists the mechanical properties induced by microstructural variations in TiAl6V4 alloys.

Challenges that are currently remaining in the area of metallic biomaterials in general, and titanium alloys in particular, focus on improving tissue and blood compatibility, decreasing toxicity and carcinogenicity, decreasing elastic modulus and stress shielding, and advancing miniaturization of devices. Corrosion resistance is also one of the very significant issues that are investigated when it comes to the performance of metallic biomedical devices. From this viewpoint, Ti alloys show improved biocompatibility and corrosion resistance compared with Co-Co alloys and SS. However, the presence of Al and V is of concern due to the inherent toxicity of these elements. New alloy formulations replace V with Nb or Fe. Thus V-free Ti-based alloys, such as Ti6Al7Nb and Ti5Al2.5Fe, are being developed, with good results in terms of mechanical properties. However, these new formulations lead to alloys with much higher elastic moduli than that of the bone, making them less appropriate for use in the areas of the body where stress shielding is a significant issue. The cortical bone has a modulus of 10 to 30 GPa, and the cancellous bone has a modulus of 0.01 to 2 GPa, while the lowest modulus achieved for a Ti alloy was 40 GPa in Ti-35Nb-4Sn.

Although for most orthopedic applications a low Young's modulus is advantageous, in some instances this could also become a disadvantage. For example, in

Table 3.2 Mechanical properties as a function of microstructure for Ti6AlV4 alloys.

Microstructure	YS (MPa)	UTS (MPa)	El. (%)	RA (%)	K_{IC} (MPa)
Equiaxed (Std)	951	1020	15	35	61
Lamellar (Std)	884	949	13	23	78
Equiaxed (ELI)	830	903	17	44	91
Equiaxed (CMG)	1068	1096	15	40	54

Oxygen content: Std: 0.15%–0.2%; ELI: 013 Max; CMG: 0.18%–0.2%.
El., Elongation; RA, reduction in area; UTS, ultimate tensile strength; YS, yield strength.
Reproduced with permission from Geetha M et al.[32]

Table 3.3 Mechanical properties of CoCrMo alloys.

Mechanical properties	Tensile strength (MPa)	Yield strength (MPa)	Elongation after fracture A (%)
Co-Cr wrought	1172	827	12
Co-Cr cast	665	450	8

spinal fixation devices, the surgeon needs to subject the device to bend to match the spine curvature. However, due to the limited space available for performing this procedure, a higher springback of the device is not desirable. Lower moduli result in higher springback, and thus an equilibrium needs to be found between the need to avoid stress shielding by lowering the Young's modulus and the need to lower the chances of springback.[31]

The modulus of elasticity for Ti6Al4V is around 110 GPa, half the value for Co-based alloys. This is significant because, as we will learn later in this book, implants with low elasticity moduli, as close as possible to that of bone (ranging between 14 and 21 GPa), perform better when in operation. The reason behind this is the tendency of the material with a higher Young's modulus to absorb most of the load placed at the implant location, which in turn results in bone resorption. The compression strength of the Ti alloys is similar to that of 316 SS and Co-based alloys. The main disadvantage of Ti alloys is their poor shear strength, which renders them unsuitable for fracture fixation devices that are exposed to torsion forces, such as bone screws, or plates.[33]

Stainless steel implants

The modern use of metallic materials started with the development of the "Sherman vanadium steel." This material was manufactured into bone screws and plates for fracture fixation.[34] SSs are alloys that contain iron (Fe), chromium (Cr), and nickel (Ni), as well as other transition metals. These transition metals are well tolerated in the human body if present in trace amounts. However, if leaching out in large amounts from implants, they can lead to serious health issues, including cancer and neurological diseases. Thus as with all other metallic materials used in implants, it is essential to avoid corrosion of SS under the aggressive chemical environment of the human body. The Sherman vanadium steel proved not to offer sufficient corrosion resistance and its use was therefore discontinued.

The next, more resistant to corrosion and at the same time mechanically stronger SS used for fracture fixation devices was the 302 SS. This material contains, in addition to Fe, 18% Cr and 8% Ni. Because the biological environment is rich in salt, Mo was later introduced, which aids in the corrosion resistance in salt water. This SS later became known as the 316 SS.

In the 1950s, the carbon content of 316 SS was reduced from 0.08 to 0.03 *w/o* for better corrosion resistance to chloride solution, and it became known as SS 316L. This is an austenitic SS, which is now used in most biomedical implants.

Cr is a major component of corrosion-resistant SS. The minimum effective Cr concentration is 11 w/o. Chromium is included in these formulations because it can form a passivating layer, providing excellent corrosion resistance. The 316 and 316L SSs can be hardened by cold working, not by heat treatment, and are nonmagnetic, which is significant for the ability to not interfere with magnetic resonance imaging (MRI) measurements. SS is mostly used in fracture fixation devices such as fracture plates, screws, or hip nails.[35]

From the different biocompatible alloys used for implantable devices, SS is the most easily manufactured in a cost-efficient manner. Although the manufacturing process offers remarkable advantages, yielding high-quality SS (surgical grade), it requires detailed attention to the melting process, carbon content, impurities, and passivation. This last step is performed to remove any free iron (Fe) atoms and increase the thickness of the surface oxide layer. The passivation strategy provides corrosion protection beyond the native resistance of the metal, ensuring its biocompatibility.[16]

Cobalt-chromium alloys

Cobalt-chromium alloys are another category of metallic materials that are widely used for orthopedic and dental applications.[36] The popularity of these alloys for biomedical applications stems from their properties of being nonmagnetic, as well as resistant to wear and corrosion. The unique properties of Co-Cr alloys originate from the effect of chromium and molybdenum inducing solid solution strengthening, as well as the formation of carbides with very high hardness and the passivating layer formed by Cr.

The high corrosion resistance induced by Cr made this alloy attractive initially for dental applications, such as the casting of dental implants. Other uses include orthopedic applications such as knee, shoulder, and hip prostheses, as well as fracture-fixation devices.[37]

The first alloys from this family were CoCrW and CoCrMo, which were developed in 1907 and named "satellites." Since then, the main elements in the composition of these alloys remained constant, and modifications were made to control the carbon content. Carbon tends to form carbides with Cr, which in turn depletes it from being available to form the Cr oxide passivation layer critical for corrosion resistance, as illustrated in Fig. 3.7A. Corrosion resistance, as discussed previously, is very important for biomedical applications, since leaching of metallic ions can severely alter biocompatibility. At the same time, corrosion eventually leads to mechanical failure. Thus a low carbon content is desirable both for Co-Cr alloys and SS. On the other hand, the inclusion of carbon in the composition is necessary due to the formation of hard carbides (generally M_7C_3 and $M_{23}C_6$), which assist with abrasion resistance. There are two main variants of Co-Cr alloys used for implants today, namely the cast and wrought versions.[38] These alloys have identical composition, with 58 to 69 wt% Co and 26 to 30 wt% Cr. The processing history of the alloys differentiates them from each other, with differences being reflected in the microstructure and mechanical properties. The effect of composition on the microstructure is illustrated by optical micrographs obtained from Co-Cr alloys. The images in Fig. 3.7B−E show the microstructural changes induced by the formation and precipitation of chromium carbides.

The crystal structure and defects in Co-Cr alloys affect their properties and processability. For example, Cr, Mo, and W, which are often included in Co-Cr alloys, stabilize the HCP-Co structure but also lead to a higher tendency to form stacking faults. This defect is associated with partial dislocations and is described as a planar

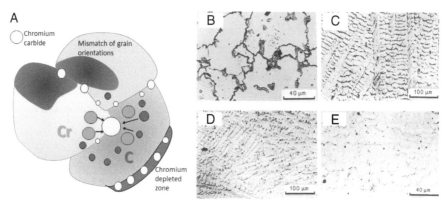

FIGURE 3.7

Co-Cr alloys: effect of carbide precipitation on microstructure. Illustration of chromium depletion at the grain boundary by carbide precipitation (A); optical micrographs of three cast Co-Cr alloys with different compositions: as cast Co-Cr alloy containing no Co, Cr, Ni, and C, with carbides along grain boundaries (B); cast alloy containing C, Cr, Mo, W, Ni, and C, with face-centered cubic (FCC)-Co structure (C); cast alloy with a similar FCC-Co structure to image C (D); and cast alloy at higher densification, showing spherical carbide particles formed (E).

Reproduced with permission from Yamanaka et al.[39] and Shi et al.[42]

anomaly in the stacking sequence of crystal planes, thereby disrupting the continuity of a perfect lattice. The formation of stacking faults lowers the ductility and workability of Co-based alloys, including Co-Cr alloys.[39]

The wrought CoCrMo alloys have a finer microstructure than the cast version. This processing route leads to superior chemical homogeneity and ductility. The formation of hard carbides in the microstructure leads to Co-Cr alloys being difficult to machine.

Co-Cr alloys are often used in dentistry, especially for the manufacturing of metallic frameworks for removable partial dentures. Other applications of Co-Cr alloys include metallic substrates for porcelain-fused-to-metal dental restorations and welded structures such as disks, rods, and screws for spine surgeries. Welding allows for the melting of materials together, which creates stronger joints in metallic assemblies, thereby improving overall fatigue resistance. Fig. 3.8 offers an example of high-frequency welding, also known as ultrasonic impact treatment (UIT), and its effect on the Co-Cr alloy microstructure. UIT is a technology for the fabrication of complex implantable metallic devices that creates residual compressive stresses as illustrated in Fig. 3.8A; the stresses introduced by this technique improve the resistance of welded structures by grain refinement, and grain size reduction, as presented in Fig. 3.8B–D.

The use of Co-Cr alloys is quickly expanding as an alternative to the Ni-Cr–based alloys, also used in dentistry, mainly due to Ni toxicity, as well as for the long-term preservation of shine in the case of removable prostheses. Co-Cr alloys also compare favorably with Ni-Cr-Mo alloys in terms of corrosion resistance, the latter being more likely to release ions.[40]

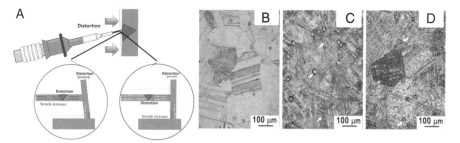

FIGURE 3.8

Ultrasonic impact treatment (UIT) effect over the Co-Cr alloy microstructure. Graphic illustration of UIT principle (A). Optical micrographs of wrought Co-28Cr-6Mo alloy before and after UIT: wrought Co-Cr alloy before treatment, (B) wrought Co-Cr alloy after UIT for 30 s (C), and after UIT of 120 s (D). UIT is employed to produce ultrafine-grained nanostructures.

Reproduced with permission from Petrov et al.[43]

Summary

The mechanical and electrochemical working conditions of dental and orthopedic implants are very complex, including stress cycles, friction, fatigue, and corrosion. Overall, the use of traditional metallic alloys (Ti alloys, SS, Co-Cr alloys) dominates in orthopedic surgery, due to their favorable benefit/risk ratio, and the amount of evidence available of their performance in vivo along with extended clinical testing. However, a great deal of research is underway trying to improve the excessive toughness of metallic alloys when compared with natural tissue, while reducing the incidence of failure rate after a prolonged implantation time (current implant life is about 15 years, which is in many situations insufficient). This chapter summarized the desired properties in a metallic biomaterial and the challenges that metallic implants undergo when implanted in harsh conditions such as physiological environments.

Problems
Identify the atomic bond in each of the following compounds:
1. H_2SO_4, FeS, NaI, $MgCl_2$, and Zn.
   a) According to your previous response, from the listed compounds which one is expected to show the highest melting point?
   b) How does metallic bond explain ductility in metals?
2. The surface of metallic materials is described as highly reactive. What is the implication of their reactive nature under physiological conditions? What would be your proposed approach to overcome this challenge and increase long-term stability?

3. Analyzing both the atomic structure and mechanical properties of metallic materials:
   a) How does the APF affect the overall mechanical properties of metals?
   b) Assuming two identical polycrystalline compositions, which material will perform with a higher strength: FCC crystal structure or BCC crystal structure? Explain.
   c) How do the presence and quantity of slip-plane systems correlate with ductility in metallic materials?
4. What are the main effects of the introduction of grain boundaries in polycrystalline metallic materials? Describe the mechanism of grain boundary strengthening.
5. Additive manufacturing has brought the possibility to print metallic materials for the fabrication of biomedical devices. Selective laser melting provides local fast heating and rapid cooling, inducing the formation of unique microstructures (fine-grain structures). However, the process also brings up the formation of defects such as unmelted powders, microcrack- entrapped gas pores, and rough surfaces. Discuss the extent of the long-term outcomes of these defects post-implantation.
6. From the stress-strain curve provided here:

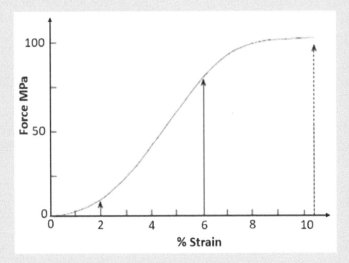

a) Identify the different regions (elastic, plastic, fracture).

b) Calculate:

- Yield stress
- Ultimate stress
- Young's modulus

c) Assuming this material is implanted as a joint replacement in bone tissue, is it expected to induce stress shielding after implantation? Explain.

7. As an R&D engineer, your input is requested to design an enhanced device that will be permanently implanted to repair the load-bearing function of a joint. Details on the candidate metals are presented here:

Material	Density (ρ) (g/cm^3)	Yield stress (MPa)	Maximum elongation (%)	Elastic modulus (GPa)
SS 316 type	8	190	40	190
Pure Ti	4.5	140	15	110
Ti6Al4V (F136)	4	795	10	114

a) What force can each of the materials carry without permanent deformation?

b) Which one offers better properties as a load-bearing implant? Explain.

c) Describe the primary consequences of choosing the wrong material; consider in your answer the effect of density, yield stress, elongation, and Young's modulus on the implant's performance.

8. You are requested to determine the degree of plastic deformation of a promising new biomedical alloy. If implanted, the alloy will be under significant tensile stress. What material properties would you analyze?

9. Ti and Ta are used for implantable biomaterials due to their high corrosion resistance. Why are these materials especially stable? Describe the mechanisms and desired conditions to maintain stability under physiological conditions.

10. From the following metallic couples, which one is more likely to corrode?

a) Mg or Ca

b) Au or Ti

c) Li or Pt

d) Fe or Zr

e) Predict the maximum oxidation states of Pb, Cr, Mo, Rh, Zr, Zn, Mg, Hg, and Fe.

f) What are the main biological factors inducing corrosion?

11. If a sample of silver and a sample of iron come into contact, the iron corrodes but the silver does not. If a sample of silver comes into contact with a sample of gold, the silver corrodes but the gold does not. Explain this phenomenon.

12. A new device will be implanted, and decisions need to be made regarding the optimal material to be used for cortical bone fixation (tissues are expected to be under an average stress of 100 MPa).

Tissue/material	Density (ρ) (g/cm^3)	Yield stress (MPa)	Maximum elongation (%)	Elastic modulus (GPa)
Cortical bone	2	105	—	10
SS 316 type	8	190	40	190
Pure Ti	4.5	140	15	116
Ti6Al4V (F136)	4	795	10	114
CoCrWNi (F90)	8.5	310	20	200

a) What alloy would you use? Explain your rationale.
b) If corrosion is a big concern and tensile stress is only 10 MPa in the implanted area, would your decision change? Why?
c) The implant is expected to be removed shortly after healing. What material would provide favorable conditions to avoid excessive tissue damage during the removal?

13. Using standard potentials and molarity for ion concentrations, calculate the open circuit potential of the following electrochemical reactions (balance the equations with water-related chemical species when necessary, i.e., H^+, OH^-, and H_2O):

a) $H_2O_2 + Ni \rightarrow H_2O + Ni^{2+}$
b) $H_2O + Mg^{2+} \rightarrow H_2O_2 + Mg$
c) $Al^{3+} + OH^- \rightarrow Al + O_2$
d) $Ni + PbO_2 \rightarrow Pb^{2+} + Ni^{2+}$

14. Propose one alloy modification that will possibly decrease the corrosion risk, and another to enhance the corrosion (oxidative response).

15. What type of corrosion is localized in a specific area of the metallic surface while the rest of the passivation layer is apparently intact? Why is this corrosion much more prone to provoke a catastrophic failure than uniform corrosion generally does? Explain two possible scenarios that might induce this corrosion in metallic implantable devices.

16. Ni-Ti alloy, also known as nitinol, displays a shape-memory behavior (ability to return to its initial shape by temperature changes).
a) What type of implants would you consider a good target to use this material instead of traditional alloys?
b) Explain the microstructure transformations occurring during heat treatment and limitations on the implantation of these alloys.

17. Clinical studies have reviewed the failure of Ti-alloy implants due to corrosion. What types of corrosion might have led to the failure of these implants? How would you prevent this from happening?

18. Titanium and iron create a galvanic cell after implantation under physiological conditions. What pH range will favor this reaction to happen? What would be the cell potential?
19. An orangutan raised in captivity falls from a high tree. It requires a hip prosthesis due to a fractured femur caused by the impact. While the anatomy of the patient is very similar to humans, laboratory analyses indicate unique metabolic processes and a different chemical environment (pH, temperature). Would you implant a material based on clinical data collected from humans? How would you go about selecting a metal or an alloy given the available knowledge of corrosion in varying environmental conditions?

References

1. Prasad K, Bazaka O, Chua M, Rochford M, Fedrick L, Spoor J, Symes R, Tieppo M, Collins C, Cao A, et al. Metallic biomaterials: current challenges and opportunities. *Materials*. 2017;10(8):1–33.
2. Tibbitt MW, Rodell CB, Burdick JA, Anseth KS. Progress in material design for biomedical applications. *Proc Natl Acad Sci U S A*. 2015;112:14444–14451.
3. Uchihara Y, Grammatopoulos G, Munemoto M, Matharu G, Inagaki Y, Pandit H, Tanaka Y, Athanasou NA. Implant failure in bilateral metal-on-metal hip resurfacing arthroplasties: a clinical and pathological study. *Journal of Materials Science: Materials in Medicine*. 2018;29(3):1–7.
4. Minnath MA. Metals and alloys for biomedical applications. In: Balakrishnan P, Sreekala MS, Sabu T, eds. *Fundamental Biomaterials: Metals*. Duxford, UK: Elsevier; 2018:167–174.
5. Nag S, Banerjee R, Fraser HL. A novel combinatorial approach for understanding microstructural evolution and its relationship to mechanical properties in metallic biomaterials. *Acta Biomater*. 2007;3:369–376.
6. Pileni MP. Impact of the metallic crystalline structure on the properties of nanocrystals and their mesoscopic assemblies. *Acc Chem Res*. 2017;50:1946–1955.
7. Beriha B, Sahoo UC, Steyn WJM. Determination of endurance limit for different bound materials used in pavements: a review. *Int J Transp Sci Technol*. 2019;8:263–279.
8. Roitsch S, Heggen M, Lipińska-Chwałek M, Feuerbacher M. Single-crystal plasticity of the complex metallic alloy phase β-Al–Mg. *Intermetallics*. 2007;15:833–837.
9. Bahl S, Shreyas P, Trishul MA, Suwas S, Chatterjee K. Enhancing the mechanical and biological performance of a metallic biomaterial for orthopedic applications through changes in the surface oxide layer by nanocrystalline surface modification. *Nanoscale*. 2015;7:7704–7716.
10. Shayan M, Padmanabhan J, Morris AH, et al. Nanopatterned bulk metallic glass-based biomaterials modulate macrophage polarization. *Acta Biomater*. 2018;75:427–438.
11. Silva-Bermudez P, Rodil SE. An overview of protein adsorption on metal oxide coatings for biomedical implants. *Surf Coating Technol*. 2013;233:147–158.

12. Hanawa T. Research and development of metals for medical devices based on clinical needs. *Sci Technol Adv Mater.* 2012;13(6):1—15.

13. Wolford ML, Palso K, Bercovitz A, Survey HD. NCHS Data Brief. *Hospitalization for Total Hip Replacement Among Inpatients Aged 45 and Over: United States, 2000—2010.* 2015;186(February):1—7.

14. Badilatti SD, Christen P, Parkinson I, Müller R. Load-adaptive bone remodeling simulations reveal osteoporotic microstructural and mechanical changes in whole human vertebrae. *J Biomech.* 2016;49:3770—3779.

15. Niinomi M, Nakai M. Titanium-based biomaterials for preventing stress shielding between implant devices and bone. *Int J Biomater.* 2011;2011(836587):1—10.

16. Walley KC, Bajraliu M, Gonzalez T, Nazarian A. The chronicle of a stainless steel orthopaedic implant. *Orthop J Harv Med Sch.* 2016;17:68—74.

17. Eliaz N. Corrosion of metallic biomaterials: a review. *Materials.* 2019;12(407):1—91.

18. Noumbissi S, Scarano A, Gupta S. A literature review study on atomic ions dissolution of titanium and its alloys in implant dentistry. *Materials.* 2019;12(368):1—15.

19. Örnek C, Leygraf C, Pan J. Passive film characterisation of duplex stainless steel using scanning Kelvin probe force microscopy in combination with electrochemical measurements. *npj Mater Degrad.* 2019;1:1—8.

20. Geringer J, MacDonald DD. Modeling fretting-corrosion wear of 316L SS against poly (methyl methacrylate) with the point defect model: fundamental theory, assessment, and outlook. *Electrochim Acta.* 2012;79:17—30.

21. Mellado-Valero A, Muñoz AI, Pina VG, Sola-Ruiz MF. Electrochemical behaviour and galvanic effects of titanium implants coupled to metallic suprastructures in artificial saliva. *Materials.* 2018;11(1):1—19.

22. Pourbaix M. Electrochemical corrosion of metallic biomaterials. *Biomaterials.* 1984;5:122—134.

23. Grzegorczyn S, Ślezak A, Przywara-Chowaniec B. Concentration polarization phenomenon in the case of mechanical pressure difference on the membrane. *J Biol Phys.* 2017;43:225—238.

24. Saini M, Singh Y, Arora P, Arora V, Jain K. Implant biomaterials: a comprehensive review. *World J Clin Cases.* 2015;3:52—57.

25. McNaney JM, Imbeni V, Jung Y, Papadopoulos P, Ritchie RO. An experimental study of the superelastic effect in a shape-memory nitinol alloy under biaxial loading. *Mech Mater.* 2003;35:969—986.

26. Balla VK, Bodhak S, Bose S, Bandyopadhyay A. Porous tantalum structures for bone implants: fabrication, mechanical and in vitro biological properties. *Acta Biomater.* 2010;6:3349—3359.

27. Kumar CSSR, Mohammad F. Magnetic nanomaterials for hyperthermia-based therapy and controlled drug delivery. *Adv Drug Deliv Rev.* 2011;63:789—808.

28. Bagot PAJ, Kruska K, Haley D, et al. Oxidation and surface segregation behavior of a Pt—Pd—Rh alloy catalyst. *J Phys Chem C.* 2014;118:26130—26138.

29. Kirmanidou Y, Sidira M, Drosou ME, et al. New Ti-alloys and surface modifications to improve the mechanical properties and the biological response to orthopedic and dental implants: a review. *BioMed Res Int.* 2016;2016(2908570):1—21.

30. Kastenhuber M, Rashkova B, Clemens H, Mayer S. Enhancement of creep properties and microstructural stability of intermetallic β-solidifying γ-TiAl based alloys. *Intermetallics.* 2015;63:19—26.

31. Ahmed W, Jackson M. In: Mark J. J, Waqar A, eds. *Surface Engineered Surgical Tools and Medical Devices*. Boston, MA: Springer; 2007:595.
32. Geetha M, Singh AK, Asokamani R, Gogia AK. Ti based biomaterials, the ultimate choice for orthopaedic implants—a review. *Prog Mater Sci*. 2009;54:397−425.
33. Liu X, Chen S, Tsoi JKH, Matinlinna JP. Binary titanium alloys as dental implant materials—a review. *Regen Biomater*. 2017;4:315−323.
34. Uhthoff HK, Poitras P, Backman DS. Internal plate fixation of fractures: short history and recent developments. *J Orthop Sci*. 2006;11:118−126.
35. Yang K, Ren Y. Nickel-free austenitic stainless steels for medical applications. *Sci Technol Adv Mater*. 2010;11(014105):1−13.
36. Kassapidou M, Franke Stenport V, Hjalmarsson L, Johansson CB. Cobalt-chromium alloys in fixed prosthodontics in Sweden. *Acta Biomater Odontol Scand*. 2017;3:53−62.
37. Kuznetsov VV, Filatova EA, Telezhkina AV, Kruglikov SS. Corrosion resistance of Co−Cr−W coatings obtained by electrodeposition. *J Solid State Electrochem*. 2018;22:2267−2276.
38. Navarro M, Michiardi A, Castaño O, Planell JA. Biomaterials in orthopaedics. *J R Soc Interface*. 2008;5:1137−1158.
39. Yamanaka K, Mori M, Sato K, Chiba A. Characterisation of nanoscale carbide precipitation in as-cast Co−Cr−W-based dental alloys. *J Mater Chem B*. 2016;4:1778−1786.
40. Wang H, Feng Q, Li N, Xu S. Evaluation of metal-ceramic bond characteristics of three dental Co-Cr alloys prepared with different fabrication techniques. *J Prosthet Dent*. 2016;116:916−923.
41. Wang J, Li JY, Zhang Y, Yu WM. Effects of the addition of micro-amounts of calcium on the corrosion resistance of Mg-0.1Mn-1.0Zn-XCa biomaterials. *J Mater Eng Perform*. 2019;28(3):1553−1562.
42. Shi L, Northwood DO, Cao Z. Alloy design and microstructure of a biomedical Co-Cr alloy. *J Mater Sci*. 1993;28:1312−1316.
43. Petrov YN, Prokopenko GI, Mordyuk BN, et al. Influence of microstructural modifications induced by ultrasonic impact treatment on hardening and corrosion behavior of wrought Co-Cr-Mo biomedical alloy. *Mater Sci Eng C*. 2016;58:1024−1035.
44. Vanysek P. Electrochemical series. In: Haynes WM, ed. *CRC Handbook of Chemistry and Physics*. Boca Raton, FL: Taylor & Francis; 2015:8−21.

Bioceramics

Learning objectives

This chapter offers a broad overview of ceramic materials and their applications in the biomedical industry, with a focus on hard tissue replacement. At the end of the chapter, the reader will:

- Become familiar with what ceramic materials are and their main structure and properties.
- Learn about the type of ceramics that have the appropriate characteristics to make them useable as part of hard tissue replacement devices.
- Understand the chemical, mechanical, microstructural, and surface properties of these bioceramics in the context of their applications in medicine.
- Gain a working knowledge of how bioceramics work in biomedical devices and what properties are significant and have to be considered in biomaterial selection for a specific application.

Ceramics in biomedical devices

Ceramics[1] are another major class of materials that are used as components of biomedical devices. Implantable bioceramics should meet several criteria to be appropriately used. They should be nontoxic, noncarcinogenic, nonallergenic, noninflammatory, and functional in vivo (in the body) for the lifetime of the implant.

Ceramic materials[2] are generally chemically stable oxides or metallic salts that have high chemical inertness in body fluids. Most ceramics used in the biomedical industry are ionic compounds, but there are also covalent ceramics (e.g., carbides, nitrides), albeit not used for biomedical implants. Ceramics are known for their high compressive strength, which is an advantage for their use in applications where high mechanical loads are expected, such as in orthopedic devices.

Another advantage of ceramics is their ease of fabrication. The most conventionally used method for fabricating ceramic prosthesis is the powder, or slurry, technique. In this method, a green body of the part intended to be used in a prosthesis is subjected to a heating regimen that will lead to the formation of a sintered body, where most pores are eliminated, and the mechanical strength is sufficient

Introductory Biomaterials. https://doi.org/10.1016/B978-0-12-809263-7.00004-4

for the intended application, as illustrated in Fig. 4.1. Steps such as surface treatment and cooling are also part of this procedure. Examples of ceramics are: silicates, metal oxides (alumina, magnesium oxide, etc.), carbides such as silicon carbide, and ionic salts such as sodium or cesium chloride and zinc sulfide.

Ceramic powders can be sintered to form a larger, solid product. Sintering[3] involves high temperatures that are used to agglomerate small particles into bulk materials. Sintering is conducted at temperatures below the melting point of the material being processed, to avoid formation of liquids. The applied heat and pressure cause adjacent particle surfaces to form a single grain boundary. Points of contact between particles form due to atomic diffusion. The process is accompanied by shrinkage and a reduction in porosity. Sintering is the common method of applying porous ceramic coatings to implants. The porosity of the as-processed material is controlled by the time and temperature of the heat treatment.

There are some disadvantages to the use of ceramics in prostheses. One of them is that they do not undergo plastic shear deformation at room temperature. This makes ceramics susceptible to crack propagation (Fig. 4.2). They tend to fracture instead of yielding or undergoing plastic deformation. Thus ceramics have low tensile strength if they are not processed to be flaw-free. However, if processed to produce a flaw-free structure, tensile strength increases substantially. For example, glass fibers are twice as strong as high-strength steel. The cause of low tensile strength in ceramics is notch sensitivity. A notch or crack is a stress-raiser. That means that the effect of the crack presence is greater than would be expected based on reduced cross-sectional area.

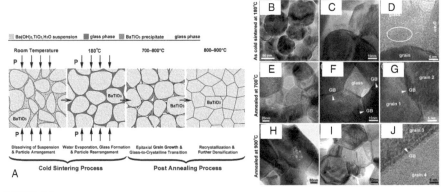

FIGURE 4.1

Cold sintering of ceramics: Illustration of the hydrothermal-assisted cold sintering of $BaTiO_3$ nanocrystal ceramics (A). Transmission electron microscopy microstructure of the evolution of $BaTiO_3$ ceramics after cold sintering at 180°C (B–D), annealing at 700°C (E–G), and annealing at 900°C (H–J).

Reproduced with permission from Guo et al.[8]

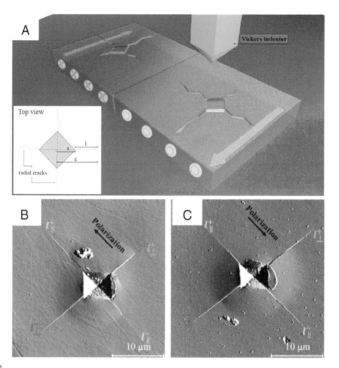

FIGURE 4.2

Schematic of a crack and crack propagation in materials. Top view of typical radial crack propagation for indentation fracture toughness measurement with corresponding crack and diagonal lengths (A). Atomic force microscopy (AFM) topography showing crack propagation and induced polarization (B and C).

Reproduced with permission from Cordero-Edwards et al.[9]

Measured stress at tip (S_m) exceeds the nominal tensile stress (s):

$$S_m = 2s(c/r)^{1/2} \qquad (4.1)$$

where c is the crack depth and r the crack tip radius (defined as the major radius of an equivalent ellipse). Sharper cracks have a smaller crack radius, thus a higher measured stress at tip, and greater stress concentration. In ductile materials, when s exceeds the yield strength, the notch tip will widen (radius increases) and the stress concentration decreases.

Notch sensitivity refers to how much a material is sensitive to developed notches under cyclic loading and is measured by the notch sensitivity factor, q.

$$q = \frac{K_f - 1}{K_t - 1} \qquad (4.2)$$

where K_f is the fatigue stress concentration factor and K_t is the theoretical stress concentration factor.

Because brittle materials have higher notch sensitivity, bioceramics will be more likely to encounter this problem compared with more ductile materials such as metals.

Example 4.1

For notch sensitivity in materials, choose among the options below and explain your answer:
a) A value of $q = 1$ indicates full notch sensitivity
b) A value of $q = 0$ indicates no notch sensitivity
c) a) and b) are both valid answers
d) None of the above

Answer Both a) and b) are correct. When the notch sensitivity factor q is equal to 1, the fatigue stress concentration factor is equal to the theoretical one, which makes the material fully notch sensitive. On the contrary, when $q = 0$, there is no difference between the sensitivity between notched and unnotched materials.

Structure of ceramics

The next topic we will discuss is a general view of the ceramic structure. Ceramics are composed of two or more elements with the general formula A_mX_n, where A is a metal and X is a nonmetallic element, while m and n are integers. The simplest ceramic structure is of the form AX (Fig. 4.3). Generally, ceramics have more complex structures than metals. They can have covalent or ionic bonds and differences in electronegativities of the composing elements. Ceramics are crystalline in nature

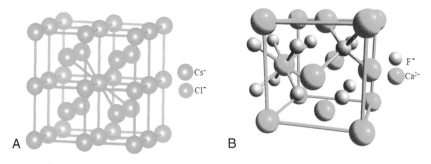

A   B

FIGURE 4.3

AX structures of ceramics. Illustration of an AX structure: CsCl crystal structure (A). Illustration of an AX_2 structure: CaF_2 crystal structure (B).

Reproduced with permission from Chaviano.[10]

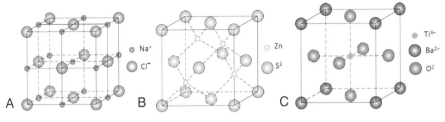

FIGURE 4.4

Examples of crystal structures of ceramic materials. The unit cell of rock salt or NaCl crystal structure (A), unit cell of zinc blende (ZnS) crystal structure (B), and unit cell of perovskite crystal structure (C).

Reproduced with permission from Callister and Rethwisch.[1]

and their crystals are electrically neutral. In AX structures, if the size of the ions is approximately equal, the structure becomes simple cubic (e.g., CsCl). However, if the sizes of the component ions are quite different, a face-centered cubic (FCC) structure is created (positive ions can be fitted in the tetragonal or octagonal spaces among larger negative ions) (Fig. 4.4).

Bioceramics in implantable devices

Properties of bioceramics

The criteria for the use of a ceramic in implanted biomedical devices include requirements for the material to be nontoxic, noncarcinogenic, nonallergenic, and noninflammatory, and have adequate tissue response and mechanical properties, as well as the ability to be biofunctional for the lifetime of the implant. Based on their properties, bioceramics can be classified as bioactive, resorbable, and bioinert.

Bioactive materials, such as bioglass and high-temperature sintered hydroxyapatite (HA), have the ability to form a direct bond with the tissue, as well as have a positive effect on postoperative healing. Some bioceramics are biodegradable, such as calcium phosphates and low-temperature sintered HA, and will disappear under physiological conditions, generally to give way for new bone formation. Finally, other ceramics such as alumina and zirconia are said to be bioinert. These do not elicit any tissue response, either positive (bioactivity, biodegradability) or negative (toxicity, inflammation, carcinogenicity). While this is a generally accepted term, in reality, under the harsh body environment, which contains a variety of ions, proteins, and blood components, no material is fully bioinert and will be impacted under normal implant operating conditions. However, bioinert ceramics tend to be affected to only a lesser extent by in vivo chemical and biological attack and usually offer good prognosis of high resistance when used as part of an implanted biomedical system, for example, in a total hip replacement device. While these ceramics do not form a direct bond with the bone, the body is encasing them into a very thin (a few molecular layers) layer of fibrous tissue, which separates them from other surrounding tissues.

As opposed to bioinert ceramics, their bioactive counterparts do not present with this fibrous separating layer at the interface with the surrounding tissues and just directly bind to them. This is a desirable effect because such direct bonding offers the advantage of good stability of the implant-tissue integration, leading to long-term adequate functioning of the biomedical implant in which these bioceramics are being used.

Resorbable ceramics, on the other hand, gradually degrade and resorb in the living host after implantation, leaving behind new tissue free of artificial materials. An example is tricalcium phosphate (TCP), which is used for repairing bone defects and is gradually resorbed at the same speed with new bone tissue formation. High surface area, high porosity, as well as low crystallinity and low grain size tend to favor an increased speed of resorption. This knowledge can be used to tune the resorption velocity when designing bioresorbable ceramics within a range of desirable resorption velocity.

Because of their brittleness and high elastic modulus, bioceramics are not generally used for load-bearing implants, such as the stem of a hip implant. Bioactive ceramics such as HA, however, are commonly used to coat metallic hip implant stems in order to promote superior tissue-implant interactions.[4] More commonly, ceramics are used in dentistry, mainly owing to their high compressive strength and wear resistance. In orthopedics, bioceramics are used in certain parts of joint replacement implants (e.g., the acetabular cup in hip replacement devices).

There is no common standard for evaluation of some ceramic properties, such as bioactivity and clinical usability. It is very difficult to translate the in vitro (i.e., in the test tube or under the laboratory conditions) behavior of ceramics to true in vivo (in the body) conditions. For example, some ceramic materials, such as alumina (Al_2O_3) and zirconia (ZrO_2), are difficult to dissolve in aqueous solutions outside the body and also normally present with high wear resistance. Wear resistance, as the name suggests, is the ability of a material to resist loss from its surface in response to the mechanical forces being applied to it. However, the very same ceramics can degrade when used in vivo in implants. A combination of the aggressive biological environment and mechanical forces being continuously applied to such ceramics during normal operation of the implant can lead to the formation of microparticles, also called wear debris. An example would be an alumina acetabular cup that is used in conjunction with a titanium alloy stem in a total hip joint replacement (also called hip implant) that could degrade and produce such microparticles. Because of the body's immune response, these microparticles can be recognized as a foreign body and elicit strong toxicity, although the material in itself may not be chemically toxic. At the same time, such wear of a ceramic component, such as alumina or zirconia, may lead to crack propagation and ultimate device failure.

We will learn in Chapter 7 (Biocomposite materials) that the main mineral component of the bone tissue is a combination of amorphous calcium phosphate, together with a crystalline calcium phosphate, with the chemical formula $Ca_{10}(PO_4)_6(OH)_2$, HA. Bioactive ceramics have the ability to form a direct bond with the living tissue and can promote bone tissue growth when part of a bone inter-facing implant intended to repair or replace hard tissue. The most used in vitro method to assess bioactivity of ceramics is the Kokubo method.[5,6] In this method, simulated body fluid (SBF), a solution of equal concentrations of inorganic minerals as found in blood plasma, is used for testing. This is termed the *Kokubo solution.*[7] When immersed in the Kokubo solution, ceramics tend to form apatite on their sur-faces, which is beneficial for osseointegration (formation of a direct bond with living bone tissue) when these biomaterials are used in orthopedic implants. Although there is still debate resulting in other solutions being proposed to mimic blood plasma, the original Kokubo solution is still the most well known and often used in practice. This solution contains sodium, magnesium, potassium, and calcium ions, as well as chorine, hydrocarbonate, phosphate, and sulfate, and is balanced at a pH of 7.4.

Some examples of bioceramics that will be discussed further in this chapter are HA, calcium phosphates, alumina, and zirconia. Their properties and therefore ap-plications differ markedly. Table 4.1 summarizes the properties of these bioceramics and can serve as a guide when reflecting on how they may be applied for a particular application in surgery.

Hydroxyapatite

One of the most used and well-known bioceramics is HA. It is a phosphate from the apatite family and its chemical formula is $Ca_5(PO_4)_3(OH)$, often written as $Ca_{10}(PO_4)_6(OH)_2$ to indicate two units in the basic crystal structure (Fig. 4.5). It is the mineral component of hard tissues such as bone, dentine, and dental enamel, which are natural composites that contain HA, as well as protein, other organic materials, and water. HA displays excellent biocompatibility, with an ability to form a direct chemical bond with hard tissues (supports bone growth and osseoin-tegration: formation of a direct bond with living bone tissue). Because of these properties, it can be used as artificial bone, in implant construction, as well as implant coatings. It can be obtained either through processing of sea corral or syn-thetic production.

Table 4.1 Mechanical properties of bioceramics commonly used in biomedical implants.

Bioceramic	Young's modulus (GPa)	Tensile strength (MPa)	Compressive strength (MPa)
Alumina	380	350	4500
Zirconia	150–200		
Calcium phosphates	40–117	69–193	510–896
Glass ceramics	22	56–83	500

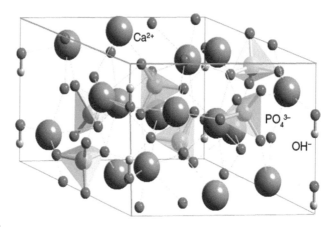

FIGURE 4.5

The crystal structure of hydroxyapatite.

Reproduced with permission from Greeves.[11]

HA has two main crystalline forms, apatite and b-witlockite depending on the Ca/P ratio, presence of water, and impurities. The apatite variety (hydroxyl- or hydroxy-) will more likely form at temperatures below 900°C. Both forms are tissue compatible, with the apatite form more closely related to the mineral of bone and teeth than b-witlockite.

Polycrystalline HA has a Young's modulus in the range of 40 to 117 GPa and a density of 3.16 g/cm³. It has, however, a lower fracture toughness than that of bone. It is highly tissue compatible (i.e., does not elicit negative tissue effects, such as inflammation) and is able to form a direct bond with hard tissues. It does not induce bone growth, however. Instead, it can act as a guide and framework for bone growth. Both the dense and porous forms of HA can be used in different ways in prostheses (Fig. 4.6).

Porous HA is used as a scaffolding for bone repair due to the suspected ability to gradually resorb and convert to natural bone tissue when implanted (Fig. 4.7). The dense HA has less tendency for conversion to natural bone than the porous version but displays better mechanical properties. This is because porosity is necessary to ensure bone cells can fit through the scaffold, and then act on producing bone and resorbing the porous ceramic scaffold.

There are related ceramic compounds that present with very high biocompatibility and also biodegradability. This is useful when temporary prosthetics are needed, or to use for filling small hard-tissue defects or damage due to disease or trauma. Another use could be found in drug delivery devices. Examples of bioresorbable ceramics include aluminum-calcium-phosphorous oxides, ferric-calcium-phosphorous oxides, TCPs, zinc-calcium-phosphorous oxides, zinc-sulfate-calcium-phosphorous oxides, and calcium sulfates.

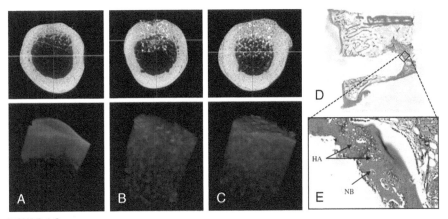

FIGURE 4.6

Osteogenesis induced by implanted hydroxyapatite *(HA)* composites. Axial and three-dimensional images of micro—computed tomography of the empty defect (without HA) (A), control HA without spheres (B), and poly(lactic-co-glycolic acid) (PLGA) microspheres immobilized on HA scaffold (10 weeks postimplantation) in beagle femur (5-mm drill hole defect) (C). Histological slides showing bone integration (D). HA particles surrounded by new bone *(NB)* (E).

Reproduced with permission from Zampelis et al.[12] and Son et al.[13]

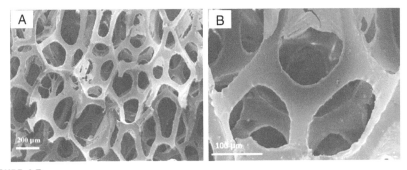

FIGURE 4.7

Porous hydroxyapatite (HA) scaffold with 90% porosity. Scanning electron microscopy (SEM) images of scaffold synthesis by polymeric sponge method using sol-gel—derived HA powder (A) and polymer slurry (B).

Reproduced with permission from Tripathi and Basu.[14]

Calcium phosphates

There are four forms of TCPs, with the chemical formula $Ca_3(PO_4)_2$, and while all are bioactive and biodegradable, they have different degrees of degradation speed in vivo, depending on their phase, level of crystallinity, and microstructure. These are the α-TCP, the β-TCP, the amorphous TCP, and the apatitic TCP. HA, with the formula $Ca_{10}(PO_4)_6(OH)_2$ when fully dense and crystalline, has a significantly lower degradation rate than that of any TCP and can be converted into calcium phosphate via hydrolysis. The crystalline forms of TCP, α and β, are high-temperature phases and have a level of crystallinity that closely mimics the calcium phosphate mineral component of bone tissue. This is one of the reasons why TCPs have been good candidates for being used as bone substitutes in skeletal repair. Other reasons include their ability to fully integrate with the living tissue owing to their bioactivity. Upon their biodegradation, TCPs promote new bone growth, a property called osteoconductivity.

The amorphous and apatitic TCP are unstable phases. Amorphous and α-TCP forms biodegrade at the highest speed among all forms of TCP. As mentioned earlier in the chapter, the lower the crystallinity of a phase, the higher the dissolution rate. Factors accelerating it also include a fine grain size in the crystalline forms and a high surface area. Thus nanopowders would degrade the fastest, while fully dense, nonporous crystalline forms of calcium phosphate or HA will degrade the slowest.

In an experiment observing the effects of implantation of β-TCP granules with diameters between 100 and 300 μm in bone tissue where a hole has been previously drilled, backscattered scanning electron microscopy (SEM) images were taken after 3 weeks and again after 12 weeks postimplantation to observe the effect of these materials on bone growth.

After 3 weeks of implantation, new bone tissue has been formed, as observed in the SEM image in Fig. 4.8. There were no observed negative tissue reactions. After

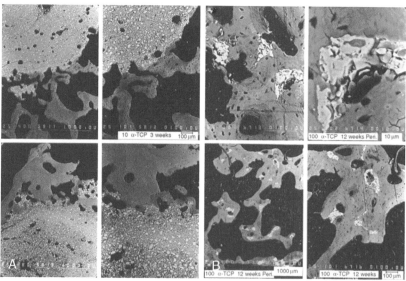

FIGURE 4.8

Backscattered scanning electron microscopy images of bone growth on α-tricalcium phosphate (α-TCP) granules after 3 weeks (A) and 12 weeks (B) post-implantation into a bone defect from bioceramics and their clinical applications.

Reproduced with permission from Oonishi et al.[15]

12 weeks post-implantation, the α-TCP granules were largely replaced by new bone (Fig. 4.8). These experiments showcase the biodegradability of TCPs, as well as their positive tissue interactions and osteoconductive properties.

Both α and β phases of TCP present with similar biodegradable and osteoconductive behavior when used for bone defect repair. Other calcium phosphate phases in addition to α- and β-TCP include tetracalcium phosphate and octacalcium phosphate. They are all bioresorbable ceramics that resorb by in vivo solution-mediated dissolution, as well as by a process called cell-mediated resorption, where cellular components of the bone tissue contribute to calcium phosphate material resorption followed by new bone deposition by bone-forming cells called osteoblasts.

The drawbacks of using calcium phosphates are their poor fatigue resistance and brittleness, which are exacerbated in systems where high porosity is required. Since pores with an average diameter of over 100 μm are necessary both for vascularization and for cells to proliferate and produce new tissue for most of the orthopedic applications, this severely limits the usability of these materials for any load-bearing applications. However, they can be used as coatings, bone defect fillers, or guiding structures (scaffolds) for tissue growth in tissue engineering. The specific applications of such bioceramics will be discussed in more detail in the following chapters.

Alumina

The chemical formula of alumina is Al_2O_3 and its crystal structure is shown in Fig. 4.9. To be able to use alumina in prostheses, a purity of over 99.5% is required. Other property requirements include compressive strength between 4 and 5 GPa and a flexural strength over 400 MPa. The elastic modulus of alumina is 380 GPa, and density is between 3.8 and 3.9 g/cm^3. Alumina is generally a quite hard ceramic, with a Mohs hardness of 9 (compare with the diamond's hardness of 10 and talc's hardness of 1).

For biomedical devices, alumina has the advantage of low friction and wear properties, making it appropriate for use in joint bearings. When the components of an artificial joint are in operation, the repetitive movements involve continuous contact and friction between them. Concerns that result are the production of wear particles, as well as damage to the components and improper functioning of the prosthesis. Alumina is used in hip ball construction (occasionally socket—acetabular cup) (Fig. 4.10), porous coatings, and dental applications such as crowns. When used as a bone interface, bone will grow right up to the alumina and seal but will not ingrow into the bulk.

Alumina can be manufactured in single crystal and polycrystalline forms. Single crystals up to 10 cm in diameter have been formed by feeding alumina powders onto a seed crystal under extreme heat. Like with other materials, the strength of polycrystalline alumina depends on porosity and grain size. Strength increases as porosity decreases and as grain size decreases (at normal temperatures). The reason behind this is that grain boundaries interfere with the progression of dislocations

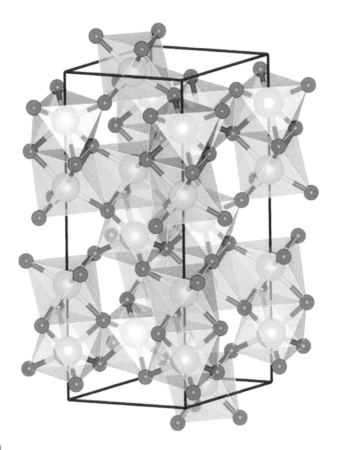

FIGURE 4.9

Crystal structure of alumina (Al_2O_3).

Reproduced with permission from Royal Society of Chemistry.[16]

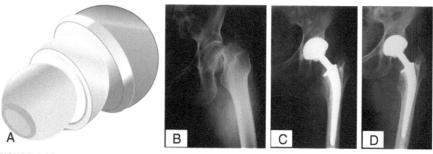

FIGURE 4.10

Alumina hip ball and socket. Illustration of a prosthetic hip ball showing a pink ceramic femoral head, a yellow polymeric liner, and a metal acetabular cup (A). Total hip arthroplasty: preoperatively (B), postoperatively (B), and at 9 years of follow-up (D).

Images B–D reproduced with permission from Solarino et al.[17]

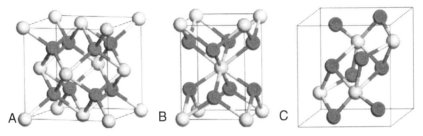

FIGURE 4.11

Crystal structures of zirconia (ZrO_2): cubic (A), tetragonal (B), and monoclinic (C). Red and blue atoms correspond to zirconia and oxygen, respectively.

Reproduced with permission from Han and Zhu.[18]

Zirconia

Yttria-stabilized zirconia is used for fabricating the femoral head of total hip prostheses. The cubic structure of zirconia belongs to the fluorite structure (Fig. 4.11A). Zirconia has some advantages over alumina, such as finer grain size and well-controlled microstructure, with no residual porosity and higher fracture strength and toughness.

The wear properties of biomaterials are usually evaluated against ultra-high-molecular-weight polyethylene (UHMWPE) and measured by the wear factor:

$$\text{Wear factor} = \left[\frac{\text{Wear volume } (mm^3)}{\text{Load(N)} \cdot \text{sliding distance (m)}} \right] \tag{4.3}$$

The wear factor has a smaller value for zirconia compared with alumina and 316L. The smaller the wear factor is, the better the wear and friction properties are. That indicates excellent wear and friction properties for zirconia, making it appropriate for wear bearings in orthopedic implants and dental applications. Although the higher the strength and toughness are, the small the wear factor is expected to be, processing conditions, geometric conditions, and microstructure can affect the wear properties of bioceramics in general.

Example 4.2

An alumina ceramic sample has been reinforced with 10 wt% SiC microparticles, thus increasing its fracture toughness. How do you expect the wear factor to be influenced? Should this composite be used in acetabular cups?

Answer Reinforcing alumina with the tougher SiC ceramic is expected to slow down crack initiation and increase the toughness of the resulting ceramic composite. It is expected that the wear volume would be smaller for Al_2O_3-SiC compared with pure Al_2O_3. However, achieving the lowest wear factor is not the only criterion for material selection for implantation. Before selecting the composite for use in implantation, testing that measures its toxicity, carcinogenicity, osseointegration properties, as well as in vivo studies would be necessary.

Glass ceramics

Glass ceramics are polycrystalline ceramics made by controlled crystallization of glasses. They have properties that place them between ceramics and glasses, with at least one glassy and one crystalline phase in their composition. Their most attractive property is their surface bioactivity, which makes them of interest for use as coatings on prostheses. Upon implantation in the host, these surface-reactive materials form strong bonds with adjacent tissue. These materials were first discovered in 1953 by researchers at Corning Inc., USA. The microstructure of glass ceramics shows a much smaller grain size than for conventional ceramics (Fig. 4.12).

To understand the crystallization of glass ceramics, one has to understand the thermal transitions that occur in response to subjecting a material to thermal treatment. A critical parameter for understanding the formation of glass ceramics is the concept of glass transition temperature, T_g. Like the name suggests, T_g is the temperature under which there is a slowing down of molecular motion and there is no structural relaxation. Below the glass transition, the previously molten liquid solidifies into an amorphous glass solid. The glass is formed by the rapid quenching of the molten liquid below T_g. Controlled heating of the solid glass material to a temperature T_c can lead to crystallization, which means the component atoms arrange into an ordered and periodic structure. Further increase in temperature leads to these component atoms falling out of their periodic crystalline arrangements and

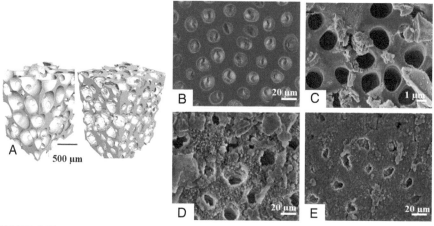

FIGURE 4.12

Bioactive glass ceramics. X-ray microtomography images of bioactive glass scaffolds (A). Induced remineralization: scanning electron microscopy images of NovaMin particles (B), human dentine untreated (C), 24 hours after application (D), and 5 days after application (E) in simulated saliva.

Reproduced with permission from Hench and Jones[19] and Jones.[20]

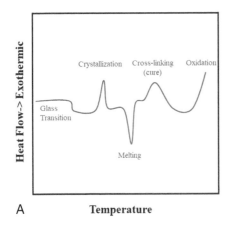
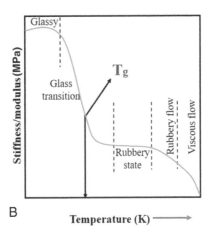

FIGURE 4.13

Example of a differential scanning calorimetry curve showing thermal phase transitions including glass transition, crystallization, and melting (A); thermal transitions encountered in glass-ceramic systems (B). T_g, Glass transition temperature.

becoming once again disordered and fluid, above the melting temperature T_m (Fig. 4.13).

Glass ceramics are of interest for applications in dentistry and orthopedics due to their ability to be processed with controlled microstructures and compositions that offer advantages of high mechanical strength and ability to form direct bonds with bone living tissue—they are bioactive.

Bioglass (Leitz, Wetzlar Co.) and Ceravital (University of Florida) are glass ceramics developed specifically for implants. Both are SiO_2-CaO-Na_2O-P_2O_5 systems with controlled compositions. As stated earlier, bioglass composition is typically manipulated to induce direct bonding with the bone. For the bonding to occur, the glass ceramics must simultaneously form a calcium phosphate- and SiO_2-rich film layer on the surface of ceramic. With correct composition, glass ceramics will bond with bone in approximately 30 days. The strength of interface between the bone and the bioglass after 6 weeks in rat femurs was found to be about the same as the strength of the bulk glass ceramic (83.3 MPa in this case).

Glass ceramics have a low thermal coefficient of expansion and their grain size can be controlled, thus allowing for stronger materials. Because they are resistant to surface damage, the tensile strength is improved compared with regular ceramics. Their resistance to scratching and abrasion is close to that of sapphire. However, the mechanical strength cannot be substantially improved due to the limited range of composition that allows induction of bone growth. That makes glass ceramics inappropriate to be used for major load-bearing implants. The best uses are for fillers, dental restorative implants, and coatings.

Summary

Ceramic materials are stable compounds with a range of properties that make some of them potentially useful for applications in surgery, for example, in dentistry or ortho-pedics. Compared with the large variety of ceramic materials in general, few of these meet the necessary requirements for mechanical properties and positive or neutral implant-tissue interactions for use in biomedical implants. Ceramics can be either inert, bioactive, or biodegradable, and the biomaterial selection is highly dependent on the specifics of the biomedical system incorporating them. These specifics include the mechanical forces expected to be acting on the implant under normal operating conditions, the level of the desired implant-tissue integration, or whether or not it is expected that they will be used in an application where dissolution of the ceramic is expected (biodegradable implant). The most used inert bioceramics include alumina and zirconia. Calcium phosphates are the class of bioceramics that are used when biodegradability is required, such as for repairing bone defects or as scaf-folds in tissue engineering. On the other hand, bioactive ceramics such as glass ce-ramics and porous HA are part of implant coatings when design with a strong implant-tissue bond is the goal.

Problems

1. What concerns would be present in a hip prosthesis where its stem would be entirely composed of HA?
2. An aluminum material is sintered in a polycrystalline form. What component would this material be most appropriate to be used for in a hip prosthesis design?
3. What tests could you design to test the tissue response of a bioceramic ma-terial when used for implantation?
4. A piece of glass has a 3-mm radius crack. When subjected to a stress of 500 MPa, calculate the stress intensity factor at the crack tip.
5. A polymeric acetabular cup was measured to have a wear factor of 37×10^{-6} mm^3/N m. An alumina acetabular cup was measured to be 6×10^{-7} mm^3/N m. Based on this information only, what material would constitute the best selection for this application?
6. Is the wear factor in alumina dependent on the grain size? Explain your answer.
7. How do you expect the toughness of a bioceramic to influence wear properties?
8. What is the most appropriate ceramic for the properties provided below (a–e): (1) Al_2O_3, (2) $Ca_{10}(PO_4)_6(OH)_2$, (3) ZrO_2, (4) diamond-like carbon, (5) pyrolitic carbon. (a) Used to make dental implants. (b) Used to coat the surface of heart valve discs. (c) Has capacity to bond directly with bone. (d) Similar to the diamond, coated on a surface. (e) Single crystal is used for making jewels.

9. A bone cement is a mixture of a solid and a liquid component. This bone cement is used in situ and shrinks upon polymerization of a monomer. The shrinkage of the bone cement is given by $S_c = (1 - P_s)S_0$, where P_s is the fraction of solid in the mixture and S_0 the shrinkage percentage of the bone cement matrix material.
    a) If the bone cement mixture undergoes a shrinkage of 0.4%, calculate the fraction of solid and liquid material in the mixture.
    b) Assume an area of bone of 5 mm × 5 mm × 5 mm is filled with bone cement with a 0.4% shrinkage. Calculate the dimensions of the bone cement after shrinkage.

10. Which properties on the list below are a characteristic of a bone cement?
    a) Viscoelasticity
    b) Absorption shock ability
    c) Osteoinductive
    d) All of the above
    e) None of the above

11. Make a comparison between alumina and zirconia in terms of microstructure and mechanical properties, including wear properties. Describe one possible implant application for each ceramic and comment on the advantages and disadvantages of each choice.

12. A Co-Cr alloy implant was coated with a bioactive glass that contained 50% SiO_2 and 2 wt% Al_2O_3. When compared with control uncoated Co-Cr implants, the bioglass coating resulted in enhanced osseointegration and bone growth. What could be the material properties that could be the root of such improved performance? Describe an experimental procedure that could be used to verify your hypothesis.

13. Is each of the following statements true or false? Explain your answer.
    a) In a ceramic fiber—reinforced ceramic matrix composite, the ceramic fiber should be inherently stronger than the matrix.
    b) In a ceramic fiber—reinforced ceramic matrix composite, the interfacial strength between the fibers and matrix should be high.
    c) The creep resistance is higher in a ductile phase-toughened ceramic matrix composite than in the ceramic matrix.
    d) The fracture toughness of a metastable zirconia particle—reinforced composite is lower than that of the matrix the composite is made up of.

References

1. Callister WD, Rethwisch DG. *Fundamentals of Materials Science and Engineering: An Integrated Approach.* 4th ed. Hoboken, NJ: Wiley; 2012; xxv, 910.
2. Chiang Y-m, Birnie DP, Kingery WD. *Physical Ceramics.* New York: J. Wiley; 1997; xiv, 522.
3. Narayan R, Colombo P, Ohji T, Wereszczak A, American Ceramic Society. *Advances in Bioceramics and Porous Ceramics: A Collection of Papers Presented at the 32nd International Conference on Advanced Ceramics and Composites, January 27-February 1, 2008, Daytona Beach, Florida.* Wiley: Hoboken, NJ, 2009; xii, 360.
4. Wu G, Huang F, Huang YP, et al. Bone inductivity comparison of control versus non-control released rhBMP2 coatings in 3D printed hydroxyapatite scaffold. *J Biomater Appl.* 2020;34(9):1254−1266.
5. Vallet-Regí M. Bio-ceramics with clinical applications. In: Vallet-Regí M, ed. *Bio-Ceramics with Clinical Applications.* West Sussex, UK: John Wiley & Sons; 2014. https://doi.org/10.1002/9781118406748.
6. Narayan R, Bose S, Bandyopadhyay A. *Biomaterials Science: Processing, Properties and Applications V.* Hoboken, NJ: John Wiley & Sons; 2015; vii, 198.
7. Kokubo T, Takadama H. How useful is SBF in predicting in vivo bone bioactivity? *Biomaterials.* 2006;27(15):2907−2915.
8. Guo HZ, Guo J, Baker A, Randall CA. Hydrothermal-assisted cold sintering process: a new guidance for low temperature ceramic sintering. *ACS Appl Mater Inter.* 2016; 8(32):20909−20915.
9. Cordero-Edwards K, Kianirad H, Canalias C, Sort J, Catalan G. Flexoelectric fracture-ratchet effect in ferroelectrics. *Phys Rev Lett.* 2019;122(13):135502.
10. Chaviano M. Structure of Solids. https://minerva.mlib.cnr.it/mod/book/tool/print/index.php?id=269%26chapterid=78; 2018.
11. Greeves N. *Hydroxyapatite $Ca_5(OH)(PO_4)_3$.* ChemTube3D; 2019. https://www.chemtube3d.com/sshydroxyapatite/.
12. Zampelis V, Tagil M, Lidgren L, Isaksson H, Atroshi I, Wang JS. The effect of a biphasic injectable bone substitute on the interface strength in a rabbit knee prosthesis model. *J Orthop Surg Res.* 2013;8:25.
13. Son JS, Appleford M, Ong JL, et al. Porous hydroxyapatite scaffold with three-dimensional localized drug delivery system using biodegradable microspheres. *J Control Release.* 2011;153(2):133−140.
14. Tripathi G, Basu B. A porous hydroxyapatite scaffold for bone tissue engineering: Physico-mechanical and biological evaluations. *Ceram Int.* 2012;38(1):341−349.
15. Oonishi H, Oonishi H, Kim SC, et al. *Clinical Application of Hydroxyapatite.* Sawston, Cambridge: Woodhead Publishing Limited; 2008. https://doi.org/10.1533/9781845694227.3.606.
16. Royal Society of Chemistry, Deutsche Bunsen-Gesellschaft für Physikalische Chemie, Koninklijke Nederlandse Chemische Vereniging & Società Chimica Italiana. *Physical Chemistry Chemical Physics: PCCP.* Cambridge: Royal Society of Chemistry; 1999.
17. Solarino G, Piazzolla A, Mori CM, Moretti L, Patella S, Notarnicola A. Alumina-on-alumina total hip replacement for femoral neck fracture in healthy patients. *BMC Musculoskel Dis.* 2011;12:32.

18. Han Y, Zhu JF. Surface science studies on the zirconia-based model catalysts. *Top Catal*. 2013;56(15-17):1525−1541.
19. Hench LL, Jones JR. Bioactive glasses: frontiers and challenges. *Front Bioeng Biotech*. 2015;3:1.
20. Jones JR. Review of bioactive glass: from Hench to hybrids. *Acta Biomater*. 2013;9(1): 4457−4486.

Polymeric biomaterials

Learning objectives

This chapter presents an overview of polymers used for a large array of applications, including orthopedic devices, dentistry-related devices, drug delivery systems, and tissue engineering, among others.

At the end of the chapter, the reader will:

- Get a general understanding of polymers' synthesis, structure, and their physical, chemical, and mechanical properties.
- Understand the relationship between polymers' properties and their applications, from the perspective of meeting the requirements for use in the biomedical industry.
- Become familiarized with the benefits of using polymeric biomaterials over the use of metals and ceramics.
- Learn the criteria involved in material selection for very specific roles in complex biomedical devices.

Introduction

By definition, a polymer is a substance composed of macromolecules. A macromolecule has long sequences of one or more species of atoms or groups of atoms linked to each other by a primary bond. Macromolecules are formed by linking monomers (repeating units) by a process known as polymerization. Polymers are composed of nonmetallic elements, having carbon as the most predominant element. These macromolecules are synthesized by primary covalent bonding of small units (mers) in the main molecular chain backbone. Fig. 5.1A illustrates a polymerization process starting with a single monomer, which forms a dimer (in the illustration represented by a heterodimer), and yielding a complex long chain of coiled polymeric networks. When the monomers are identical, the synthesis results in homopolymer chains, whereas the presence of monomers of different nature yields copolymers that can follow different configurations or arrangements. Fig. 5.1B illustrates a typical example of natural biopolymers: proteins. In this case, the repeating units

Introductory Biomaterials. https://doi.org/10.1016/B978-0-12-809263-7.00005-6

77

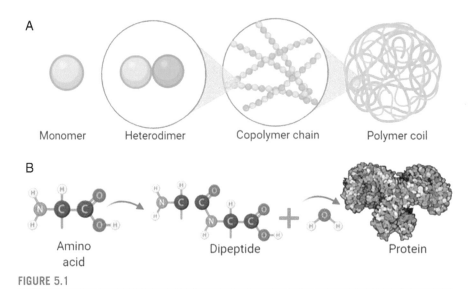

Monomer   Heterodimer   Copolymer chain   Polymer coil

Amino
acid

Dipeptide

Protein

FIGURE 5.1

Polymerization. (A) High-level illustration from monomers to amorphous random coils.
(B) From aminoacids to quaternary structure of proteins.

(amino acids) are linked to each other via peptide bonds, forming dipeptides, to yield
a three-dimensional (3D) arrangement of polypeptide chains known as proteins.

Polymers are the most versatile materials that can be synthesized and carefully
tailored to offer appropriate chemical, physical, interfacial, and biocompatible performance, which permits the exploration of multiple applications. This statement is
supported by the widely expanded use of these materials across different fields,
bringing promising applications in medicine, biotechnology, food, biosensors,
implantable devices, and drug delivery systems, among others. This versatility is
a consequence of a number of advantageous properties offered by this group of materials, such as high strength or modulus-to-weight ratio (in other words, these materials are light weight while stiff and strong), toughness, resistance to corrosion,
transparency, and low cost.[1]

Many of these properties are indeed distinctive to polymers and are explained by
the long-chain molecular structure characteristic of these materials. Every polymer
presents unique attributes; however, most polymers share some general traits:

- **Chemical resistance:** This refers to the ability of a polymeric material to
  maintain its original properties after exposure to a specific chemical agent,
  environment, or an extended period of time. The main factors affecting this
  behavior are the molecular structure of the polymer, type and concentration of
  the chemical agent, environmental conditions (temperature, time), and the
  mechanical stress applied.[2]
- **Light weight with significant degrees of strength:** Polymeric structures usually
  have very long and branched molecular arrangements. This configuration

prevents close packing, resulting therefore in a low density. However, their high strength-to-weight ratio gives to polymers such as polyurethane (PU) a strength equivalent to steel or concrete despite their significantly lower weight.

- **Electrochemical corrosion resistance:** Linked to their chemical resistance, organic nature (carbon-based composition), and poor conductivity, polymers are not susceptible to electrochemical corrosion, improving their overall biocompatibility in physiological environments.

Biomedical applications: material design rationale

When considering biomedical applications, additional factors must be assessed in order to fulfill the design requirements for optimal performance in biological/physiological environments. These preliminary principles are the foundation to produce safe, effective, durable, and compliant devices.

- Stability and corrosion resistance (under the aggressive conditions of physiological environments)
- Plasticity and easy processing
- Strength and fatigue resistance
- Coating activity
- Good cell adhesion and integration to surrounding tissue
- Low immune response
- Extended lifetime

The technical requirements are highly specific and dependent on device application, with the material selection being the critical cornerstone of the overall performance. Fig. 5.2 presents general criteria for material selection and examples of polymers traditionally used for specific applications.

In general, certain criteria should be followed in the intent to rationalize material selection:

1. **Surfaces characteristics:** For many applications, the success of the device depends on the proper identification and control of critical surface properties.
   Hydrophilicity and surface energy: These properties govern most of the device-host interactions postimplantation. When the material is in contact with blood, a highly charged, nonthrombogenic surface is desired to reduce protein adsorption, offering a protective effect. This is achieved by introducing polar groups that create negatively charged interfaces.
   Modulation of protein adsorption: Strong water-biomaterial interactions modulate cell adhesion, hence biocompatibility.
   Water adsorption: This has a key role in strength and durability of biomaterials that might be impacted by hydrolytic degradation.

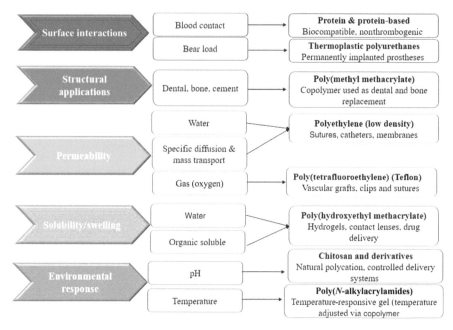

FIGURE 5.2

Rationalized polymer selection, presenting primary criteria, specific capabilities, and examples of commonly used polymers and applications.

Water sorption: When designing drug delivery systems and sutures, a controlled degradation rate and optimum diffusion are regulated by introducing polar groups (grafted polymers and hydrophilic polymeric coating).

2. **Bulk properties:** Certain properties such as permeability, diffusion, degradation rate, hydrophilicity, and mechanical strength must be assessed on a case-by-case basis.

3. **Surface modification:** In general, physical and chemical strategies of polymeric surface modification (e.g., polymerization or grafting of poly(ethylene oxide) [PEO], heparinization, or the design of highly specific bioactive surfaces) may enhance the biocompatibility of the implanted material.

Classification of polymers

Type of polymers by composition (monomeric species)

As stated earlier, polymers are organized in long chains linked through covalent bonding. They can also be classified based on the type and organization of the monomeric species in their chain, having therefore two main categories: homopolymers and copolymers (Table 5.1).

Table 5.1 Examples of FDA-cleared homopolymers and copolymers.

Trade name	Type of polymer	Properties	Validated applications
Propylux HS2	Homopolymer (polypropylene)	Withstands high temperatures with low moisture absorption. Allows sterilization (steam autoclaving and cold treatment)	Cranio-maxillofacial trays
Formolene	Homopolymer (polypropylene)	Unique formulation to provide excellent low water carryover	Food packaging and other food contact uses
Gore-Tex	Homopolymer (expanded polytetrafluoroethylene [ePTFE])	Inert material, high strength, low water adsorption. Waterproof while allowing gas passage	Vascular grafts, surgical meshes, ligament and tendon repair
Dacron	Homopolymer (poly(tetrafluoroethylene) [PTFE])	Low immune response	Nonabsorbable sutures
Natrelle	Homopolymer (poly(dimethylsiloxane) [PDMS])	Chemical stability, stable under sterilization, good elastomeric properties	Breast implants
ELVAX EVA	Block copolymer (ethylene vinyl acetate)	Toughness, puncture resistance, and good adhesion to paper and nonwoven substrates	Medical packaging foaming web
OsteoFab Patient Specific Cranial Device (OPSCD)	Block copolymer (polyetherketoneketone [PEKK])	Semicrystalline thermoplastic with high heat and chemical resistance. Withstands high mechanical loads	Replacement of bony voids in the cranial skeleton
Tecophilic HP-60D-35	Random copolymer (polyvinylpyrrolidone/vinyl acetate [PVP/VA])	Thermoplastic polyurethane medical grade. Absorbs equilibrium water contents of up to 100% of the weight of dry resin	Medical tubing applications (catheters)

FDA, *US Food and Drug Administration.*

- **Homopolymers:** These polymers are derived from only one molecular species and have a structure that looks like this: (-A-A-A-A-A-A-)$_n$ (Fig. 5.3 (1)melt). They show good thermoplastic properties and permit classic processing by melt-spinning into fibers.

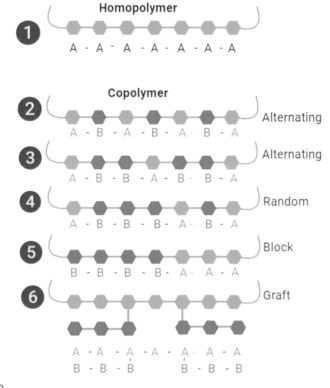

FIGURE 5.3

Schematic representation of homopolymer and typical copolymer configuration.

- **Copolymers:** On the other hand, these polymers are derived from different monomers and can follow various configurations as illustrated in Fig. 5.3 (2−6). Copolymerization is commonly used to modify and/or improve the performance of homopolymers by blending their properties toward a tailored modification of a polymeric material. It is important to note that copolymerization does not yield a mixture of individual homopolymers, but instead both monomers are covalently incorporated in the individual molecules.[3] While alternating and statistical copolymers can be produced by simultaneous polymerization of two different monomers together, the syntheses of block and graft copolymers entail particular methods. In an effort to differentiate the products, there is a nomenclature for copolymer as follows:

    Block copolymers: These are linear. In this type of polymer, repeat units exist only in long sequences. Two common types are AB and ABA (Fig. 5.3 (5)).
    Graft copolymers: These are branched polymers with branches that have a different chemical structure than that of the main chain. In branched co-polymers, the main chain or the side chains can be either homopolymers or copolymers.

Statistical copolymers: These copolymers are arranged in certain sequences that follow some statistical model. However, if it obeys Bernoulli statistics a random copolymer is synthesized. In such case the monomers are distributed randomly (and usually unevenly) along the polymer chain, as illustrated in Fig. 5.3 (4).

Alternating copolymers: This arrangement results in monomers distributed in a regular alternating fashion. Typically, the molar concentration of co-polymers is equitable in the polymeric chain.

Both block and graft copolymers usually have behaviors that are attributable to each homopolymer. However, since the chemical bonds between homopolymer sequences are unique to each polymer, these will also influence the overall properties.[4] Block copolymers differ from statistical copolymers by the monomer distribution. In the simplest block polymer configuration, a longer sequence of repeating units (-A-A-A-) is linked to a sequence of different repeating units (-B-B-B-); such products are referred to as AB block copolymers. This polymeric structure often shows multi-phase morphologies, thus displaying attributes of all elements that conform to the polymer.[3] By contrast, statistical copolymers display properties following a reasonable average approximation of the corresponding homopolymer properties.

Type of polymers by degree of crystallinity

Polymers can be also classified according to their degree of organization, which is defined in terms of crystallinity. By definition, highly organized structures with symmetrical orientation are described as crystalline, while the ones following no or limited order are known as amorphous. Polymers are never completely crystalline; therefore when categorizing morphology, we will refer to crystalline regions and amorphous regions. In crystalline polymers, the chains still might fold; however, they form organized folded chains known as *lamellae*. In some cases, these lamellar structures show some short chains that loop out from the ordered structure. This creates amorphous regions.

Degree of crystallinity holds a significant role on the melting temperature (T_m), which is described as the temperature at which ordered long-chain structures transition to random disorganized arrangements. Amorphous solids do not melt, instead they display a transition from glassy (brittle, hard, and rigid) to rubber (soft and flexible), which is described as the glass transition temperature (T_g). With greater degree of crystallinity the polymer will have higher T_m, become rigid, and show less sensitivity to solvent penetration, as shown by common thermoplastics such as nylon, polypropylene (isotactic), polyethylene (PE), and polystyrene (syndiotactic), as presented in Fig. 5.4A. By contrast, polymers that tend to have a high concentration of amorphous regions are softer, have glass transition, and show the tendency to be penetrated by solvents. Some examples include polypropylene (atactic), poly(methyl methacrylate) (PMMA), and polybutadiene (Fig. 5.4B).

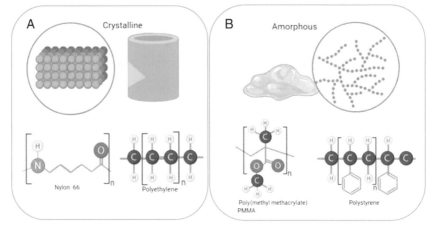

FIGURE 5.4

Schematic representation of crystalline (A) and amorphous polymers (B).

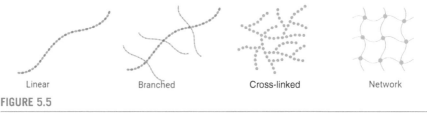

FIGURE 5.5

Configuration of polymeric chains after polymerization. Linear, branched, cross-linked, and network structures.

Type of polymers by physical structure

As a polymeric chain grows it folds back randomly, generating different structures (as illustrated in Fig. 5.5) and can be summarized as:

a) **Linear polymers:** These polymers align horizontally and have no side chains. The long chains are typically held together by weak Van der Waals or hydrogen bonds, which makes them easy to break with temperature. These polymers are typically thermoplastic.

b) **Branched polymers:** These polymers have branches or side chains of significant length. The side (shorter) chains can interfere with efficient packing, therefore branched polymers tend to show a lower density when compared with linear polymers. Also, since the short side chains do not bridge from one long backbone to another, they are typically highly susceptible to be broken down by heat, which allows the polymer to be thermoplastic.

c) **Cross-linked polymers:** These polymers have 3 to 10 structures in which all chains are connected with each other by junction points. Unlike linear polymers, these polymers are held together by strong covalent bonds, which make them thermoset polymers in most cases.

d) **Network polymers:** These polymers are structures with intertwined connections that display a higher density of cross-linking (3D linkages). These polymers are virtually impossible to soften without resulting in degradation of the underlying structure. Thus they behave as thermoset polymers.

Variation in skeletal structure gives rise to major differences in properties. To illustrate the last point, linear PE has a melting point of about 20°C higher than branched, while network polymers do not melt upon heating.[5]

Polymer synthesis

Polymerization is traditionally categorized according to the type of reactions occurring in the synthesis. Overall, there are three main reactions governing polymerization, the most common being condensation and addition.[6] Fig. 5.6 illustrates the classic example of a linear homopolymer: PE. The monomers of ethylene are sequentially linked to yield PE polymer, which typically contains long chains with about 75,000 repeating units connected together.[7]

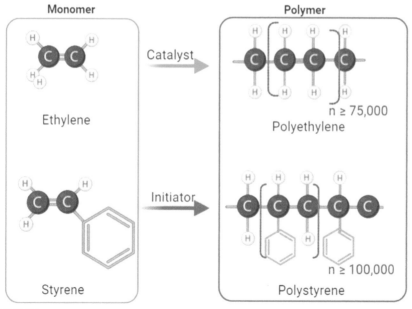

FIGURE 5.6

Schematic representation of the addition (chain-growth) polymerization reaction, occurring during polyethylene and polystyrene polymerization.

Step-growth polymerization

A step-growth polymerization is a stepwise reaction between bi- or multifunctional monomers. All monomers are reactive. This synthesis can be driven by two different reactions: condensation or addition.

Condensation

Condensation polymerization is a synthesis reaction that involves the conventional functional group transformation of polyfunctional reactants via a step-growth process. This reaction often occurs with loss of a small byproduct (e.g., water, ammonia), and generally fuses two different components in an alternating fashion. A classic example is presented by the natural synthesis of proteins (Fig. 5.1B). Following a classic condensation reaction, a carboxylic acid from one amino acid reacts with an amine available from the second amino acid. A substitution leads to the formation of an amide group (peptide bond) and water is produced as a byproduct. This condensation reaction leads to polyamides as well, also known as nylons. Nylon was the first commercially available polymer, which was produced in the 1930s. Condensation reactions are also the basis for the formation of polyesters (Dacron, Mylar), polyamide (Kevlar), Polyamide (Nylon 6), and polycarbonates (Lexan). Examples of naturally occurring condensation polymers include cellulose, polypeptides, and poly(β-hydroxybutyric acid).

Polyaddition

On the other hand, polyaddition is a reaction between functional groups without the generation of byproducts. A typical step-growth polyaddition example is the synthesis of PU, which is produced by reaction of a polyol with a diisocyanate.

Some characteristic features of step-growth polymerization are:

- All molecules (monomer, oligomer, polymer) can react among them.
- Large quantities of monomers are consumed early in the reaction.
- There is no termination step.
- The reaction is rapid at the beginning and slows down as the polymeric chain increases molecular weight (MW).
- Long reaction times are required to synthesize long (high MW) polymers.
- Coexistence of molecular species of any length.

Chain-growth polymerization

Chain-growth polymerization forms polymers in a similar fashion to adding links to a chain.

Addition

Addition reactions take place by combination of monomers with two reaction sites; the monomers are added to the growing polymeric chain. A typical example of addition of PE polymerization is illustrated in Fig. 5.6. PE is synthesized by linking ethylene ($CH_2=CH_2$) monomers together through covalent bonding. Carbon atoms share electrons with two other hydrogen and carbon atoms to form the polymer. The chemical formula for PE is $-CH_2-(CH_2-CH_2)_n-CH_2-$. In this case, the addition reactions involve breaking down of a covalent double bond between monomers with an identical structure to form a polymer containing only single C-C bonds.

Now, how do we get the addition reaction to proceed: does it start spontaneously? An initiator, or a catalyst, is needed. Free radicals are usually used as initiators, which are activated chemically, by heat, or by ultraviolet (UV) light. Examples of such free radicals are benzoyl peroxide and 2,2′-azo-bis-isobutyrylnitrile.

By contrast to step-growth polymerization, chain-growth polymerization is characterized by:

- During propagation, only monomers react to the active sites at the end of the growing chain.
- Monomers exist during the reaction, and their concentration decreases steadily with time.
- The initiator splits into two radicals and each radical adds a monomer, converting it into a new radical that propagates.
- There are two mechanisms during this polymerization: initiation and propagation. In most cases there is a termination step as well.
- The reaction increases velocity depending upon the initiator concentration and the polymer MW.
- Long reaction times have high degrees of conversion.
- The mixture contains mainly monomers and polymers (only a minor amount of growing polymeric chains).

Material properties of polymers
Thermal properties

Thermal characterization is especially important to determine the melting temperature (T_m), crystallization temperature, and glass transition temperature (T_g). Although these properties are not directly linked to material performance post-implantation (biocompatibility), they provide key insights on processing conditions. Crystalline/lamellar phases are characterized by their melting temperatures, T_m. Amorphous polymers and phases are characterized by their glass transition temperatures, T_g. At the glass transition temperature, these polymers go abruptly from a hard-glassy state to a rubbery, soft state, which corresponds to the onset of chain motion. Both T_g and T_m increase with increasing chain stiffness and increasing strength of intermolecular forces.[9] The polymeric degree of crystallinity displays a direct positive effect on other properties such as density, sharp melting point, strength, hardness, and brittleness. The temperature that the polymer is submitted to will greatly affect its response to stress, as illustrated in Fig. 5.7.

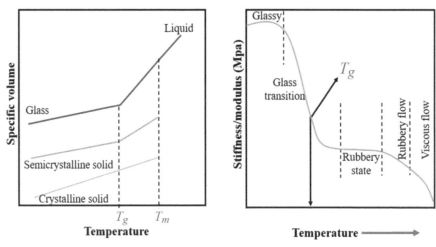

FIGURE 5.7

Glass transition temperature: Polymeric response to temperature as a function of specific volume (A); influence of temperature on mechanical properties of polymers (B).

- **Thermoplastic polymers:** This category of materials can be either crystalline or amorphous, and shows several common characteristics with other polymer classes.[10] They can be either linear or branched, but they are not networked polymers. Thus they can be melted and molded (and remolded) into virtually any shape. Due to these desirable traits, thermoplastics constitute by far the largest proportion of polymers that found uses in industry. They do not crystallize easily upon cooling to the solid state since their usually highly coiled and entangled chains are difficult to organize in a crystalline structure. Some thermoplastics do crystallize but generally form semicrystalline materials.[11]
- **Thermoset polymers:** Thermosets are network polymers whose chain motion is restricted by a high degree of cross-linking. Because of the restrictions imposed by this type of structure, thermosets are usually rigid and intractable once formed. These polymers are irreversibly hardened by curing; they do not melt under heating, rather they degrade at high temperatures.[12]
- **Elastomeric polymers:** Another important class of polymers is known as elastomers. These were first developed for making synthetic rubbers for the military. Since World War II there has been a spike in the research and understanding of properties of polymers and their various synthesis routes.[13] Polymers can stretch to high elongations, between 3 and 10 times their original dimensions. Their elastic nature allows the original dimension to recover when stress is released.[14]

Mechanical properties

The mechanical properties of polymers are generally inferior to those of metals and ceramics, especially when high loads are applied,[15] which makes them less suitable for load-bearing devices. However, their ability to modify the chemical structure and their ability to degrade or change conformation in the presence of various stimuli, such as pH or temperature changes, have led to polymers being the best choice for some applications, including soft-tissue replacement, biosensors, and drug delivery systems, among others. Mechanical parameters of key relevance for successful biomedical applications can be summarized as:

- Strength
- Elastic modulus
- Compressive strength
- Wear
- Creep resistance

When elastic (not elastomeric or long-range) materials are stressed, there is an immediate and proportional strain (extension changes) response. Fig. 5.8A illustrates a comparison among different types of polymers. The load (stress) is calculated by accounting for the cross-sectional area of the testing material as follows:

$$\sigma = F/A \tag{5.1}$$

where F is the applied force (N), and A is the average cross-sectional area of the samples (mm^2). The elastic modulus represented in Fig. 5.8A by a straight red line is the slope of the curve in the linear region and provides information on the stiffness of the tested material. This property is calculated by:

$$E = \sigma/\varepsilon \tag{5.2}$$

Therefore,

$$E = \frac{F/A}{\Delta L/L} \tag{5.3}$$

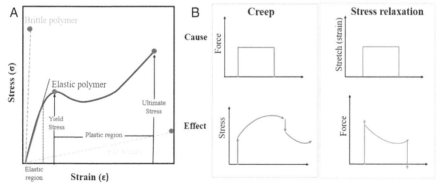

FIGURE 5.8

Mechanical properties of polymers. Illustration of tensile stress-strain curves for polymers (A) and viscoelastic behavior of polymers (B).

where $\Delta L / L$ is the material extension per unit length.

According to the illustration in Fig. 5.8A, brittle polymeric materials, such as polystyrene (PS), MPPA, or polycarbonate, endure significant mechanical stress, but they will not withstand much elongation before reaching failure. Thus brittle polymers are strong but not very tough. Flexible (elastic) polymers like PE and polypropylene display a limited tolerance to deformation but tend to not break. This ability to deform is described by the modulus of elasticity. The initial modulus is high, but with enough stress eventual plastic (permanent) deformation is achieved. These polymers might be less strong than brittle polymers but perform much better in terms of toughness. Finally, elastomers display a different mechanical response. They have low moduli, allowing them to stretch easily within the elastic region, performing with not only high elongation, but superior reversible elongation (recovery). Some common examples of elastomers are polyisoprene, polybutadiene, and polyisobutylene.

Now, let us introduce viscoelasticity. Polymers display viscoelastic behavior mainly explained by their unique structure that combines both solid and fluid behaviors. When stresses are lifted from polymeric materials (before failure), they undergo strain recovery. Both the deformation and the following recovery are time dependent. There are two characteristic properties of polymeric materials. The first one is creep, which refers to the increasing deformation under a constant load. The second significant behavior is stress relaxation, which refers to a decrease of the stress under constant deformation (strain), as illustrated in Fig. 5.8B.

The tensile (stretching) response of polymers is nonlinear and depends on the strain rate, exhibiting steady deformation and recovery when subjected to loading and unloading; this *time-dependent material response* is called *viscoelastic behavior*.[16] For viscoelastic materials, the relationship between stress and strain can be expressed as:

$$\sigma = \sigma(\varepsilon, \dot{\varepsilon}) \tag{5.4}$$

$$\sigma = \mu \, \dot{\varepsilon} \tag{5.5}$$

Thus,

$$\dot{\varepsilon} = d\varepsilon/dt \tag{5.6}$$

where σ = stress, ε = strain, $\dot{\varepsilon}$ = strain rate, μ = viscosity, and t = time.

When thinking of material design, it is crucial to pay careful attention to the specific application conditions and the environment that the polymer must withstand when implanted. The great news about polymers is that unlike other materials, their microstructure is easy to tailor, allowing the possibility of "pooling strengths" by combining two polymers with different/complementary properties. This operation yields a new polymer with the potential to perform with a blend of properties while mechanical properties are controlled. Such result can be obtained by copolymerization, blending, and fabrication of composite materials.

Molecular weight and its effect on material properties

MW is the cornerstone of most polymer properties. This might not come as a surprise, since we understand the essential role of MW dictating the properties of any molecular compound. However, in polymers, this statement holds added significance. This is because during synthesis, the yielded polymeric chains (with the exact same composition) coexist along with a distribution of chains displaying different MWs (variation among the length of monomer enchained), which introduces an obvious impact on the material properties. This particular variation creates a characteristic unique to polymer MW. It is to be noted here that when we refer to polymer MW, we are always describing an average value. Therefore calculating MW requires measuring the individual chains (number and specific MW).

The degree of polymerization (DP) is the average length (number of enchained monomers) in an average polymer chain, as illustrated in Fig. 5.9A. The number-average is defined as the number of repeating units in the polymeric chain and can be calculated using the number-average MW.

$$DP = \frac{\text{Number-average MW } (M_n)}{\text{MW of repeat unit}} \qquad (5.7)$$

Therefore,

$$\text{Number-average MW } (M_n) = DP * (\text{MW of repeat unit}) \qquad (5.8)$$

Considering the MW value as a pool of measurements, it is always helpful to have an insight on how widely distributed the individual values are. In polymers, this distribution is described by the *dispersity* (Đ), also known as the polydispersity index (PDI). Most of the polymers will display a PDI falling in a range between 1.0

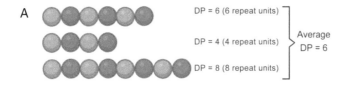

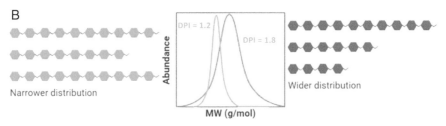

FIGURE 5.9

Graphic illustration of degree of polymerization *(DP)* (A) and polydispersity index *(DPI)* (narrow and wide distributions) (B).

and 2.0 (with some exceptions being higher than 2). A PDI of 1.0 would refer to an ideal distribution where all polymeric chains are exactly the same length (same MW), while higher values will represent a *polydisperse* nature of the polymer (Fig. 5.9B).

$$PDI = \frac{M_w}{M_n} \tag{5.9}$$

The PDI is measured as the ratio of the weight-average MW (M_w) to the number-average MW (M_n). M_n (defined in Eq. 5.10) is determined experimentally from the mole fraction distribution of different sized molecules in a sample and is defined as the statistical average MW of all the polymer chains in the sample.

$$M_n = \frac{\sum M_i N_i}{\sum N_i}, \quad M_w = \frac{\sum M_i^2 N_i}{\sum M_i N_i} \tag{5.10}$$

M_w is the weight-average molar mass calculated from the weight fraction distribution of different sized molecules. It accounts for the MW of a chain contribution to the overall MW average. Since larger chains weigh more than the short ones, M_w is skewed to higher values, and always greater than M_n. As the weight dispersion of molecules in a sample narrows, M_w approaches M_n, and in the unlikely case that all the polymer molecules have identical weights (a pure monodisperse sample), the M_w/M_n ratio (PDI) becomes unity.

Many polymer properties indicate a strong dependence on the size of the chains. For example, when thinking about the molar mass and its influence on properties, it is true that for network polymers the only molar mass that has any significance is that of the chains between the cross-links. In turn, for a homopolymer, the molar mass is dependent on the DP.[17]

The properties of the polymers can be fine-tuned by varying their synthesis and structure.[8] For example, a DP of well over 1000 and a high MW are necessary if the goal is to synthesize solid polymers. For PE the MW is higher than 28,000 g/mol. This explains why polymers in use in most industries are giant molecules.[18] Low MW polymers generally behave like wax (e.g., paraffin wax to produce candles). If the MW is even lower, the polymer can be an oil, while the parent compounds of monomers are often gases.

Example 5.1

A batch of PS $[C_8H_8]_n$ is synthesized with a degree of polymerization (DP) of 30,000.
a) Calculate the MW of the repeat unit.
b) Compute the number-average MW.

Solution

a) $[C_8H_8]_n = 8(12.01 \text{ g/mol}) + 8(1.008 \text{ g/mol})$

MW $[C_8H_8]_n = 104.14 \text{ g/mol}$

b) From Eq. (5.7):

$$\text{Number-average MW } \left(\overline{Mn}\right) = \text{DP} * (\text{MW of repeat unit})$$

$$\overline{Mn} = 30{,}000\left(104.14 \text{ g/mol}^{-1}\right)$$

$$\overline{Mn} = 3.12 \times 10^6 \text{g/mol}^{-1}$$

MW very much affects polymer properties. For example, low-molecular-weight polyethylene (LMWPE) is used to make soda bottles. On the other hand, ultra-high-molecular-weight polyethylene (UHMWPE) is used for load-bearing applications such as for the plastic bearings in artificial hips.[19]

The nature of the synthesis reaction also has a significant effect on polymer properties. Heat and catalysts are used to synthesize ethylene polymers, for example, and nylon is formed by a condensation reaction where water is released while bathtub caulk is a silicone polymerization reaction where acetic acid is released.[20] The condensation byproducts are typically trapped in the resulting material and leach out over time, which is of significance for biomedical applications, where any toxic leaching products are not acceptable. Slight chemical modifications in a polymer structure can also lead to big changes in properties. For example, if we substitute the hydrogen atoms in PE with fluorine (F), the resulting material is well known as Teflon (polytetrafluoroethylene [PTFE]), which has very different properties than PE. Polyvinyl chloride (PVC) is rigid because the chlorine atoms are large and tend to prevent molecules from sliding over one another. Thus PVC is stronger, stiffer, and much more brittle than UHMWPE.[21]

In PVC, larger side groups than in PE make chain sliding and molecular alignment much more difficult. Because the material is amorphous, it is also optically transparent (Fig. 5.10).

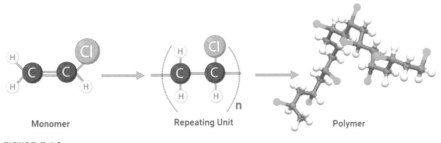

Monomer   Repeating Unit   Polymer

FIGURE 5.10

Schematic representation of polyvinyl chloride (PVC) polymerization. The PVC repeating unit is illustrated in brackets.

Polymer size (chain length [DP]) plays a critical role in polymer conformation and material architecture. The full extended length of a polymer might be described when both bond angle and length are known. L_{max} describes the length of a "fully extended" polymer and is calculated using the following equation:

$$L_{max} = n\ell \, \sin\frac{\theta}{2} \qquad (5.11)$$

where θ is the bond angle, ℓ is the bond length, and n is the total number of chain bonds.

On the other hand, the average end-to-end distance (r) for a random coiled polymer can be approached by:

$$r = d\sqrt{n} \qquad (5.12)$$

Note that this is root-mean-squared end-to-end distance for a random section of a freely jointed chain.

Example 5.2

Calculate the fully extended length (L_{max}) and the average end-to-end distance (r) of a polymeric chain of PE with an $\overline{Mn} = 500,000 \text{ g/mol}^{-1}$.

Solution

PE has an idealized structure, with bond angles (θ) of 112 degrees and C-C bond lengths (ℓ) of 0.154 nm.

Repeating unit MW [C_2H_4]:

$[C_2H_4]_n = 2(12.01 \text{ g/mol}) + 4(1.008 \text{ g/mol})$

MW $[C_2H_4]_n = 28.05$ g/mol

Applying Eq. (5.2)

$$DP = \frac{\text{Number-average MW}}{\text{MW of repeat unit}} = \frac{500,000 \text{ g/mol}^{-1}}{28.05 \text{ g/mol}^{-1}}$$

$$DP = 1782.53$$

Now we know that our polymer has 1783 repeat units along the average chain. The PE repeat unit has two carbon bonds, therefore $n = (\text{Carbon chain per repeat unit}) * (DP) = 2(1783) = 3566$ total chain bonds.

Using Eq. (5.4):

$$L_{max} = 3566(0.154 \text{ nm})\sin\frac{112}{2} = 455.28 \text{ nm}$$

Finally, approaching the conformation of a coiled polymer the average end-to-end distance is given by Eq. (5.5):

$$r = d\sqrt{n} = 0.154\sqrt{3566} = 9.20 \text{ nm}$$

Polymers in drug delivery

Historically, polymers have played an integral role in the growth and expansion of drug delivery technology by allowing controlled release of therapeutic cargos. For more than 50 years, polymers such as cellulose derivatives, poly(ethylene glycol) (PEG), and poly(N-vinyl pyrrolidone) have been used in the pharmaceutical industry to deliver bioactive agents. PEG is widely used as the gold standard in nanomedicine to increase the circulation time and improve drug efficacy. Despite all the advantages this polymer displays, the extended use of PEG has triggered the appearance of anti-PEG antibodies,[22] which threatens to accelerate the blood clearance, thereby leading to low efficacy. New materials need to be explored in order to overcome PEG immunogenicity.

From a device perspective, delivery systems can be categorized by their delivery mechanisms, as diffusion controlled, solvent activated (swelling or osmotically tuned), chemically controlled (biodegradable), or externally triggered (temperature, pH). Polymers have improved pharmacokinetics compared with small-molecule drugs with a longer circulation time. The ability of polymers to respond to stimuli makes them useful in a different manner, such as polymeric drugs, in combination with small-molecule drugs or biopolymers.[23] Thus we can say that polymers can be used in polymer therapeutics as implantable drug delivery devices,[24] polymeric drugs,[25,26] polymer-drug conjugates,[27] polymer-protein conjugates,[28] polyplexes (a combination of nucleic acids and synthetic polymers),[29] or polymeric micelles.[30]

Material selection criteria

For a polymer to be safely applied for therapeutics or as an agent in tissue regeneration and repair, there are several criteria that it needs to meet. If the polymer is the drug itself, it has to be water soluble, nontoxic, nonimmunogenic, biodegradable, and safe for all stages of drug delivery (the degradation products need to be nontoxic at all times). If a polymer is not biodegradable (e.g., polymethacrylates), it could still be used in some applications; however, its size needs to be below the renal threshold to ensure it is not accumulated in the body.[31] For biodegradable polymers (e.g., polyesters), toxicity and/or the immune response of the degradation products must be considered. These systems must overcome harsh conditions and allow specific targeting, intracellular transport, and recognitive feedback control.

Implantable drug delivery devices

Since many drugs are not suitable for oral delivery due to potential side effects or first-pass metabolism, among other challenges, implantable devices are alternative systems to achieve an effective delivery of low-concentration drug cargo, minimizing the side effects while increasing patient compliance. Currently we find passive polymeric implants and dynamic or active polymeric implants. The first are relatively simple devices that rely on passive diffusion for drug release. Generally they are made of packed drugs within a biocompatible polymer molecule employing common synthetic polymers such as silicones (e.g., poly(dimethyl siloxane), poly-(urethanes), poly(acrylates)) or copolymers such as poly(ethylene vinyl acetate) as a polymer matrix where the drug is dispersed. In this case the release rate is controlled by the membrane/matrix thickness and the permeability of the drug passing through the membrane. This approach has been widely used to develop contraceptive delivery devices.[24] On the other hand, the dynamic polymeric implants have a positive driving force to control the drug release. Most of the devices in this category are electronic systems (e.g., polymeric pump-type implants). Implantable devices have demonstrated a versatile use for a variety of clinical applications, including women's health, oncology, ocular diseases, pain control, and central nervous disorders, as presented in Table 5.2.

Polymeric micelles

Micelles are defined as aggregates of molecules in colloidal dispersions. In the presence of a dispersed phase and a continuous phase, micelles form spontaneously in response to the dispersed molecule and their amphiphilic nature (they display both hydrophilic and lipophilic properties). The main driving force of micelle formation is the overall reduction in the free energy of the system.

As illustrated in Fig. 5.11A, the polar heads of the amphiphilic molecules face the aqueous environment, leaving the tails (nonpolar) to interact with the nonpolar fraction (oil-like phase). By contrast when the media is nonpolar, inverse micelles will form by trapping the water-like phase. Finally, polymeric micelles are

Table 5.2 Examples of commercial implantable drug delivery devices.

Brand name	Type	Polymer	Drug delivered	Use
Norplant	Subcutaneous	Silicone	Levonorgestrel	Contraception
Estring	Intravaginal	Polyethylene vinyl acetate (PEVA)	Etonogestrel	Menopausal symptoms
Zoladez	Subcutaneous	Poly(lactic-co-glycolic acid) (PLGA)	Goserelin	Prostate cancer
Vantas	Subcutaneous	Methacrylate-based hydrogel	Histrelin	Prostate cancer
Ocusert	Intraocular	PEVA	Pilocarpine, alginic acid	Open-angle glaucoma
LiRIS	Intravesical	Silicone	Lidocaine	Interstitial cystitis/pain
Probuphine	Subcutaneous	PEVA	Buprenorphine	Opioid abuse
Med-Launch	Subcutaneous	PLGA	Risperidone	Schizophrenia

Adapted with permission from Stewart et al.[24]

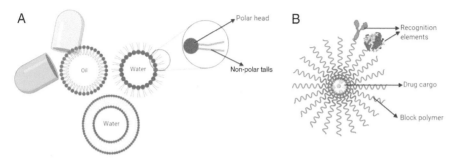

FIGURE 5.11

Graphic illustrations of a selection of drug delivery systems. Micelle, reverse micelle, and liposome with a zoom-in of a monomer of fatty acid (amphiphilic *mer*) (A); polymeric micelles featuring classical components (B).

presented as nanocarriers (Fig. 5.11B) by arranging amphiphilic block copolymers in aqueous solutions.[32] These particles have a hydrophobic core and a hydrophilic shell. This configuration provides protection for the cargo and facilitates the drug loading and transport through different tissues; also it allows for highly specific targeting by immobilizing recognition biomolecules on the particle surface. Interestingly their small size and hydrophilic shells render them invisible to macrophages, increasing the circulation time of the drug cargos. The outer hydrophilic shell usually is composed of polyesters such as PEO, PEG, poly(ε-caprolactone) (PCL), poly(D,L-lactide) (PDLLA), and poly(glycolide) (PGA). Recent advances have demonstrated the potential of poly(L-amino acids) (PAAs) as potent self-adjuvanting delivery platforms, having poly(L-glutamate and aspartate) as promising natural polymers for drug delivery systems.[33]

Delivery mechanisms

Diffusion-controlled systems

In these systems a drug is dispersed in a non-swellable or fully swollen matrix that does not degrade during its therapeutic life. In dissolved systems ($C_0 < C_S$), C_0 is the initial loading concentration and C_S is the saturation concentration.[32] Assuming the concentration of diffusing species is a function of both time and position $C = C(x,t)$; the modeling of drug release from swellable polymers is governed by Fick's second law[34]:

$$\frac{\partial C}{\partial t} = D \frac{\partial^2 C}{\partial x^2} \tag{5.13}$$

For solving, we need boundary and initial conditions:

$$\text{At } t = 0, \ C = C_0 \text{ for } 0 \leq x \leq \infty$$
$$\text{At } t > 0, \ C = C_s \text{ for } x = 0$$
$$C = C_0 \text{ for } x = \infty$$

Derived from Fick's second law, drug release is described in the case of controlled diffusion models by the following equation:

$$\frac{M_t}{M_0} = 4\left(\frac{D_t}{\pi h^2}\right)^2 \tag{5.14}$$

where M_t is the amount of drug released at time t, M_0 is the total mass of drug loaded into the device, D is the diffusion coefficient of the drug within the polymer matrix. π is 3.14 and h the thickness of the device. This approximation accounts for the release of the first 60% of the loaded drug.

Thus the equation that describes the late release (40%) is as follows:

$$\frac{M_t}{M_0} = 1 - \left(\frac{8}{\pi^2}\right)\exp\left[\frac{-\pi^2 D_t}{h^2}\right] \tag{5.15}$$

This model is based on approximations that assume both the unchanged dimension and physical properties of the materials during release, and release from a slab.[35]

Chemically controlled: biodegradable polymers

Some polymers can degrade under certain conditions, and this property can be useful in applications such as drug delivery, temporary orthopedic pins and screws, or surgical sutures. Some of these polymers are biodegradable, meaning that they are degraded by a biological agent, often an enzyme.[36] Biodegradable polymers have the characteristic of being able to dissolve under physiological conditions, generally by chemical attack. An example of a degradable polymer is PU. These are generally stable polymers, which are used, for example, in foam cushions.[37,38] However, in the presence of peroxides, the superoxide's chains can break, and the polymer degrades.

By definition, biodegradation differs from degradation as follows:

- **Biodegradation:** Susceptibility to degradation by biological activity, followed by a lowering of polymer molecular mass.
- **Degradation:** Chemical modifications in a polymeric material that lead to changes in material properties.

The origin of this property is that some covalent bonds that are present in polymers are prone to hydrolysis, which can occur either nonenzymatically or enzymatically. Examples of degradable polymers include polyhydroxybutyrate (PHB),[39] polyhydroxybutyrate-co-hydroxyvalerate (PHBV),[40] polyglycolic acid (PGA),[41] polylactic acid (PLA),[42] and PCL,[43] among others presented in Fig. 5.12. On the basis of origin, biodegradable polymers can be classified as synthetic (mostly polyesters) and natural, having an impact on the specific mechanisms of degradation followed. Synthetic condensation polymers, such as polyesters, PUs, polyamides, polyanhydrides, and polyureas, degrade by hydrolysis under aqueous conditions, microbial, or enzymatic action. By contrast, polymers with additives like the case of pro-oxidant or photosensitizer added to polymers (e.g., addition of Mn^{2+}/Mn^{3+}

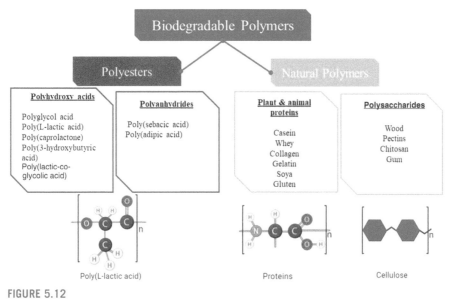

FIGURE 5.12

Selection of biodegradable polymers used as biomaterials.

to polyolefins) degrade by oxidation due to heat or light exposure.[44] Natural polymers (e.g., chitosan, cellulose, chitin, PLA, and hyaluronic acid) degrade by hydrolysis, enzymatic, or microbial action.[44]

Biodegradable polymers have been widely explored for medical products, such as surgical sutures and tissue ingrowth materials, or for controlled drug release devices (explained in the next sections), plasma substitutes, etc. Their biocompatibility is outstanding due to the fact that the degradation reactions usually involve hydrolysis (either enzymatically induced or by nonenzymatic mechanisms) to nontoxic small molecules that can be metabolized by or excreted from the body. However, a number of additional applications have been shown to be successful such as food packaging, agriculture, and personal care, as shown in Table 5.3.

Continuous advancement in polymer chemistry yields an increasing number of potential applications. Degradable materials come into play as pivotal solutions for sustainable use of polymers in biomedical, agriculture, personal care, and packaging technologies. Despite their great potential, the high production cost and complex manufacturing processes still pose restrictions for extended use and adoption of these materials.

Hydrogels

Hydrogels are 3D hydrophilic polymeric networks. These materials have the ability to absorb large amounts of water (or physiological fluids). Hydrogels can be

Table 5.3 Examples of commercial applications of biodegradable polymers.

Application	Use	Brand name	Polymer	Degradation mechanism
Agriculture	Biodegradable mulch	Ecoflex	Low-density polyethylene (LDPE)/starch composite	Photo-initiated chemical degradation
Agriculture	Agrochemical delivery	Bifender FC	Natural polysaccharides (alginates, starch, cellulose, and chitosan)	Enzymatic and microbial hydrolysis
Personal care	Thickeners	TEGO	Natural polysaccharide, vinyl acetate, vinyl pyrrolidone	Enzymatic and microbial hydrolysis
Personal care	Emulsion-based products	SEPIGEL 304	Alkylene oxide-base homo-, copolymers: acrylic acid−based polymer, and polyacrylamide	Mineralization to carbon dioxide
Packaging	Food product protection and storage	Green Cell Foam	PVA, PCL, PLA, and PHA/starch composites	Microbial amylases (*Bacillus subtilis*)
Biomedical	Internal fixation devices, vascular grafts, surgical meshes, stents, scaffolds	Monocryl sutures (Ethicon)	Polyglyconate, polyglycolic acid (PGA), PLA	Simple hydrolysis mechanism
Biomedical	Nerve repair, wound dressing, stents, surgical meshes	Neurowrap	Polyhydroxybutyrate (PHB)	Bacteria and fungi enzymatic hydrolysis
Biomedical	Drug delivery systems	Atridox	Poly(L-lactide) (PLLA), poly(D,L-lactide) (PDLLA), poly(lactide-coglycolide) (PLGA), chitosan	Hydrolytic mechanisms
Biomedical	Shape memory: suture, surgical meshes	Ventrio Hernia Patch (BD Bard)	Polydioxanone (PDO)	Bulk erosion

PCL, *Poly(ε-caprolactone)*; PHA, *polyhydroxyalkanoates*; PLA, *polylactic acid*; PVA, *poly(vinyl alcohol)*.

stabilized via chemical or physical interaction. There exist "reversible" or "physical" gels since the molecular entanglement is governed by secondary forces (ionic, H bonding, or hydrophobic forces) to form the network structure. By contrast, "chemical" or "permanent" gels are formed by covalent bonds linking different macromolecules. The latter might be charged or uncharged depending upon the nature of the functional groups available in the structure.

Physical hydrogels are often reversible and dissolve under certain environmental triggers (pH, temperature, or ionic strength). On the other hand, chemical-charged hydrogels usually display changes in swelling under variation of pH and are able to undergo shape changes under electric fields. The first hydrogel was pioneered by Wichterle and Lim in 1960, when they described the use of poly-2-hydroxyethylmethacrylate (PHEMA) as a biocompatible material to be used for contact lens applications.[45] There are numerous reports of widely diverse applications for hydrogels, going from temperature-triggered release of antimicrobial agents for periodontal tissue (chitosan/gelatin/β-glycerol phosphate),[46] injectable thermosensitive poly(lactic-co-glycolic acid) (PLGA) hydrogels for cancer treatment,[47] to pH-sensitive hydrogels (hydroxypropyl methyl cellulose [HPMC]) for the treatment of schizophrenia.[47] An additional selection of US Food and Drug Administration (FDA)−cleared hydrogels is presented in Table 5.4, while additional details on hydrogels used for soft-tissue replacement can be found in Chapter 10.

Table 5.4 Examples of FDA-cleared hydrogels.

Brand name	Polymer	Drug loaded	Application
Atridox	Poly(D,L-lactide) (PLA)	Doxycycline	Bioabsorbable flowable—periodontal treatment with subgingival delivery
Eligard	Poly(lactic-co-glycolic acid) (PLGA)	Leuprolide acetate	Prostate cancer treatment
Timoptic-XE	Purified anionic heteropolysaccharide derived from gellant gum (GELRITE)	Timolo malate	Glaucoma
Relday[a]	Sucrose acetate isobutyrate	Risperidone	Schizophrenia and bipolar disorder
Oncogel[a]	PLGA/polyethylene glycol (PEG) copolymer (PLGA/PEG)	Pacitaxel	Anticancer therapy

[a] Under clinical trials (phase II).
FDA, US Food and Drug Administration.

Stimuli-responsive polymers (externally triggered)

Some polymers can respond to stimuli, such as changes in pH or temperature. These are properties of interest for pharmaceutical applications or tissue targeting. Stimuli-responsive polymers can encapsulate or bind certain molecules, such as drugs, peptides, or nucleic acids, and subsequently transport them to the desired location upon application of an external stimulus. They mimic biological systems in the sense that a change in pH or temperature results in a change in properties. Examples include changes in conformation, changes in solubility, alteration of the hydrophilic/hydrophobic balance, or release of a bioactive molecule (e.g., drug molecule).[48]

pH-responsive polymers

For polymers that are pH responsive, a change in acidity is a particularly useful environmental stimulus to exploit in the development of polymeric drug carriers owing to the numerous pH gradients that exist in both normal and pathophysiological environments. For example, it is well documented that the extracellular pH of tumors is slightly more acidic than normal tissues, with a mean pH of 7.0 in comparison with 7.4 for blood and normal tissues.[49] Most pH-responsive polymers are ionizable, with pK_a between 3 and 10. A change in pH results in conformational changes for soluble polymers and in swelling behavior for hydrogels. pH-responsive swelling and collapsing behavior can be used to induce controlled release of model compounds like caffeine, drugs like indomethacin, and cationic proteins like lysozyme.[50] Micellar polymers that are pH responsive can enter the cells via an endocytosis process. While the endocytic pathway begins near a physiological pH of 7.4, it drops to a lower pH (5.5–6.0) in endosomes and approaches pH 5.0 in lysosomes. Therefore polymeric micelles that are responsive to these pH gradients can be designed to release their payload selectively in tumor tissue or within tumor cells.[51] When weak bases (generally, amines) are unprotonated, the block of the copolymer that they comprise has a relatively hydrophobic character, so the polymer is not water-soluble. Upon protonation (pH decrease), charges are introduced, thus imparting water solubility to the block and triggering the disintegration of the micelle into monomers (Fig. 5.13).

Conversely, for example, a poly(2-vinylpyridine)-*block*-poly(ethylene oxide) (P2VP-*b*-PEO) copolymer undergoes spontaneous and reversible micellization in aqueous solutions as the pH is increased from acidic to neutral or basic. P2VP has also been incorporated into triblock copolymers that form micelles having pH sensitivity. Triblock copolymers such as poly(L-lactide) (PLLA)-*b*-PEO-*b*-polysulfadimethoxine that are pH sensitive (owing to the weakly acidic nature of the sulfonamide groups) have been shown to perform a phase transition around pH 7.0 and are expected to be useful as anticancer drug carriers since tumor pH is known to be close to 7.0.[52]

Micelles formed from these copolymers have block copolymers with weak acidic groups such as carboxylic acids, which tend to be insoluble (and aggregate into micelles) at acidic pH, where the carboxylic acids are uncharged and more hydrophobic. Disruption of micelle formation occurs at neutral or basic pH when the

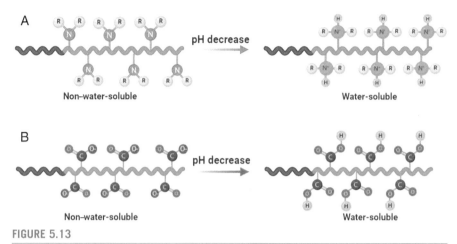

FIGURE 5.13

Illustration of the change in polymer solubility upon protonation. A copolymer having weakly basic amines (A) and weakly carboxylic acids (B) in one of the blocks.

carboxylic acids become ionized. This can be problematic for biological applications since it is usually desirable to have stable micelles during circulation in the blood at pH 7.4. For example, poly[sodium 2-(acrylamide)-2-methylpropanesulfonate-*block*-poly(sodium 6-acrylamidohexanoate)] and poly(sodium 4-styrene sulfonate)-*block*-poly(sodium 4-vinyl benzoate) form micelles at acidic pH values less than 5, but exist as monomers at and near-neutral pH.

Temperature-responsive polymers

The repulsive ionic interaction determines the transition temperature and the volume change at the transition. Polymers, which become insoluble upon heating, have a so-called lower critical solution temperature (LCST). Systems, which become soluble upon heating, have an upper critical solution temperature (UCST). LCST and UCST systems are not restricted to an aqueous solvent environment, but only the aqueous systems are of interest for biomedical applications. The change in the hydration state, which causes the volume phase transition, reflects competing for hydrogen bonding properties, where intra- and intermolecular hydrogen bonding of the polymer molecules are favored compared with solubilization by water.[53] Polymers that are cationic can be used in gene delivery applications. Gene delivery introduces foreign DNA into host cells, which can be used for gene therapy and genetic modification of agricultural products. Depending on how the foreign DNA is introduced into the host cells, gene delivery can be viral or nonviral. Viral gene delivery is based on the fact that viruses have the ability to transfer their DNA into cells. For the desired gene to be delivered to a host cell, it is included in a virus particle, which in turn can enter the host cell and deliver it.[54]

Temperature-responsive polymers show a phase transition, which exhibit a change in physical properties upon changes in temperatures. Hydrophobic gels, such as *n*-isopropyl acrylamide (NIPAM) gels, undergo a phase transition in pure water, from a swollen state at low temperatures to a collapsed state at high temperatures.[55] In other cases, the basis for the property changes is hydrogen bonding with a change in ionic interaction, for example, gels with cooperative hydrogen bonding, such as interpenetrating polymer networks that are temperature responsive. Interpenetrating polymers of poly(acrylic acid) and poly(acrylamide) undergo a phase transition in pure water (the swollen state at high temperatures).[56] Poly(NIPAM) (PNIPAM) is the most prominent candidate as a thermo-responsive polymer, even though the second polymer in this class has a nearly identical transition temperature, poly(*N,N*-dimethyl acrylamide) (PDEAM) (30°C−34°C). PNIPAM copolymers have been mainly studied for the oral delivery of calcitonin and insulin.[57] P(NIPAM-co-BMAco-AAc) was studied for the intestinal delivery of human calcitonin.

Poly(*N*-ethyl oxazoline) (PEtOx) has a transition temperature around 62°C, which is too high for any drug delivery application. However, a double thermo-responsive system by graft polymerization of EtOx onto a modified PNIPAM backbone was prepared. Currently these systems are being explored for their potential in drug delivery because they tend to aggregate into micelles above the LCST.[58] This property can be used for the preparation of drug-loaded micelles.[59]

Polypeptides can also show LCST behavior when hydrophilic and hydrophobic residues are well balanced. A polymer made of pentapeptide (GVGVP) as the repeating unit exhibits a volume phase transition at 30°C, which is the transition temperature for hydrophobic folding and assembling.[57] Below the phase transition, water molecules are structured around the polymer molecule. The attractive forces weaken upon heating and finally go into the bulk phase. Above the phase transition temperature, there is the stabilization of the secondary supramolecular structure, that is, a twisted filament structure of β-spirals, which have type II β-turns. The phase transition of these protein-based polymers can be described in terms of an increase in order. This occurs due to hydrophobic folding and assembly.[60]

Polyamidoamine (PAMAM) dendrimers represent an exciting new class of macromolecular architecture called "dense star" polymers. Unlike classical polymers, dendrimers have a high degree of molecular uniformity, narrow MW distribution, specific size and shape characteristics, and a highly functionalized terminal surface.[61] The manufacturing process is a series of repetitive steps starting with a central initiator core. Each subsequent growth step represents a new "generation" of polymer with a larger molecular diameter, twice the number of reactive surface sites, and approximately double the MW of the preceding generation.

To summarize, there are several polymeric candidates for nonviral gene therapy including dendritic linear hybrid polymers based on PAMAM dendrimers and PEG, which contain a targeting moiety for a cell surface receptor, and hybrid polymers that self-assemble with DNA to nanoparticles of 200 nm diameter with a PEGylated outer shell with surface receptor targeting moieties.[62] The idea of employing colloidal assemblies that can simultaneously deliver multiple therapeutic and diagnostic agents, however, remains an elusive goal.

Polymers in implantable prosthesis

As discussed in previous chapters, implantable prostheses hold the potential to enhance human quality of life and extend longevity. These prostheses are devices that require surgical procedures to be placed inside the human body. Due to the unique requirements, these devices involve the use of diverse materials including metals, ceramics, and of course polymers. Historically, synthetic polymers have been used for orthopedic implants, due to their ability to provide structural support while reducing friction between contacting surfaces. Among the most studied materials, we find nylon, polyesters, PTFE, polymethylmethacrylate (PMMA), PE, and silicones.[63] In the context of long-term functionality, as in the case of prosthesis, biocompatibility (blood interactions, hypersensitivity, fibrous encapsulation) and mechanical performance must be carefully analyzed. In general, selection of the mechanically optimal polymeric biomaterial for the development of prosthetic implants will depend upon the endpoint application.

Orthopedic applications

Polymers and copolymer composites have rapidly gained attention as suitable materials for joint replacement. Two of the most common joint replacement surgical procedures performed are knee and hip, and to date, polymeric biomaterials are the first choice for low-friction hip replacement. The general criteria for material selection intended for orthopedic applications can be summarized as follows:

- **Biocompatibility:** Under the specific physiological environment where the device will be implanted, and in combination with the other materials forming the device (How is the material/device affecting the environment? Namely body tissue and fluids).
- **Sterilization compatibility:** The material must remain unchanged after sterilization.
- **Mechanical requirements:** Withhold the maximum load, toughness, wear resistance.

Historically, highly stable polymeric systems such as UHMWPE and PTFE have been analyzed and tested due to their outstanding mechanical properties and high wear resistance. However, when implanted, the acetabular cup of PE produces debris that is recognized and attacked by the immune system, triggering inflammatory responses, along with a series of undesired side effects and complications. In recent years a different approach was developed to stabilize PE by blending it with vitamin E, which results in the interruption of the oxidation cycle and therefore decreased oxidation damage, improved biocompatibility, and lower bacterial adhesiveness (Fig. 5.14). PTFEE was initially used but soon discontinued as a consequence of

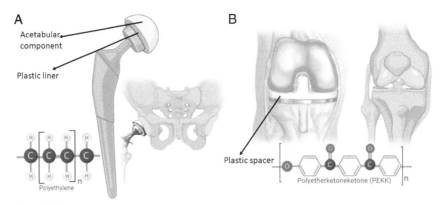

FIGURE 5.14

Illustration of orthopedic implantable prosthesis using polymers. Plastic liner: functional component of a hip replacement implant (A); plastic spacer holding the space between the two contacting surfaces (B).

unpredicted high wear and distortion that resulted in granuloma (localized areas of inflammation). To date, polymeric biomaterials are used to fabricate components of hip and knee implants. New bearing materials have been introduced in an effort to substitute PE for the overall performance of the implants, specifically for wear resistance to improve long-term outcomes after knee replacement implantation. Today we have FDA-approved commercial polymers such as Zeniva PEEK (polyetherketoneketone), manufactured by Solvay, which offers a longer service life and exhibits 50% less wear when compared with metallic implants. In addition, the modulus of elasticity of this material is remarkably similar to that of the cortical bone, improving fixation, reducing wear, and protecting the surrounding natural tissue. More polymers for replacement of hard and soft tissue are discussed in Chapters 9 and 10, respectively.

Dental applications

Dental treatments can range from simple fillings to complete teeth implants. A great majority of the materials used among all different procedures involve polymeric materials (especially the ones related to restorative and prosthetic purposes) (Fig. 5.15).

The material selected for this application must resemble the physical, mechanical, and aesthetic properties of natural teeth. On top of the good mechanical properties and wear resistance of polymers, PMMA is tasteless, odorless, nontoxic, and resistant to most microbial populations interacting with teeth and surrounding tissue.[63] Recently the development of artificial teeth has been accelerated having ultra-high-molecular-weight polyethylene (UHMWFE), PTFE, and polyethylene terephthalate (PET) as the major candidates due to their outstanding mechanical profiles. Other applications of polymeric materials such as PMMA can be found in dental bridges and full prosthetic dentures.

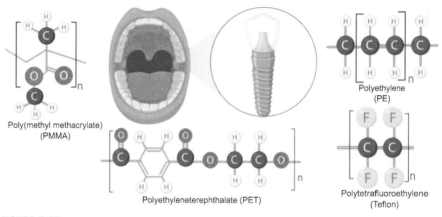

FIGURE 5.15

Illustration of a full dental implant and common materials used in dental prosthesis.

Polymers in tissue replacement

The goal of tissue engineering is to achieve regeneration of a tissue or organ by artificially growing a new tissue or organ that will behave in a functional way to the parent tissue or organ. Strategies to achieve this goal include harvesting donor tissue and dissociating it into individual cells first. The cells can be either adult or embryonic. Adult cells can be either differentiated, which means they are already defined as belonging to a specific tissue such as skin, muscle, and bone, or they can be stem cells. The cells can theoretically directly be implanted or induced to proliferate in the desired manner in tissue culture. However, the question arises: Once these cells proliferate, how do they organize into the desired shape to produce a specific tissue or organ? The key to this is guiding their organization on a so-called scaffold. Scaffolds are generally materials that resemble the geometrical, mechanical, and chemical properties of the tissue to be created. Once the cells occupy the scaffold and proliferate, a tissue such as bone or skin can be created.[64]

The entire process of starting from donor tissues and seeding scaffolds to create tissues or organs requires a significant amount of interdisciplinary knowledge, found at the intersection of cell biology, chemistry and biochemistry, and materials science. Numerous challenges stand in the way of achieving success, many of which will be discussed next.

The main components of tissue engineering approaches include either (1) cells alone, (2) cells with scaffolds, and (3) scaffolds alone. The biomedical behavior of each one of these components can be enhanced by in vitro microenvironmental factors before their application into strategies that can be successful in creating successful tissues or organs.

Cells that are used in tissue engineering include autologous parenchymal cells, allogeneic parenchymal cells, and marrow stromal stem cells. Scaffold materials, cells, and soluble cell regulators have to work together to get optimal regeneration of tissues and organs.

As discussed earlier, for a tissue engineering approach to lead to the successful fabrication of feasible tissue, scaffolds are absolutely necessary to mimic the shape and mechanical properties of said tissue.[65] The scaffolds sometimes can be cell-free, which means it is not necessary to seed them with cells. An example from this category are scaffolds intended for creating small amounts of bone-like materials. These are generally calcium phosphate minerals, which can be used to repair bone defects, have a similar mineral structure as the bone but are resorbable and induce growth of new bone in their place.

In other types of tissues, the scaffolds must include cellular components that will in turn work to create the new desired tissue. For example, collagen tubes can be seeded with endothelial cells (ECs) to fabricate blood vessels. In this case, the collagen serves an adhesive function, helping to increase the EC adhesion to the polymeric scaffold structure, which results in the coverage of the inner surface by a single layer of ECs (Fig. 5.16).

All scaffolds must have high porosity and adequate pore size to facilitate cell seeding and diffusion of both cells and nutrients throughout the entire structure. The scaffolds must be absorbed by the surrounding tissues after the new tissue is regenerated, without a surgical procedure. The rate of scaffold degradation has to be as close as possible to the rate of new tissue formation. This means that while cells are

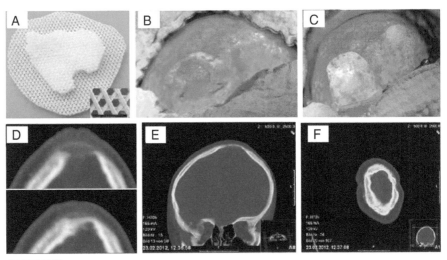

FIGURE 5.16

Example of a cell-free clinical application of tissue engineering: Calvarial reconstruction using polycaprolactone-calcium phosphate scaffolds. Scaffold designed from medical computed tomography (CT) imaging data and fabricated by fused deposition modeling; inset showing scanning electron microscopy scaffold (A). Calvarial defect (B), implanted scaffold (C), CT images showing bony consolidation of the defect after 6 months (D), and complete filling of the scaffold with the bone after 24 months (E and F).

Reproduced with permission from Probst et al.[79]

fabricating their own natural matrix structure around themselves, the scaffold is able to provide structural integrity while needed and will eventually be resorbed and disappear.[64]

Cells and scaffolds are not, however, sufficient for success. In addition, regulatory molecules, called growth factors, may be included in these strategies. Growth factors have cellular regulatory actions that guide cell migration, cell morphogenesis from one type to another, as well as cellular proliferation. These factors are diffusible signaling proteins that induce and promote cascades of events that influence tissue fate and behavior and act as growth stimulators, such as cell differentiation, survival, inflammation, and tissue repair.[66,67]

Examples of applications of tissue engineering include the fabrication of artificial skin, cartilage, bone, blood vessels, as well as nerve and skin regeneration.[68,69] Examples of scaffolds used for these applications include natural polymers such as collagens, collagen-glycosaminoglycan copolymers, fibrin, poly-(hydroxybutyrate), sodium alginate, chitin, and chitosan. Synthetic polymers used as scaffolds are PLA, PGA, PCL, polyanhydrides, or poly(orthoesters).[70]

Polymeric scaffolds need to be proven to be nontoxic and nonimmunogenic and include all components of biocompatibility. These include chemical, mechanical, pharmacological, or even optical properties, as required by the function of the newly created tissue. When designing polymeric scaffolds for vascular tissue engineering, additional evidence of low thrombogenicity is needed to avoid platelet adhesion and activation of coagulation cascade. The scaffolds should also display affinity to cells at a molecular level, so they can be compatible with cell survival and proliferation.

Growth factors are produced both locally by cells and systemically from other sites. In mediating extracellular communications, growth factors act directly on the very cell that produced them (autocrine effect), act on neighboring cells surrounding the growth factor producing cell (paracrine effect), relay a single growth factor communication signal received by one cell to neighboring cells due to direct cell-to-cell interaction (juxtacrine effect), and act on cells distant from the site of growth factor production by traveling through the bloodstream (endocrine effect). Growth factors reside in the interstitial fluid, on the cell surface, and in the extracellular matrix (ECM). These growth factors are not only important for growth, development, and day-to-day maintenance of bone tissues but are mobilized during times of tissue remodeling and injury.

This brings us back to polymers. Because the in vivo environment has a high level of complexity and is aggressive chemically and biologically, most polymers will degrade. However, this does not mean any polymer would be amenable to be used as a tissue engineering scaffold since the rate of degradation will make all the difference from this point of view.

Cell-based therapeutics have largely been unsuccessful for both clinical and financial reasons. Even if a new tissue engineering construct matched the clinical performance requirements, often the increases in costs were not justified by the performance.[71] Scaling up tissue fabrication and getting away from costly manual cell seeding and culturing of artificial scaffolds have not proved economically and logistically feasible to achieve. The most significant roadblocks standing in the way of success are the inability to (1) mimic the cellular organization of natural tissues, (2) scale up to affordable fabrication strategies for clinical application, and (3) induce vascularization throughout the construct before the most inner cells die.

Even if cells are distributed throughout a large-scale scaffold, there is a need for a vascular supply to nourish the cells in the interior of the scaffold. While a vascular supply can grow into an implanted scaffold from surrounding vascularized tissue (Figure 5.17A–D), this process takes time; cells in the interior may die before it occurs.

Scaffold-based tissue engineering methodologies could benefit from a manufacturing process that could fabricate scaffolds that not only have controlled distributions of cells but also spatial gradients of cells (Fig. 5.17A). For example, the idea of building up complete systems of seeded scaffolds, like a functioning arm with interconnected muscle, tendon, cartilage, nerve, skin, and vasculature,

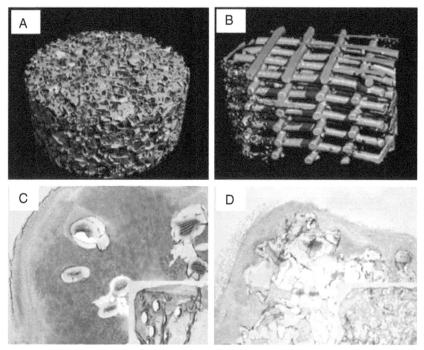

FIGURE 5.17

Cartilage regeneration by chondrocyte delivery on designed bioplotter-fabricated scaffolds. Poly(ethylene glycol) (PEG)/polybutylene terephthalate (PBT) scaffolds fabricated by porogen leaching (A) and bioplotter fabrication (B). Cartilage matrix *(red areas)* generation in bioplotter scaffold after 21 days in a mouse (C) and generation after 21 days in the porogen-leached scaffold (D).

Reproduced with permission from Hollister.[72]

has been put forward. However, the creation of such heterogeneous systems, as well as whole organs, will require capabilities not only for spatial control over cells but also for scaffold materials and microstructure (see Fig. 5.17B).

It may be desirable to have a scaffold design where one portion of the scaffold incorporates a material or microstructure optimized for strength, while another portion has another material composition or microstructure optimized for initial tissue ingrowth. The ability to create concentration gradients of growth factors is also key. Concentration gradients of growth factors are required for cellular responses, that is, the preferential cellular migration from lower to higher concentrations of a mediator.

With additive manufacturing (AM) techniques, objects from 3D model data sets can be constructed by joining material in a layer-by-layer fashion, as opposed to a subtractive way, as most traditional manufacturing methodologies operate. In terms of tissue and organ manufacturing, the additive nature ensures minimal waste of scarce and expensive building material, namely cells, growth factors, and biomaterials. The use of 3D model data enables the fabrication of customized tissues, which is a *condition sine qua non* for patient-specific treatment concepts. Further, AM techniques offer a high level of control over the architecture of the fabricated constructs, guarantee reproducibility, and enable scale-up and standardization. The first step to produce a 3D object through AM is the generation of the corresponding computer model either by the aid of 3D computer-aided design (CAD) software or imported from 3D scanners.[64] There are a large number of imaging methods for data acquisition of human or animal body parts, such as X-ray computed tomography, magnetic resonance imaging, ultrasound echoscopy, single-photon gamma rays (SPECT), and bioluminescence imaging. The CAD model is then tessellated as a Standard Template Library (STL) file, which is currently the standard file for facetted models. Before manufacturing, the STL model is mathematically sliced into thin layers (sliced model), which are reproduced into a physical 3D object by the AM device. Several well-developed and commercially available AM techniques have been employed to design and fabricate scaffolds for tissue engineering applications.[72]

A challenge in additive tissue manufacturing using cell-laden hydrogels is to develop a polymer along with processing conditions that are appropriate for both accurate printing and cell culture. Often these criteria impose opposing requirements. For accurate printing of form-stable structures, high polymer concentrations and cross-link densities are desired, whereas for cell migration and proliferation and subsequent ECM formation, both need to be maintained low. For example, a currently used naturally derived printable biopolymer, namely calcium-cross-linked alginate, has only a small processing window in which both printing and cell culture are possible: the bioprinting window. This bioprinting window can be defined for other hydrogel systems by varying the polymer concentration and cross-link density and assessing the influence on printability and support for cell culture. Often the bioprinting window will be small, if at all present. After several years of predominantly proof-of-principle studies demonstrating the (bio)printability of a gel with a particular AM system, researchers are increasingly optimizing gel parameters and processing conditions in systematic and quantitative ways (Fig. 5.18).

The incorporation of cells into a computer-controlled fabrication process creates living cell/material constructs rather than cell-free scaffolds. Most attempts of additive tissue manufacturing so far have utilized hydrogels designed for purposes other than AM. However, the development of polymers specifically for AM of cell-laden constructs has been explored to a limited extent and may help overcome the limitations of current gels and expand the bioprinting window. One of the few examples of a hybrid gel tailor-made for AM is based on a PEG-polyphenylene oxide (PPO)-PEG block copolymer. The thermosensitive block copolymer conveniently allows for dispensing a cell suspension at ambient temperature, which solidifies upon collecting at 37°C. However, although most cells remain viable during the plotting process, the gel does not support cell viability in culture; all cells die within a few days, while the thermogel slowly dissolves into the culture medium. By functionalizing the terminal hydroxyl units of PEG-PPO-PEG with a peptide linker followed by a methacrylate group, a mechanism for covalent cross-linking as well as biodegradability has been introduced, resulting in increased viability over 3 weeks of culture.[73] A similar

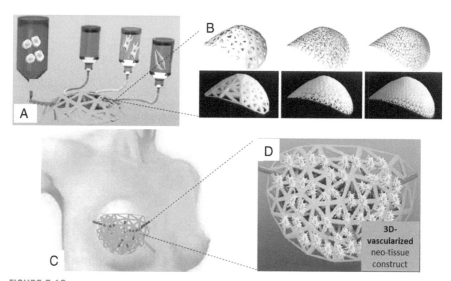

FIGURE 5.18

Schematic representation of the additive tissue manufacturing principle. Concurrent additive manufacturing of scaffolding structure (biodegradable thermoplastic) and cells suspended in gels: pre-adipocytes in adipose-mimetic extracellular matrix (ECM) gel and smooth muscle cells in gel mimicking their native ECM (A). Personalized scaffolds for breast reconstruction (B). Implantation after mastectomy (C), and manufactured 3D neo-tissue construct (D).

Reproduced with permission from Melchels et al.[80]

approach of a synthetic gel that allows for both thermal gelation as well as UV-initiated chemical cross-linking was recently demonstrated by the same group. The polymer is an ABA block copolymer composed of poly(N-(2-hydroxypropyl) methacrylamide lactate) A-blocks and hydrophilic PEG B-blocks with MW of 10 kDa. The hydrophobic A-blocks not only induce LCST behavior employed for printing but are also partly derivatized with methacrylate groups that allow for photopolymerization for increased strength and shape stability.[74]

Polymer network chains give hydrogels their mechanical stability, but at the same time restrict the mobility for cells to migrate and proliferate. Therefore it is important to match the kinetics of degradation with firstly the cell migration and proliferation and subsequently tissue formation, such that the newly deposited ECM can take over the load to a certain extent from the partially degraded polymer network. Moreover, the rate of tissue formation and remodeling depends on many factors and is different for various tissues. Hence it is of utmost importance to study those in vitro and/or in vivo mimetics in great detail.

So far, fabrication of cell-laden PEG structures by AM techniques has mostly employed off-the-shelf PEG-diacrylates in conjunction with tripeptide Arg-Gly-Asp (RGD)-PEG-acrylate, resulting in gels that support cell viability but are nondegradable and therefore of limited use in tissue engineering. However, it can be easily envisioned that the aforementioned strategies of introducing hydrolyzable links, tethered grow factors, and enzyme-sensitive cleavage sites will also be used with alternative hydrogel platforms such as thermosensitive PEG-PPO-PEG, allowing application in AM technologies.

Sterilization of polymers

We refer to sterilization when the goal is to remove all forms of life that would be present on a biomedical component. This includes any biological agents located on a medical device. Sterilization is always included as part of the preparation of biomedical devices as it is necessary to prevent any infections in the patient. Sterilization destroys all microorganisms from the surface of a material. In any biomedical device, pores, cracks, and crevices can hide and be sites for pathogen presence. The most common sterilization techniques include radiation, the use of gas ethylene oxide (EtO), and steam autoclave.[75]

For sterilization that includes radiation, gamma radiation is used. This radiation kills most microorganisms and is compatible with most polymers. However, the dosage rate needs to be optimized based on the type of polymer. It is a technique generally used in disposable biomedical devices. Gamma radiation is now a standard industry procedure and is often preferred to chemical sterilization with EtO. The advantage is that it is free of any chemical residues and does not subject the device to high temperatures. The typical dosage is 2.5 Mrad. One possible challenge is that gamma radiation is ionizing radiation that can lead to polymer discoloration and affect some mechanical properties, depending on the polymer being irradiated.[76] Polymers with high MW and narrow MW distribution are compatible in gamma irradiation sterilization. Among these, the presence of aromatic rings in the polymer structure as well as an amorphous structure are favorable traits of polymers compatible with this method. Sometimes, in addition to discoloration, effects can include increased strength and decreased elongation via induced cross-linking or loss of strength and elongation due to chain scission.

EtO sterilization uses a gas that is flammable, toxic, and colorless. It is used to sterilize polymers because it can kill microorganisms on a medical device and is highly compatible with most polymers. Although it has been the most-used early method of sterilization, it is now being replaced with gamma irradiation due to the potential health hazard involved by exposure to EtO and the high number of steps required for its completion. The benefits include the use of low temperatures and no effects on thermoplastics properties.[77] Steam sterilization uses water vapor to remove microorganisms from medical devices. However, most polymers cannot safely be exposed to steam sterilization multiple times without their properties being negatively affected.[75] Common steam sterilization conditions include exposing the device at 121°C for 30 minutes, or at 134°C for 20 minutes.

Autoclave steam sterilization is an often-used sterilization technique for reusable medical devices. It is based on the pressure vessel of the autoclave producing steam that kills the cells of the microorganisms on medical devices. The same temperatures and times of operation as above apply. However, like steam sterilization, this method is not compatible with most biomedical polymers. Some polymers are compatible with steam sterilization. These include polypropylene, polyphenylene oxide (PPO)/polyphenylene ether (PPE), polycarbonate, polyamides, polyetherimide (PEI), liquid crystal polymer (LCP), polyphenylene sulfide (PPS), and PEEK.

A certain sterilization technique cannot be employed by considering only one polymeric component. Instead, the entire device needs to be tested to determine if a sterilization technique is appropriate and successful. Biomedical devices can contain different materials, often metals, ceramics, and different types of polymers at the same time. The condition of "sterile" is defined by the probability of sterility for each item to undergo sterilization. This probability is commonly referred to as sterility assurance level (SAL), and is defined as the probability of having a single microorganism on the surface after sterilization (expressed as 10^{-n}). Recommended sterility levels for biomedical devices are typically 10^{-6} (which translates into a probability of a cell surviving sterilization of one in one million).[78] Medical devices that come into contact with sterile body tissues or fluids, such as surgical instruments, wound dressing, and implanted medical devices are considered critical. For example, catheters and wound dressings need a 10^{-6} SAL. The recommended sterility for devices and materials in contact with skin is allowed to be 10^{-3} SAL.

Summary

The continuous advancement in engineering polymeric materials along with the understanding of molecular and physiological interactions of tissue-implantable materials continues to push forward the development of new biomaterials. The introduction of biodegradable polymers and hydrogels has proven to be a promising avenue to control specific cell routes with many potential applications. This chapter covered the versatility of polymeric materials and their successful incorporation in a wide variety of medical devices. These materials go from grafted coatings enhancing tissue-implant interactions, scaffold materials for tissue replacement, to highly engineered drug delivery systems with specific target recognition. Despite the ability to synthesize tailored polymers for new applications, the challenges related to degradation, leaching, and potential toxicity still impose limitations when considering long-term implantation of synthetic polymers.

Problems

1. What is DP? How is it estimated and how it relates to the MW of polymers?
2. "p" is the functional conversion (consumption of monomers). How is it calculated and why is it important to know for a given polymerization reaction?
3. Three values of p are given below:
   - $p = 0.10$
   - $p = 0.75$
   - $p = 0.995$

   For each value of p, answer the following questions:
   a) Is it a step-growth reaction or a chain-growth reaction? Why?
   b) Which would form the highest value of DP? Explain.
   c) Which would form the greatest total concentration of polymer? Explain.

M fractions (g/mol)		# (Moles) of molecules in each fraction
0	5000	0.1
5000	10,000	0.3
10,000	15,000	0.8
15,000	20,000	0.9
20,000	25,000	1
25,000	30,000	1.1
30,000	35,000	0.9
35,000	40,000	0.6
40,000	45,000	0.8
45,000	50,000	1.3

4. You are a gifted polymer scientist and made a batch of PS at a laboratory scale using free radical polymerization. After running some tests you concluded that it contains the following MW fractions (presented in the table). Using this information:
   a) Calculate M_n, M_w, and M_z.
   b) Plot the data showing the full distribution of MWs.
   c) Label the three different MW averages.
5. How is the degree of crystallinity in polymeric materials quantified? Explain in detail.
   a) What are the main factors determining the degree of crystallinity in a polymer?
   b) How does % crystallinity relate to polymeric density and MW? Explain.
   c) What will be the effect of heat sterilization on the MW of a polymer?
6. What is the fully extended length of PE molecule that has an MW of 75,000 g/mol?
   a) What would be the extended length of the same polymer if it has a DP of 350?
7. A poly(acrylic acid) molecule has a DP of 550. What is its absolute MW? Explain the role that molecular structure plays on the hydrolysis rate.
8. Why is the reported MW of a polymer (e.g., 85,000 g/mol) not always a good approximation of its actual size? Hint: Is a polymer that is 20,000 g/mol always smaller in size than a polymer that is 50,000 g/mol? Explain.
9. What specific characteristics should a polymeric material attain to be safely implanted? Contrast the benefit/risk of synthetic versus natural polymers.
10. Identify the following polymers by their repeating unit:

e)

f)

g)

11. The repeat unit MW of an unknown polymeric material is 230 g/mol and has the following composition. With the information provided below calculate:

-mers	%
80	5
90	25
100	30
120	25
150	15

   a) The number-average MW
   b) The weight-average MW
   c) Dispersity

12. Describe the differences between:
   a) Monomer and repeating unit
   b) Polycondensation and polyaddition
   c) Polymer configuration and polymer conformation

13. PMMA in an implantable device has a molar mass of 3.42×10^5 g/mol. Assume that carbon-carbon distance in the backbone is 0.154 nm, the angle between bonds is 109.5 degrees, ratio C_N is equal to 9.85, and M_w (repeating unit) $= 100.12$ g/mol. Estimate:
   a) End-to-end distance (in fully extended conformation).

14. Label with the respective name the following copolymer arrangements.

ABABABABABABABABABABABAB

AAAABBBBAAAABBBBAAAABBBB

AABABABBAABAAABBAABABBBA

AAAAAAAAAAAAAAAAAAAAAAAAA
B	B	B
B	B	B
B	B	B
B	B	B
B	B	B
B	B	B
B	B	B
B	B	B

15. A polymeric chain is composed of 350 repeating units of identical length. Calculate the following parameters:
    a) Assuming a free chain with an end-to-end distance of 12.5 nm, calculate the repeating unit length (in angstroms).
    b) Assuming a freely rotating chain (same number of repeating units: 350) with a repeating link length of 0.930 nm and a fixed bond angle of 109.5 degrees, calculate the end-to-end distance of the polymeric sample.

16. Two fractions of PE are mixed as follows:

Fraction	A	B
Weight fraction	0.3	0.7
(g/mol)		

   a) If the average MW of the mixed PE is 2.3 mg/mL, what is the weight-average MW of fraction A?
   b) Calculate the MW of the repeating unit of PE.
   c) Calculate the DP of the mixed PE.

17. You are given the task to design a vascular graft. List the advantages and disadvantages of using polymers as a structural material. Consider the possibility of introducing a drug-eluting system: what type(s) of the polymer will you test as potential candidates?
    a) List the main characteristics with which the material(s) you selected need to comply in order to be successfully implanted. Explain.
    b) What type of delivery systems can be used?
    c) What will be the best modulation approach in order to achieve controlled delivery of the encapsulated drug? Explain your rationale.

18. The following illustration shows the typical configuration of a polymeric micelle. Label the selected layers and explain the role of each one within a targeted controlled drug delivery system.

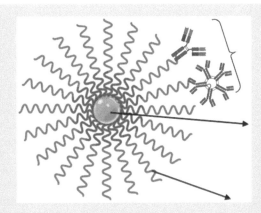

19. What are the key properties of hydrogels for bioengineering applications? What is the main motivation for their use as tissue scaffolds?

20. A polyacrylamide molecule tested for tendon replacement has a DP of 1000.000. What is its absolute MW? Based on the M_w calculated, what is your insight on the mechanical performance of this material?

References

1. Brinson HF, Brinson LC. Polymerization and classification. In: *Polymer Engineering Science and Viscoelasticity: An Introduction*. New York: Springer Science Business Media; 2008:99–157.
2. Alfredo Campo E. Microbial, weather, and chemical resistance of polymeric materials. In: *Selection of Polymeric Materials*. Norwich, NY: William Andrew Publishing; 2008:205–225.
3. Koltzenburg S, Maskos M, Nuyken O. Copolymerization. In: *Polymer Chemistry*. Berlin, Heidelberg: Springer; 2017:584.
4. Prange R, Reeves SD, Allcock HR. Polyphosphazene-polystyrene copolymers: block and graft copolymers from polyphosphazene and polystyrene macromonomers. *Macromolecules*. 2000;33(16):5763–5765.
5. Sakamoto Y, Tsuji H. Crystallization behavior and physical properties of linear 2-arm and branched 4-arm poly(L-lactide)s: effects of branching. *Polymer*. 2013;54(9): 2422–2434.
6. Moad G, Rizzardo E, Thang SH. Radical addition–fragmentation chemistry in polymer synthesis. *Polymer*. 2008;49(5):1079–1131.
7. Rice GN. Polymers and materials chemistry. In: *Encyclopedia of Materials: Science and Technology*. London, UK: Elsevier; 2001:7184–7188.
8. Deleted in review.

9. Dos Santos WN, De Sousa JA, Gregorio R. Thermal conductivity behaviour of polymers around glass transition and crystalline melting temperatures. *Polymer Testing*. 2013; 32(5):987−994.

10. Coiai S, Passaglia E, Pucci A, Ruggeri G. Nanocomposites based on thermoplastic polymers and functional nanofiller for sensor applications. *Materials*. 2015;8(6):3377−3427.

11. Andrady AL, Neal MA. Applications and societal benefits of plastics. *Philos Trans R Soc Lond B Biol Sci*. 2009;364(1526):1977−1984.

12. Nakka JS, Jansen KMB, Ernst LJ. Effect of chain flexibility in the network structure on the viscoelasticity of epoxy thermosets. *J Polym Res*. 2011;18:1879−1888.

13. Zhai J, Shan Z, Li J, Li X, Guo X, Yang R. Study on influence of terminal structure on mechanical properties of GAP elastomers. *J Appl Polym Sci*. 2013;128(4):2319−2324.

14. Lee HH, Lee JH, Yang TH, et al. Evaluation of the flexural mechanical properties of various thermoplastic denture base polymers. *Dent Mater J*. 2018;37(6):950−956.

15. Mijailovic AS, Qing B, Fortunato D, Van Vliet KJ. Characterizing viscoelastic mechanical properties of highly compliant polymers and biological tissues using impact indentation. *Acta Biomater*. 2018;71:388−397.

16. Özkaya N, Nordin M, Goldsheyder D, Leger D. *Fundamentals of Biomechanics: Equilibrium, Motion, and Deformation*. 3rd ed. New York: Springer-Verlag; 2012:1.

17. Shrivastava A. Polymerization. *Introduction to Plastics Engineering Plastics Design Library*. London, UK: Elsevier; 2018:17−48.

18. Kaligian KL, Sprachman MM. Controlled polymers: accessing new platforms for material synthesis. *Mol Syst Des Eng*. 2019;4:144−161.

19. Sobieraj MC, Rimnac CM. Ultra high molecular weight polyethylene: mechanics, morphology, and clinical behavior. *J Mech Behav Biomed Mater*. 2009;2(5):433−443.

20. Ouellette RJ, David Rawn J. Alkanes and cycloalkanes: structures and reactions. In: *Organic Chemistry Structure, Mechanism, and Synthesis*. London, UK: Elsevier; 2014:993−1020.

21. Meyer RW, Pruitt LA. The effect of cyclic true strain on the morphology, structure, and relaxation behavior of ultra high molecular weight polyethylene. *Polymer*. 2001;42: 5293−5306.

22. Thi TTH, Pilkington EH, Nguyen DH, Lee JS, Park KD, Truong NP. The importance of poly(ethylene glycol) alternatives for overcoming PEG immunogenicity in drug delivery and bioconjugation. *Polymers*. 2020;12(2):298.

23. West JL. Drug delivery: pulsed polymers. *Nat Mater*. 2003;2(11):709−710.

24. Stewart SA, Domínguez-Robles J, Donnelly RF, Larrañeta E. Implantable polymeric drug delivery devices: classification, manufacture, materials, and clinical applications. *Polymers*. 2018;10(12):1379.

25. Eldar-Boock A, Miller K, Sanchis J, Lupu R, Vicent MJ, Satchi-Fainaro R. Integrin-assisted drug delivery of nano-scaled polymer therapeutics bearing paclitaxel. *Biomaterials*. 2011;32(15):3862−3874.

26. Xu JP, Ji J, Chen WD, Shen JC. Novel biomimetic polymersomes as polymer therapeutics for drug delivery. *J Control Release*. 2005;107(3):502−512.

27. Pang X, Jiang Y, Xiao Q, Leung AW, Hua H, Xu C. pH-responsive polymer−drug conjugates: design and progress. *J Control Release*. 2016;222:116−129.

28. González-Valdez J, Rito-Palomares M, Benavides J. Advances and trends in the design, analysis, and characterization of polymer−protein conjugates for "PEGylaided" bioprocesses. *Anal Bioanal Chem*. 2012;403(8):2225−2235.

29. Pereira P, Jorge AF, Martins R, Pais AACC, Sousa F, Figueiras A. Characterization of polyplexes involving small RNA. *J Colloid Interface Sci*. 2012;387(1):84−94.

30. Gao D, Lo PC. Polymeric micelles encapsulating pH-responsive doxorubicin prodrug and glutathione-activated zinc(II) phthalocyanine for combined chemotherapy and photodynamic therapy. *J Control Release*. 2018;282:46−61.

31. Vroman I, Tighzert L. Biodegradable polymers. *Materials*. 2009;2:307.

32. Liechty WB, Kryscio DR, Slaughter BV, Peppas NA. Polymers for drug delivery systems. *Annu Rev Chem Biomol Eng*. 2010;1:149−173.

33. Skwarczynski M, Zhao G, Boer JC, et al. Poly(amino acids) as a potent self-adjuvanting delivery system for peptide-based nanovaccines. *Sci Adv*. 2020;6(5):1.

34. Lin CC, Metters AT. Hydrogels in controlled release formulations: network design and mathematical modeling. *Adv Drug Deliv Rev*. 2006;58(12-13):1379−1408.

35. Kung H, Chen H. *Packaging Technology*. 2009;7:429.

36. Song R, Murphy M, Li C, Ting K, Soo C, Zheng Z. Current development of biodegradable polymeric materials for biomedical applications. *Drug Des Devel Ther*. 2018;12: 3117−3145.

37. Mills NJ, Fitzgerald C, Gilchrist A, Verdejo R. Polymer foams for personal protection: cushions, shoes and helmets. *Compos Sci Technol*. 2003;63(16):2389−2400.

38. Koohbor B, Kidane A, Lu WY, Sutton MA. Investigation of the dynamic stress−strain response of compressible polymeric foam using a non-parametric analysis. *Int J Impact Eng*. 2016;91:170−182.

39. Priyanka U, Nandan A. Biodegradable plastic—a potential substitute for synthetic polymers. *Res J Eng Technol*. 2014;5(3):158−165.

40. Rivera-Briso AL, Serrano-Aroca Á. Poly(3-hydroxybutyrate-co-3-hydroxyvalerate): enhancement strategies for advanced applications. *Polymers*. 2018;10(7):1.

41. Rokutanda S, Yanamoto S, Yamada SI, Naruse T, Inokuchi S, Umeda M. Application of polyglycolic acid sheets and fibrin glue spray to bone surfaces during oral surgery: a case series. *J Oral Maxillofac Surg*. 2015;73(5):1017.e1-e6.

42. Tverdokhlebov SI, Bolbasov EN, Shesterikov EV, et al. Modification of polylactic acid surface using RF plasma discharge with sputter deposition of a hydroxyapatite target for increased biocompatibility. *Appl Surf Sci*. 2015;329:32−39.

43. Sisson AL, Ekinci D, Lendlein A. The contemporary role of ε-caprolactone chemistry to create advanced polymer architectures. *Polymer*. 2013;54(17):4333−4350.

44. Gnanasekaran D. *Green Biopolymers and Their Nanocomposites*. Singapore: Springer Nature Singapore; 2019.

45. Bellucci R. An introduction to intraocular lenses: material, optics, haptics, design and aberration. *Cataract*. 2013;3:38−55.

46. Pakzad Y, Ganji F. Thermosensitive hydrogel for periodontal application: in vitro drug release, antibacterial activity and toxicity evaluation. *J Biomater Appl*. 2016;30:919−929.

47. Fan DY, Tian Y, Liu ZJ. *Frontiers in Chemistry*. 2019;7:1.

48. Wei M, Gao Y, Li X, Serpe MJ. Stimuli-responsive polymers and their applications. *Polym Chem*. 2017;8:127−143.

49. Anemone A, Consolino L, Arena F, Capozza M, Longo DL. Imaging tumor acidosis: a survey of the available techniques for mapping in vivo tumor pH. *Cancer Metastasis Rev*. 2019;38(1):25−49.

50. Rizwan M, Yahya R, Hassan A, et al. pH sensitive hydrogels in drug delivery: brief history, properties, swelling, and release mechanism, material selection and applications. *Polymers*. 2017;9(6):225.

51. Jhaveri AM, Torchilin VP. Multifunctional polymeric micelles for delivery of drugs and siRNA. *Front Pharmacol*. 2014;5:77.
52. Han SK, Na K, Bae YH. Sulfonamide based pH-sensitive polymeric micelles: physico-chemical characteristics and pH-dependent aggregation. *Colloids Surf A Physicochem Eng Asp*. 2003;214:49−59.
53. Taylor M, Tomlins P, Sahota T. Thermoresponsive gels. *Gels*. 2017;3(1):4.
54. Bodoki AE, Iacob B-C, Bodoki E. Perspectives of molecularly imprinted polymer-based drug delivery systems in cancer therapy. *Polymers*. 2019;11(12):2085.
55. Li B, Thompson ME. Phase transition in amphiphilic poly (N-isopropylacrylamide): controlled gelation. *Phys Chem Chem Phys*. 2018;20(19):13623−13631.
56. Zhao Y, Kang J, Tan T. Salt, pH and temperature-responsive semi-interpenetrating polymer network hydrogel based on poly(aspartic acid) and poly(acrylic acid). *Polymer*. 2006;47(22):7702−7710.
57. Gandhi A, Paul A, Sen SO, Sen KK. Studies on thermoresponsive polymers: phase behaviour, drug delivery and biomedical applications. *Asian J Pharm Sci*. 2015;10(2):99−107.
58. Li J, Mooney DJ. Designing hydrogels for controlled drug delivery. *Nat Rev Mater*. 2016;1:16071.
59. Schmaljohann D. Thermo and pH responsive polymers in drug delivery. *Adv Drug Deliv Rev*. 2006;58:1655−1670.
60. Pena-Francesch A, Demirel MC. Squid-inspired tandem repeat proteins: functional fibers and films. *Front Chem*. 2019;7:1.
61. Jang JG, Park HB, Lee YM. Molecular thermodynamics approach on phase equilibria of dendritic polymer systems. *Korean J Chem Eng*. 2003;20(2):375−386.
62. Bolu BS, Sanyal R, Sanyal A. Drug delivery systems from self-assembly of dendron-polymer conjugates. *Molecules*. 2018;23(7):1570.
63. Dang TT, Nikkhah M, Memic A, Khademhosseini A. *Polymeric Biomaterials for Implantable Prostheses*. Elsevier; 2014.
64. Pina S, Ribeiro VP, Marques CF, et al. Scaffolding strategies for tissue engineering and regenerative medicine applications. *Materials*. 2019;12(11):1824.
65. Dzobo K, Thomford NE, Senthebane DA, et al. Advances in regenerative medicine and tissue engineering: innovation and transformation of medicine. *Stem Cells Int*. 2018;2018. https://doi.org/10.1155/2018/2495848.
66. Bonnans C, Chou J, Werb Z. Remodelling the extracellular matrix in development and disease. *Nat Rev Mol Cell Biol*. 2014;15:786−801.
67. Caballero Aguilar LM, Silva SM, Moulton SE. Growth factor delivery: defining the next generation platforms for tissue engineering. *J Control Release*. 2019;306:40−58.
68. Ikada Y. Challenges in tissue engineering. *J R Soc Interface*. 2006;3(10):589−601.
69. Metcalfe AD, Ferguson MWJ. Tissue engineering of replacement skin: the crossroads of biomaterials, wound healing, embryonic development, stem cells and regeneration. *J R Soc Interface*. 2007;4(14):413437.
70. Shim JB, Ankeny RF, Kim H, Nerem RM, Khang G. A study of a three-dimensional PLGA sponge containing natural polymers co-cultured with endothelial and mesen-chymal stem cells as a tissue engineering scaffold. *Biomed Mater*. 2014;9(4):045015.
71. Williams DF. Specifications for innovative, enabling biomaterials based on the principles of biocompatibility mechanisms. *Front Bioeng Biotechnol*. 2019;7:255.

72. Hollister SJ. Porous scaffold design for tissue engineering. *Nat Mater*. 2005;4(7): 518–524.
73. Hacker MC, Nawaz HA. Multi-functional macromers for hydrogel design in biomedical engineering and regenerative medicine. *Int J Mol Sci*. 2015;16(11):27677–27706.
74. Vermonden T, Jena SS, Barriet D, et al. Macromolecular diffusion in self-assembling biodegradable thermosensitive hydrogels. *Macromolecules*. 2011;43:782–789.
75. Dai Z, Ronholm J, Tian Y, Sethi B, Cao X. Sterilization techniques for biodegradable scaffolds in tissue engineering applications. *J Tissue Eng*. 2016;7:2041731416648810.
76. Haim Zada M, Kumar A, Elmalak O, Mechrez G, Domb AJ. Effect of ethylene oxide and gamma (γ-) sterilization on the properties of a PLCL polymer material in balloon implants. *ACS Omega*. 2019;4:21319–21326.
77. Norwich DW. Fracture of polymer-coated nitinol during gamma sterilization. *J Mater Eng Perform*. 2012;21:2618–2621.
78. Bernhardt A, Wehrl M, Paul B, et al. Improved sterilization of sensitive biomaterials with supercritical carbon dioxide at low temperature. *PLoS One*. 2015;10(6):e0129205.
79. Probst FA, Hutmacher DW, Müller DF, Machens H-G, Schantz J-T. Calvarial reconstruction by customized bioactive implant. *Handchir Mikrochir plast Chir*. 2010;42(6): 369–373.
80. Melchels FPW, Domingos MAN, Klein TJ, Malda J, Bartolo PJ, Hutmacher DW. Additive manufacturing of tissues and organs. *Prog Polym Sci*. 2012;37(8):1079–1104.

Hard tissues and orthopedic soft tissues

Learning objectives

For the purpose of this book, a natural material is one that has its provenance from the ground, plants, or animals. Although the basics of the structure and function of such natural materials are presented in this chapter, the reader is advised to consult chemistry textbooks for more information on types of bonding, basic amino acid structures, and similar general organic chemistry concepts, which are not within the scope of this biomaterials textbook.

Below are the chapter-related learning objectives that will be attained by the readers:

- Acquire a basis of understanding of tissue composition, structure, and properties, starting with a basic knowledge of the structure and properties of the biological molecules that are likely to be part of tissue composition and relevant to integration of biomaterials into devices that repair or replace tissues and organs.
- Become familiarized with the structural and mechanical properties of hard and soft tissues, which include bone and teeth, as well as orthopedic soft tissues such as tendons, ligaments, and cartilage.
- Understand the mechanisms associated with tissue healing.

Proteins and proteoglycans of the extracellular matrix of tissues

Present in all tissues, the extracellular matrix (ECM) is the material that surrounds cells; it has a three-dimensional (3D) organization that is composed of molecules that are produced by cells, such as proteins, sugar molecules, or enzymes. Although many of the same components (e.g., collagen, elastin, fibrillin, etc.) may be part of many different tissues and organs, the exact composition and structural organization of the ECM will strongly depend on the type of tissue. The ECM has the function of providing structural support (e.g., cushion for cells, mechanical strength, or adhesion), as well as playing biochemical roles (e.g., signal transduction—transmission of molecular signals by cells—and tissue repair following injury, to name only two).

Introductory Biomaterials. https://doi.org/10.1016/B978-0-12-809263-7.00006-8

Part of the ECM of tissues, proteins are complex molecules that are critical for the effective functioning of the human body. Proteins are large molecules and are essentially natural polymers, called polypeptides, where the monomers are amino acids held together by peptide bonds. There are 21 amino acids that make up the structure of proteins. Depending on the organization of these amino acids into the protein chain, the 3D structure (Fig. 6.1) and ultimately the function of a specific protein is determined.

Proteins serve an array of functions in the human body that are critical for life. For example, proteins have a structural role, that is, they provide structural support for cells and permit body movement. Transport proteins are those that can bind different small molecules as needed for various physiological processes. Messenger proteins, such as hormones, help with signaling between cells. Antibodies protect against disease by binding to bacteria or viruses and initiating the body's defense mechanisms. Plasma proteins are those that are part of the blood plasma, and include albumin (creates and maintains osmotic pressure), globulin (has a role in the immune response), fibrinogen (has a role in blood coagulation), clotting factors (have a role in the conversion of fibrinogen into fibrin, another protein that has a role in tissue healing), and regulatory proteins (have a role in the regulation of gene expression). Enzymes have a critical influence on the immense number of

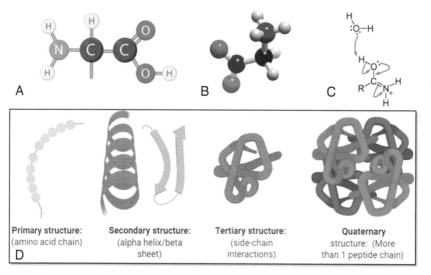

FIGURE 6.1

Proteins: synthesis and structure. Amino acid backbone structure (A). Chemical structure of the simplest amino acid: glycine (Gly) (B). Structural formula of the peptide bond occurring during extension of peptide chains (protein synthesis) (C). Illustration of the structural conformation of proteins (D).

biochemical reactions occurring in cells and take part in the formation of new proteins.[1] Understanding the structure and function of proteins is paramount to understanding not only the structure and functions of tissues, but also understanding their role in processes related to the progression and diagnosis of disease, healing after injury, as well as in finding treatment options. Proteins can have different organizations, from the α-helix and β-sheet (Fig. 6.1) to the triple helix (Fig. 6.2B) and β-strands (Fig. 6.2C). Further, these basic structures can continue to assemble into larger biological structures in tissues, such as mesh networks, filaments, fibrils with micron size diameters to larger fibers with diameters in the 100-μm range, as well as protofilaments and filamentous plaques (Fig. 6.2).

The primary structure of a protein starts with the amino acid chain (Fig. 6.1), which then can assemble into one of the secondary protein structures, the right-handed α-helix structure (Fig. 6.1). The α-helix structure is held together by

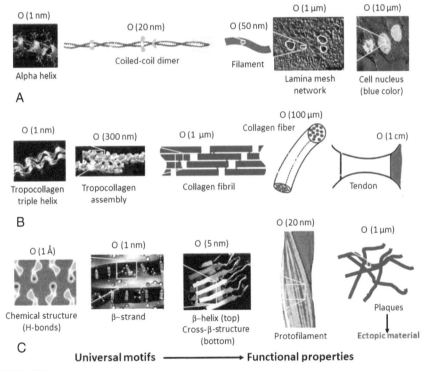

FIGURE 6.2

Schematics of three example biological protein materials. Intermediate filaments (A), collagenous tissues (B), and amyloid proteins (C) revealing their hierarchical makeup, illustrating potential material candidates benefiting from a "materiomic" perspective.

Reproduced with permission from Buehler and Yung.[31]

hydrogen bonds that form between the hydrogen atom in the amino group of the amino acids and the oxygen in the carboxyl group of the same amino acid chain, leading to the right-handed twist structure of the α-helix. The secondary structure of proteins also includes β-strands, which then can assemble into β-sheets (Fig. 6.1). The β-strand is formed when the primary structure of a protein has alternating polar and hydrophobic amino acids, which leads to a self-assembly of a structure of alternating amino acid side chains. Further, β-sheets are formed via the self-assembly of β-strands via hydrogen bonding, with hydrophobic and hydrophilic side chains pointing in different directions (Fig. 6.1). The tertiary structure of proteins refers to their 3D shape (Fig. 6.1). A single polypeptide chain acts as the backbone of the tertiary structure, from which side chains display secondary structures that interact with each other, leading to the 3D organization of each protein. Finally, some but not all proteins have a quaternary structure, which means these proteins have subunits containing multiple polypeptide chains.

All of these structures are part of the ECM of various natural tissues and impart their properties. Such an understanding of the structural building blocks of tissues is helpful for the design of successful biomedical devices that interface with tissues and organs.[2]

Collagen is the most abundant protein in the mammalian body, being the main organic component of bone, skin, tendons, ligaments, cartilage, cornea, and other tissues, most of which will be discussed later in this chapter. There are many forms of collagen, among which collagen I is one of the most abundant and found in the composition of tissues such as skin or bone. Collagen I is a fibrillar protein that is heavily deposited during tissue healing and is a large part of the structure of scar tissue. While there are other types of collagen, the fibrillar collagen I often acts to mechanically reinforce tissues and ensure they present with adequate mechanical performance as required by their function. However, there are around 20 types of collagen present in the body, depending on the type of tissue, and not all of them are fibrillary in microstructure. As mentioned previously, type I collagen is mainly found in skin, bone, tendons, and ligaments. Type II collagen is found in the cornea. Type III collagen is part of the blood vessels and arteries, as well as organs such as lungs and liver. Type V collagen is the type that is found in hair, placenta, and on the cellular surfaces.[3]

All proteins in the ECM are produced by specialized cells; collagen is produced by cells called fibroblasts. It acts together with other proteins, such as elastin and glycoproteins, to keep tissues cohesive. Other properties that collagen imparts to tissues are tensile strength, flexibility, and some level of elasticity.

The collagen type I molecule has a structure that includes a single triple-helical domain that extends across around 95% of the molecule (Fig. 6.3). However, other collagen types have different structures that have multiple triple-helical α-domains.[4] In general, we can state that the collagen I structure includes a right-handed bundle of three parallel, left-handed polyproline II-type helices (Fig. 6.3). As an example, Fig. 6.4 shows a scanning electron microscopy (SEM) image of collagen I fiber organization in cartilage tissues.

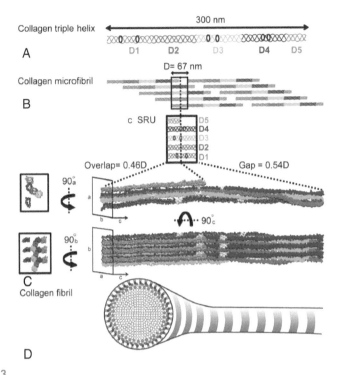

FIGURE 6.3

Structural hierarchy of collagen I in tissues. Collagen monomer (A). Microfibril, containing five monomers packed in parallel (B). The smallest repeating unit that contains the entire sequence of all five D-segments of the monomer in the microfibril configuration (C). Fibril, with concentric packing of collagen monomers with a single fibril for the overlapped region (cross-section) (D).

Reproduced with permission from Zhu et al.[32]

The fibrous collagens are organized in a primary, secondary, tertiary, and finally a quaternary structure. In the primary structure of the fibrous collagens, a glycine amino acid hinges every third amino acid, which imparts the ability to rotate (Fig. 6.3A). At the same time, the amino acid hydroxyproline provides the chain with stiffness. The primary collagen structure is specific to each tissue. For example, collagen type I is present in tendons, while collagen type II is in the cartilage composition. The function of the secondary collagen structure (Fig. 6.3B, collagen microfibril), as well as the tertiary structure (Fig. 6.3C), is in wound healing and tissue remodeling (discussed in Chapter 8). They act as substrates for the collagenase enzyme, which degrades collagen fibers. In the quaternary structure of fibrous collagens, collagen fibril is created by the side-by-side packing of several collagen molecules, which gives the banded structure that is well recognized by microscopy (Fig. 6.3D). The quaternary collagen structure has a function in the formation of blood clots by inducing the aggregation of blood platelets.

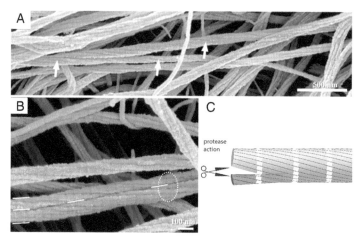

FIGURE 6.4

Scanning electron microscopy (SEM) images of grade 0 knee and hip articular cartilage showing the breakdown at the individual fiber level. Knee cartilage showing a fiber that splits to form two smaller fibers indicated by *arrows* (A); hip cartilage exhibit prototypic arranged fibrils (B), and schematic illustration of protease-induced fibril breakdown (C).

Reproduced with permission from Gottardi et al.[33]

Proteases are enzymes that have the ability to break down proteins into smaller polypeptide fragments and even down to individual amino acids (Fig. 6.4). They act by catalyzing the hydrolysis (a reaction of breakdown of a polymer when it reacts with water) of the peptide bonds in the protein structure.

Elastin is another key ECM protein that is present in the human body. It is also a structural protein with a fibrous structure but displays much higher elasticity, around 1000 times higher than that of collagen type I. Elastin has in its composition the amino acids called desmosine, isodesmosine, and lysinonorleucine (see Fig. 6.5 for their chemical structure). This protein imparts elasticity and resistance to organs and tissues. Thus organs that present with high flexibility have a high composition of elastin. These include aorta, skin, and lungs.[5]

Along with collagen and elastin, another elastic glycoprotein called fibrillin is also part of the architecture of several tissues and organs, including, but not limited to, skin, arteries, and lungs. While there are four types of fibrillin, fibrilin-1 is significant because it organizes into microfibrils, which in turn surround elastin in a sheath-like arrangement. Fibrillin proteins present with critical roles in tissue formation and repair.

Fibronectin is yet another fibrillar glycoprotein, which, along with laminin, is part of the ground substance of the ECM. Fibronectin and laminin play important roles in cellular processes such as adhesion, migration, differentiation, and growth.

Fibrin, on the other hand, although similar sounding in name as fibronectin and fibrillin, has a completely different role in the body and is not part of the ECM. We

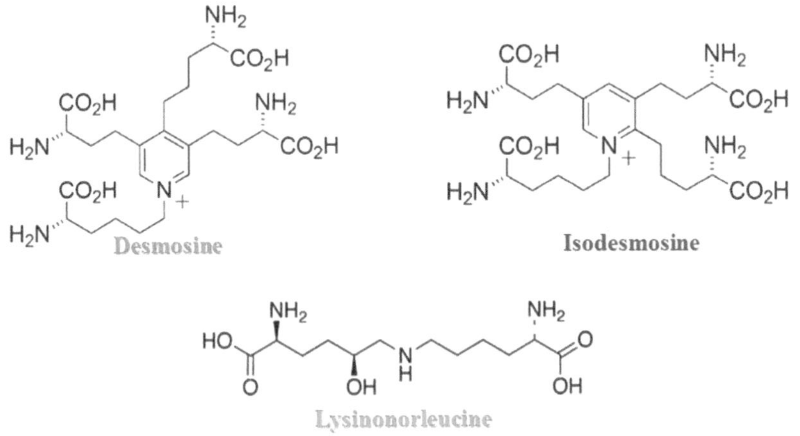

FIGURE 6.5

Elastin components: Structures of desmosine-d_4, desmosine-CH_2, and lysinonorleucine.

Adapted with permission from Murakami et al.[34]

are mentioning it here because it is a significant protein from the point of view of biomedical devices because it is involved in the mechanisms of tissue healing and repair, which will be discussed later in the textbook. It is an insoluble fibrillary protein that is the first to be produced at a site of a tissue injury and is the first protein to be formed and also the major component of blood clots. It is structurally organized into long fibrous threads and it is formed at the site of injury via the enzymatic conversion of fibrinogen. The enzyme that catalyzes this transformation is called thrombin, which is the enzyme that plays the major role in blood clotting. Fibrinogen, in turn, is a protein found in blood plasma, which is only converted into fibrin upon injury, when thrombin catalyzes its transformation.

In addition to these proteins, the ECM also contains proteins with covalently attached sugar molecules, named proteoglycans. In other words, proteoglycans are composed of glycosaminoglycans (GAGs), which are attached to core proteins via covalent bonds (Fig. 6.6). They are part of connective tissues, the ECM, and at the surface of cells.[6] GAGs are hetero-polysaccharides that are composed of a repeating disaccharide unit and are the most abundant sugar molecules in the human body. They play an important role in tissue regeneration, especially in skin and bone. They have a linear organization. The GAGs present in the ECM are hyaluronic acid, heparin, chondroitin sulfate A and B, and heparin sulfate. With the exception of hyaluronic acid, most other GAGs that are part of the proteoglycan structure have sulfate groups in their chemical composition.[7] Fig. 6.7 shows the chemical structures of two proteoglycans, chondroitin sulfate C and hyaluronic acid.

Because GAGs are highly negatively charged, they have the ability to perform their biological function of occupying space, bind water, and repel any other negatively charged molecules, which aids in the hydration of some tissues they are part of (e.g., cartilage). Their structural rigidity aids in the structural integrity of the cells

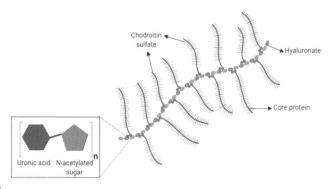

FIGURE 6.6

Schematic representation of the structure of proteoglycans.

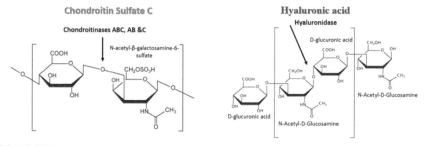

FIGURE 6.7

Proteoglycan components: Structures of chondroitin sulfate (isomers A, B, and C) and hyaluronic acid.[35])

Adapted with permission from Trowbridge and Gallo.[35]

and provides channels of intercellular communication. We will discuss in Chapters 9 and 10 the significance of some GAGs in tissue-device integration and modulation of inflammation and tissue repair. One example is the anticoagulation properties of heparin, which is used in coatings on the interior of cardiovascular devices such as stents.

Chapters 9 and 10 will present an overview of the structure and properties of both hard tissues and soft tissues that are most often the subject of biomedical device development using biomaterials.

Hard tissues

In this section, we will discuss the composition and properties of mineralized tissues (tissues that contain inorganic minerals into soft biological material matrices) such as bones and teeth. The function of such tissues is mainly load bearing and protection of other organs.[8] For example, the skull protects the brain, whereas the ribs

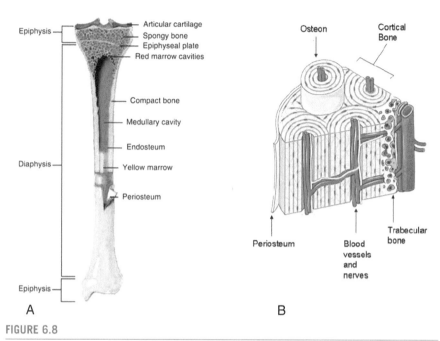

FIGURE 6.8

Graphic illustration of long bones. This classification includes the femur (thigh), tibia and fibula (calf), humerus (upper arm), and radius and ulna (lower arm), represented by the illustration (A). Three-dimensional representation of the structure and organization of trabecular and compact bone (B).

Reproduced with permission from Wegst et al.[36]

protect the heart and lungs. Another function of these tissues is to produce red blood cells, termed hematopoiesis. Bones are also the tissues that allow movement and withstand mechanical load during movement.

The structure of bone at the macroscopic level is divided into cortical bone, trabecular bone, and marrow. The exact shape and distribution of these components depend on the type of bone. However, most of the time, the cortical bone is the dense version and is present at the outermost layer, while the trabecular bone is porous and is found at the bone interior.[9]

Long bones, for example, contain mostly cortical bone, also called compact bone (Fig. 6.8). These bones are cylindrical with two rounded, larger ends and are the most common bones in the human body. Long bones are not necessarily long, but their length is greater than their width. Bone marrow is found in the medullary cavity, in the center of the tubular shaft. The bone ends contact other bones at joints.[10] These enlarged, curved ends are called epiphyses, and to prevent bone-bone friction and damage, they are covered with articular cartilage. Inside the epiphysis, trabecular (spongy) bone is present. The shaft of the long bone is called the diaphysis and is separated from the epiphysis by the growth plate in children and the epiphyseal

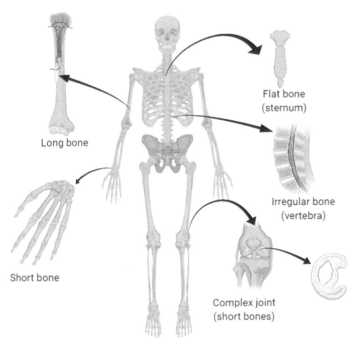

Flat bone
(sternum)

Long bone

Irregular bone
(vertebra)

Short bone

Complex joint
(short bones)

FIGURE 6.9

Bone classification, location, and main examples.

line in adults. The lining of the medullary cavity is called the endosteum and consists of connective tissue. The outer surface of the bone is called the periosteum and it is through it that blood supply reaches the bone.[11]

While long bones are the most widely found bones in the human body, there are other types of bones present (Fig. 6.9). These include short bones, which are cuboidal and have a shell of cortical bone that is filled with trabecular bone. Short bones are only found in the wrist and foot (carpal and tarsal bones). Flat bones also consist of a thin cortical shell filled with trabecular bone, as is also the case for irregular bone. Flat bones are generally not flat, but rather they are curved. They can be found in the skull (calvaria), sternum (breastbone), scapula (shoulder blade), and ribs. Irregular bones have various shapes and can be found, for example, in the facial bones or the bones of the vertebrae.[12]

Bone composition consists of 69 wt% mineral phase, 22 wt% organic phase, and 9 wt% water. The mineral phase is mainly hydroxyapatite (HA), which is a form of calcium phosphate, with citrate, fluoride, and hydroxyl ions in much smaller proportions. The organic phase is mainly collagen, up to 96 wt%, water (9 wt%), and cells (osteoclasts, osteoblasts, and osteocytes). The HA crystals reinforce the collagen fiber matrix with thin needles in the shape of lamellar sheets.[13]

Bone-forming cells (osteocytes) are present within small spaces called lacunae, which are in turn arranged in layers of concentric circles that surround a central

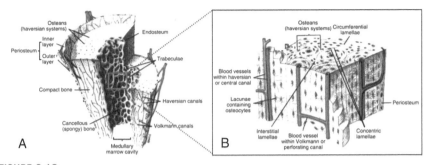

FIGURE 6.10

Compact bone vs. spongy bone. (A) Longitudinal section of a long bone showing both compact and spongy or cancellous bone. (B) Magnified view of compact bone.

Reproduced with permission from Elgazzar.[37]

canal (Haversian canal), running through the diaphysis. Lamellae are the cylindrical sections of these layers. The lamellae and the Haversian canals form osteons. Canaliculi serve to connect the osteons. Mature bone cells are called osteoblasts, whereas bone-resorbing cells are called osteoclasts. Fig. 6.10 depicts these bone components and shows a comparison between compact and spongy (also called cancellous or trabecular) bones.

Bone composition is never constant. Bone always models and remodels, and deposits and resorbs, and this activity is typically in response to the mechanical forces being applied upon it. Wolff's law of bone remodeling states just that: bone is a living tissue that models and remodels in response to the stresses applied upon it.[14] In support of this law, experiments on rats performed in zero-gravity environments showed bone resorption due to the absence of load on the bones. In the absence of gravity, there are no stresses applied to the bone, which in turn leads to no new bone being produced and bone atrophy. By contrast, in hypergravity experiments a larger amount of bone deposition has been noted in response to higher stresses endured by the bone. This is why when parts of bone are replaced with artificial components that are stiffer than the bone tissue, there is a danger of bone atrophy. When a stiffer component is placed in the proximity of the bone, it absorbs most of the stresses applied at that location. Thus the surrounding bone does not receive the required mechanical stimulation that is needed for healthy bone remodeling and is resorbed. This phenomenon is called stress shielding[15] and it is always a concern to take into account anytime a biomaterial is selected to be part of a bone-interfacing implant.

The mechanisms through which bone cells produce or resorb tissue in response to mechanical loading (mechanotransduction) are not well understood. There are several hypotheses being put forward, however. One of these theories states that bone is piezoelectric and transmits electrical signals to the cells in response to mechanical loading.[16] Other theories are related to the shear stresses produced by fluid flowing from zones of high compressive loading and affecting bone cells, osteocytes being the most responsive to it.[17] Whatever the mechanisms behind it, bone remodels only in response to cyclic loading, not static conditions.

Going back to the cortical bone, there are two types, depending on their microstructural organization. Woven bone is bone that has the collagen fibers oriented randomly and loosely packed, which results in lower density than other bone types. It is present at the growth plate in long bones and found in fetuses, infants, and toddlers. It forms de novo, that is, does not need previous tissue to form on (does not need previous hard tissue or cartilage), and its role in adulthood is skeletal defense and repair. In adults it can be found at sites of injury when new bone grows rapidly during tissue remodeling.[18]

As opposed to woven bone, the other type of cortical bone, lamellar bone, needs previous tissue to grow on. Its structure consists of lamellae organized in circular rings around the inner (endosteal) and outer (periosteal) circumference of a whole bone. A bone that forms without previous bone having been present at that particular site is called primary osteonal bone. This type of bone is composed of osteon, which has a vascular channel passing through up to 20 collagen and HA lamellae organized into cylinders, as well as bone cells associated with it. Lamellae can be reservoirs for calcium ions.[19] Osteonal bone is also called Haversian bone. Secondary osteonal bone forms as part of the bone remodeling process, where old bone tissue is resorbed and new bone forms at the same site. Secondary osteons have similar structure to primary osteons but they have a larger vascular channel and correspondingly increased number of lamellae. They are around 200 to 300 μm in diameter in adults. Osteons and lamellae are aligned mostly along the axis of the long bone and along the stress trajectories in other types of bones.

Trabecular bone, also called cancellous or spongy bone, has a similar composition and structure as the cortical bone, being organized into osteons; however, the bone tissue is further organized into plates, beams, and struts, which have marrow in the open spaces (Fig. 6.11).

Depending on the anatomical location and the stresses being applied at that site, the exact organization of the beams and struts will vary. The structure of beams and

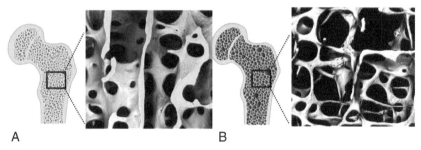

A   B

FIGURE 6.11

Healthy bone vs. osteoporotic bone. Electron micrographs of trabecular bone structure from a healthy adult showing plate and strut structure (A), and from an aging, osteoporotic individual showing beam and strut structure (B).

Adapted with permission from Hertz et al.[38]

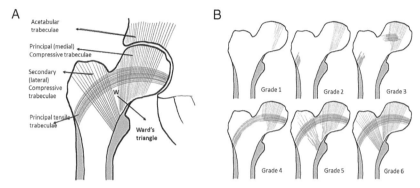

FIGURE 6.12

Sketch of the typical femoral neck and head showing the compressive and tensile trabeculae as designated by Singh. Ward's triangle, an area of reduced trabecular density, is indicated with a W (A). The six grades of the Singh index (B).

Reproduced with permission from Lasanianos et al.[39]

struts depends on the location and loading situation (Fig. 6.12). Usually, the trabecular struts are around 150 to 300 μm in diameter and form a 3D structure that has properties that are optimum for the type of bone. For example, the trabecular bone in the vertebrae of young adults is organized into vertical plates and horizontal struts, while in older adults they are organized into vertical beams and horizontal struts. In older adults, bone resorption is faster and the vertical plates become thinner. In the femoral neck, trabeculae form tensile and compressive bands that are directed along with the force trajectories experienced by the femoral neck.[20]

Trabecular macrostructure can be defined in terms of quantities such as predominant orientation, mean trabecular plate separation, mean trabecular thickness, trabecular density, and connectivity.

Bone is a viscoelastic material and the mechanical properties of both cortical and trabecular bone change with changes in the strain rate.[21] Young's modulus of trabecular tissue is estimated to be correlated to the strain rate by:

$$E = E_{static} * (de/dt)^{0.06} \tag{6.1}$$

Cortical bone also has a creep fracture response. Laboratory tests typically are quasi-static and conducted at strain rates of 0.01 to 0.001 per second. Table 6.1 gives an overview of the main mechanical property ranges for cortical bone, as well as for trabecular bone and dental tissues (these mineralized tissues will be discussed later in this chapter).

Notice in Table 6.1 that some of the values have quite a wide range, indicating the large variation in the mechanical property requirements for any biomedical implant aiming to repair or replace them, depending on the anatomical location.

Table 6.1 Mechanical properties of mineralized (hard) tissues.

	Young's modulus (GPa)	Shear modulus (GPa)	Compressive strength (MPa)	Tensile strength (MPa)	Shear strength (MPa)	Density (g/cm³)
Cortical bone	4–27	2–9	10–160	45–175	50–70	1.8–2.2
Trabecular bone	1–11		7–180[a]			1.5–1.9
Enamel	13.8	6–10	140–280	40–275	10–140	1.9
Dentine	20–84	29	95–386	30–35	6	2.2

[a] *Estimated based on regression of strength vs. structural density. Data from Park and Lakes.*[43]

The properties of hard tissues are not static and can change with time, age of the patient, activity levels, pathologies, and other similar factors.

The other hard mineralized tissues of great importance in the body are teeth (see Fig. 6.13). The term mineralized tissue refers to those tissues that contain minerals (such as HA) that usually act to reinforce a soft, organic, or biological component (e.g., collagen in bone and teeth). Teeth have a complex composition, containing four different phases plus the gums (gingiva) and the jaw bone (alveolar bone). The four phases present in the teeth are the enamel, dentine, pulp, and cementum. The enamel is composed of 97 wt% HA. This high amount of calcium phosphate makes enamel the hardest substance in the human body.

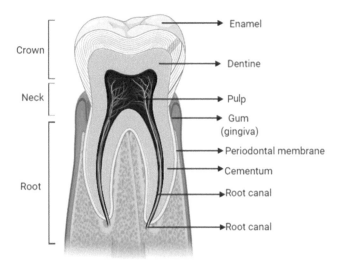

FIGURE 6.13

Illustration of the cross-section of a healthy molar tooth showing its components.

Dentine has a structure and composition that is very similar to the cortical bone and therefore the properties are also similar. More specifically, dentine contains 70 wt% HA; the remaining 30% is made of collagen, cellular components, and water. As opposed to enamel, which does not have the ability to remodel, dentine contains small dentinal tubules that allow cells, which are present in the pulp, to act to produce and resorb dentine tissues, similar to the mechanisms of bone remodeling. The dentine-producing cells are called odontoblasts. The dentinal tubules also contain nerves, which results in pain being transmitted through this tissue. The enamel, by contrast, does not contain nerves or cellular components.

The dental pulp is a soft-tissue substance that is similar to the bone marrow and serves to provide cells and blood supply to dentine, as well as containing nerve cells and thin collagen-like fibers. The pulp is contained in a chamber called the pulp chamber and is located directly under the dentine layer. In addition to odontoblasts, the pulp also contains neurons, fibroblasts (cells producing connective tissue and collagen), and macrophages (large cells that recognize, engulf, and eliminate foreign particles, including, but not limited to, bacteria and viruses). The pulp chamber also contains blood vessels.

Cementum covers the tooth root, is a phase that is similar to bone, and contains thick collagen fibers but no blood vessels, canaliculi, or osteons.[22] Its role is to attach teeth to the surrounding bone tissues in the tooth sockets with ligaments called periodontal ligaments. The mineral content of cementum is approximately 45 wt% (mostly calcium salts), while the rest is organic material (e.g., collagen and polysaccharides). There are two main types of cementum, one of which contains cementoblasts—cementum-forming cells (cellular cementum), and one which does not (acellular cementum). The acellular cementum tends to be found at the bottom two-thirds of the root, and cellular cementum at the top part. Because of the lower mineral content, cementum is softer than dentine. Cementum, like bone, is not a static tissue and remodels in response to the functional requirements when in use.

The alveolar bone supports the teeth and consists of two plates of cortical bone, with spongy (also called cancellous or trabecular) bone in the interior. The structure of the alveolar bone differs depending on location. There are areas where it is composed just of thin cortical bone with no cancellous bone. In other areas, such as in the mandible, the alveolar bone is thick and contains spongy bone, in which pores are filled with marrow. Depending on the stresses encountered at each particular site, the trabeculae vary in dimension, with thicker ones being observed in areas subjected to the highest cyclical stresses and thinner ones in areas less subjected to mechanical stress.

Orthopedic soft tissues

Orthopedic soft tissues are load-bearing connective tissues. Connective tissue is a type of tissue that provides support and protection to other tissues and organs in the human body. It is composed of cells and the intracellular matrix. The most common cells in the connective tissues are fibroblasts, which produce collagen, proteoglycans, and GAGs, which are components of the intracellular matrix. The three primary orthopedic soft tissues are the tendon, ligament, and cartilage. All these three tissues have three basic components. The first component, collagen, is as shown previously one of the main structural proteins of the human body. It is a fibrillar protein with a triple-helical structure, where individual collagen fibrils assemble in bundles that are 0.2 to 1.2 μm in diameter. Another component is elastin, also a fibrillar protein, which as opposed to the stiffer collagen is highly elastic. The tissue matrix (intracellular matrix) is composed of sugar polymers with negatively charged structures called proteoglycans. Due to the charged structure, they impart compressive strength to tissues. Orthopedic soft tissues have few nerves or blood vessels.[23]

The physical and mechanical properties of orthopedic soft tissues highly depend on their composition and microstructural arrangement. For example, the way the collagen fibers are arranged in these tissues will affect their mechanical properties. The fiber arrangement can be either parallel, in a felt-like arrangement, or in arrays of crossing fibers.[24] If the fibers are not organized in a parallel orientation, under tensile forces the fibers will need first to align and strengthen and then they will stretch inelastically, imparting more flexibility to that tissue.

Orthopedic soft tissues, like all other soft tissues, are viscoelastic, which means the stress-strain relationship is rate dependent and has hysteresis due to internal energy losses. Further, soft tissues have the properties of stress relaxation under constant strain and creep under constant stress. Creep is the property of the length of the tissue increasing in time upon application of a constant strain. Stress relaxation is the ability of tissues to need gradually less stress to maintain the same elongation over time.

All orthopedic soft tissues contain both elastin and collagen fibers, which impart good tensile properties. However, since elastin and collagen vary significantly in properties, so do each of the three tissues, depending on the proportion of these components. Tissues that have a higher proportion of collagen have a maximum tensile modulus that is close to that of individual collagen fibers.[25]

The tendon acts to connect muscles with bones and to transmit contractile forces from the muscle to the bones. As opposed to other tendons, those located in areas of the body subjected to sharp bending, such as fingers, have a tendon sheath that directs their path and protects it against friction during the repeated bending.

Tendons are composed of a large proportion of collagen fibers (86% dry weight), a low amount of elastin, and few cells. The collagen fibers are arranged in a uniaxial orientation but the actual cross-sectional appearance varies between the type of tendon and its location.[26] Fig. 6.14 shows the structural organization of the tendon, with details of the organization of collagen into three types of fiber bundles and their component collagen fibers, fibrils, and finally collagen molecules.

Because of the high collagen content, tendons have one of the highest tensile strengths of all soft tissues (50−150 MPa) and a high elastic modulus (1.2−1.8

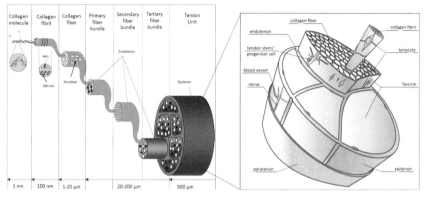

FIGURE 6.14

Structural organization of tendon.

Reproduced with permission from Wang[40] and Docheva et al.[41]

GPa). The properties vary slightly between the type of tendon and the location in the body. Because collagen is stiff, this leads to small strains occurring upon tensile loading. Tendons in load-bearing areas such as those in the legs or spinal area can exhibit strains of up to 10%, while those in non−load-bearing areas allow maximum strains of around 4%. As shown previously, tendons are viscoelastic materials but in comparison with ligaments and cartilage, its viscoelasticity is much smaller. Tendons experience a small hysteresis with 90% to 96% of the energy being recovered upon cycles of loading and unloading. The creep property of tendons leads to the tendon lengthening in time under the same stress, leading to muscles being able to contract more while maintaining the same length. This is a positive property that affords less fatigue in the muscles.[26]

The stress-strain curve for tendon shown in Fig. 6.15 is nonlinear and has three different regions. In the initial region, the so-called "toe region," the fibers align to

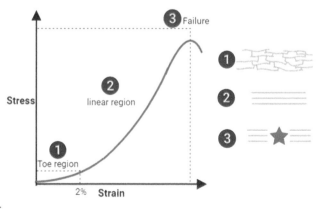

FIGURE 6.15

Typical stress-strain curve for the tendon.

the direction of the applied stress. In the middle region, the "linear region," the fibers get stretched along their long axes, and finally, in the third region, the "failure region," individual collagen fibers begin to break before the tissue fails under the applied stress.

Ligaments act to connect bone to bone and to stabilize joints. They, as other orthopedic tissues, also have a high collagen content, but to a lesser extent than the tendon (70% dry weight). Conversely, they have a higher elastin content than tendons and contain cellular components. The arrangement of the collagen fibers in the ligament and the higher elastin content make it less stiff and less strong than the tendon. Collagen fibers in the ligament are, although mostly straight, somewhat more crimped than in the tendon's exclusive parallel orientation.

Ligaments connect to bone in different ways. They can directly insert into bone when fibers cross over the mineralization front and transform gradually into fibrocartilage, mineralized fibrocartilage, and then bone. There is also indirect insertion when short collagen fibers go at an angle into the periosteum of the bone. In the latter case the fibers can be either oriented in parallel or as branching fibers. The stress-strain curve looks very similar to tendon and presents the same three regions attributed to fiber straightening, stretching, and ultimately failure.

Cartilage's main function is to maintain shape, and it also acts as a bearing surface for articular joints. Cartilage can be found in various areas such as the nose, ear, or rings of the trachea. Its structure has a high amount of dense proteoglycan matrix and a few tangled collagen and elastin fibers, much less than both the tendon and the ligament. The proteoglycan matrix being dense and charged hinders the movements of the collagen fibers. The cartilage-producing cells are called chondrocytes and are found within the spaces of the matrix; however, there are few cells available to produce additional cartilage tissue once any injury occurrs. Therefore cartilage repair is a notoriously difficult problem to solve.[27]

There are three different types of cartilage: hyaline, fibrous, and elastic. Hyaline cartilage has a proportion of 50 to 80 wt% small, evenly distributed collagen fibers and very little collagen content (Fig. 6.16). It is therefore not very elastic and has a smooth, glassy presentation. This type of cartilage is found in areas such as the ribs, growth plate, nasal septum, the windpipe, and the joints. Fibrous cartilage is more abundant in collagen, up to about 90 wt%, and is found in areas of the body such the meniscus of the knee and the outer zone of the intervertebral disc. Elastic cartilage, like hyaline cartilage, has a majority of its composition made of short collagen fibers, but contains a larger amount of elastin and, as the name suggests, has a higher elasticity.[28] Therefore it is present in areas that are bent repeatedly such as the eustachian tubes or the epiglottis. However, it also contains elastin fibers, therefore it is more flexible.

The mechanical properties of cartilage vary greatly depending on the anatomical location and the age of the subject. Because the collagen fibers have an anisotropic structure, cartilage also presents with anisotropy in properties. Most cartilage is load bearing to some extent, therefore a replacement biomaterial would need to consider all the mechanical stresses present at the site of implantation; thus it is not possible

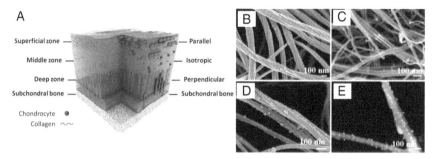

FIGURE 6.16

Articular cartilage stratification (A). Scanning electron microscopy (SEM) image of the collagen fiber meshwork from knee surface articular cartilage with hierarchical organization of 5–7 threads of prototypic fibrils forming an individual collagen fiber (B). SEM image of hip surface articular cartilage with untwisted fibers *(white arrows)* (C). Collagen fibers labeled with collagen II antibodies and inspected in cartilage by SEM (D), and imaging using the backscattering electron (BSE) in the SEM on extracted collagen type II fibers (E).

Reproduced with permission from Gottardi et al.[33] and Camarero-Espinosa and Weder.[42]

to design one single cartilage repair biomaterial. Large variation in mechanical properties is based on location and age of the subject. The tensile behavior of cartilage presents with a stress-strain curve that has the same general areas as the ligaments and tendon, that is, an initial toe region where the collagen and elastin fibers straighten, followed by a linear region where stretching occurs. Compared with ligaments and tendons, cartilage has a lower tensile strength. The compressive behavior is not dependent on the collagen content. Instead, it is influenced by the concentration of the proteoglycans. The typical mechanical property ranges include tensile modulus between 1 and 10 MPa and a compressive modulus of approximately 1 MPa. Due to changes in the collagen organization and structure, these values become smaller as the depth of the cartilage increases. The highest collagen content is at the superficial zones where the fiber orientation is parallel to the surface and it decreases in the middle zone where the fibers are more randomly oriented and oblique. In the deep zone, the collagen fibers are perpendicular to the surface and connect to the bone tissue.[28]

Cartilage has a low coefficient of friction (<0.01) because the synovial fluid, which bathes the joints for lubrication, is squeezed out under compressive loading and resorbed under tension. The meniscus is a critical component of the knee joint, has a cartilage structure, and contains a medial and a lateral area constituent, located between the tibial plateau and the femoral condyle. Ligaments cross the meniscus and help stabilize the knee joint.[29] Each meniscus component has a glossy appearance and a composition made of cells, collagen, and specialized ECM molecules. Vascularization is also present and critical to the health of the meniscus. In utero and in infancy the meniscus has full vascularization, but this starts to recede, and at age 10 only 10% to 30% of the meniscus is vascularized. At maturity the vascularization level is even lower, with only 10% to 25% vascularization.[30] The lack of vascularization leaves the interior, nonvascularized areas of the meniscus susceptible to permanent damage, as tissue repair cannot occur.

The compressive modulus of the meniscus is around 0.4 MPa and it is less permeable to fluid than the articular cartilage, although they have same fundamental function, that is, providing bearing surface and shock absorption.[28,30] Meniscus and articular cartilage have different types of mechanical forces loading them. For example, the meniscus is subjected to tensile hoop stresses that result from extrusive forces during loading. If the meniscus of the knee is lost, the articular cartilage becomes overloaded and can be damaged as well.

Summary

Understanding the structure and function of tissues present in the human body is critical for an engineer to be able to design implantable biomedical devices and successfully select the right biomaterials they are composed of, but also regulate the immune response, repair, and remodeling after injury (intentional such as in surgery, or unintentional such as in the case of accidents or disease). In future chapters we will use the knowledge about how hard and soft tissues are organized and their properties to understand the structure and properties of implantable devices that replace their functions, either partially or completely. Understanding target parameters that a biomaterial included in an implantable device needs to meet, such as porosity and pore size, surface property, chemical composition, and mechanical property requirements, is necessary for the design and fabrication of such devices.

Problems

1. List and explain the specific physiological functions of primary, secondary, tertiary, and quaternary structures of fibrillary collagens.
2. What is the main function of fibrillin?
3. What is the structure of glycosaminoglycans (GAGs)? What is the relationship between the structure and the function of GAGs?
4. From the list provided, which one is not a plasma protein?
   a) Albumin
   b) Globulin
   c) Fibrinogen
   d) Fibronectin
5. The combined density of bone is calculated by accounting for the effective volume of each of its components.
   a) Calculate the percentage volume of the main component of wet bone.

b) Using the information provided below, what is the density of 100 g of wet bone?

c) If the calculated density of dry bone is 0.12 g/cm^3, what is the percentage volume of the mineral phase?

Element	Wt%	Density (g/cm^3)
Water	9	1
Mineral phase	69	3.2
Organic phase	22	1.05

6. Explain Wolff's law for bone remodeling. List and explain at least two hypotheses that support the proposed interdependency between physiological activation mediated by mechanical signals.

7. Contrast the structures of collagen and elastin and explain the implications for the properties of tissues where they are abundant.

8. An average bone in the human body is made from natural composites that provide a tensile strength of approximately 150 MPa and a strain to failure below 2%. For a structural material, this performance would be considered poor when compared with engineered alloys that are several times stronger. Considering the complexity of the biological environment, what other conditions must the bone material meet in order to achieve its physiological function?

9. Explain which dental tissues have the ability to remodel and why.

10. Cartilage is known as a tissue that is very difficult to heal after injury. Explain the reason behind this behavior, as related to the cartilage structure.

11. Explain the difference between the structure of the trabecular bones in a young vs. an elderly individual.

12. What is the most common cell type found in connective tissue? Describe the function of these cells.

References

1. Mitrea DM, Kriwacki RW. Phase separation in biology; functional organization of a higher order. *Cell Commun Signal*. 2016;14:1.

2. National Research Council. *Beyond the Molecular Frontier: Challenges for Chemistry and Chemical Engineering*. Washington, DC: The National Academies Press; 2003.

3. Kanungo I, Fathima NN, Jonnalagadda RR, Nair BU. Go natural and smarter: fenugreek as a hydration designer of collagen based biomaterials. *Phys Chem Chem Phys*. 2015;17: 2778–2793.

4. Miranda-Nieves D, Chaikof EL. Collagen and elastin biomaterials for the fabrication of engineered living tissues. *ACS Biomater Sci Eng*. 2017;3(5):694−711.
5. Schräder CU, Heinz A, Majovsky P, et al. Elastin is heterogeneously cross-linked. *J Biol Chem*. 2018;293(39):15107−15119.
6. Elfenbein A, Simons M. Auxiliary and autonomous proteoglycan signaling networks. In: *Methods in Enzymology*. Oxford: Elsevier; 2010:3−31.
7. Liu D, Sasisekharan R. Role of heparan sulfate in cancer. In: Garg HG, Linhardt R, Hales C, eds. *Chemistry and Biology of Heparin and Heparan Sulfate*. Amsterdam: Elsevier Science; 2005:699−725.
8. U.S. Department of Health and Human Services. *Bone Health and Osteoporosis: A Report of the Surgeon General*. Rockville, MD: U.S. Department of Health and Human Services, Office of the Surgeon General; 2004.
9. Pathria MN, Chung CB, Resnick DL. Acute and stress-related injuries of bone and cartilage: pertinent anatomy, basic biomechanics, and imaging perspective. *Radiology*. 2016; 280(1):21−38.
10. Fratzl P, Weinkamer R. Nature's hierarchical materials. *Prog Mater Sci*. 2007;52(8): 1263−1334.
11. Gupta HS, Krauss S, Kerschnitzki M, et al. Intrafibrillar plasticity through mineral/collagen sliding is the dominant mechanism for the extreme toughness of antler bone. *J Mech Behav Biomed Mater*. 2013;28:366−382.
12. Clarke B. Normal bone anatomy and physiology. *Clin J Am Soc Nephrol*. 2008;3(suppl 3):S131−S139.
13. Dorozhkin SV. Calcium orthophosphates (CaPO$_4$): occurrence and properties. *Prog Biomater*. 2016;5:9−70.
14. Pearson OM, Lieberman DE. The aging of Wolff's "law": ontogeny and responses to mechanical loading in cortical bone. *Yearbk Phys Anthropol*. 2004;47:63−99.
15. Steffi C, Shi Z, Kong CH, Wang W. Modulation of osteoclast interactions with orthopaedic biomaterials. *J Funct Biomater*. 2018;9(1):1−15.
16. Ahn AC, Grodzinsky AJ. Relevance of collagen piezoelectricity to "Wolff's Law": a critical review. *Med Eng Phys*. 2009;31(7):733−741.
17. Yavropoulou MP, Yovos JG. The molecular basis of bone mechanotransduction. *J Musculoskelet Neuronal Interact*. 2016;16(3):221−236.
18. Shapiro F, Wu JY. Woven bone overview: structural classification based on its integral role in developmental, repair and pathological bone formation throughout vertebrate groups. *European Cells and Materials*. 2019;38:137−167.
19. Kolb C, Scheyer TM, Veitschegger K, et al. Mammalian bone palaeohistology: a survey and new data with emphasis on island forms. *PeerJ*. 2015;2015(10):1−44.
20. Khurana JS. Bone pathology. In: *Bone Pathology*. 2nd ed. London, New York: Springer Science Business Media; 2009:416.
21. Lee T, Chen W-M, Li L, Nathan SS, Pereira BP. Investigation of mechanical and viscoelastic properties of bovine bone using rus. *J Biomech*. 2008;41(1):S187.
22. Goldberg M, Kulkarni AB, Young M, Boskey A. Dentin: structure, composition and mineralization: the role of dentin ECM in dentin formation and mineralization. *Front Biosci(Elite Ed)*. 2011;3 E:711−E735.
23. Chanda A, Callaway C. Tissue anisotropy modeling using soft composite materials. *Appl Bionics Biomechanics*. 2018;2018:7−12.
24. Mijailovic AS, Qing B, Fortunato D, Van Vliet KJ. Characterizing viscoelastic mechanical properties of highly compliant polymers and biological tissues using impact indentation. *Acta Biomater*. 2018;71:388−397.

25. Akhtar R, Sherratt MJ, Cruickshank JK, Derby B. Characterizing the elastic properties of tissues. *Mater Today*. 2012;14:96−105.
26. Ebrahimi M, Ojanen S, Mohammadi A, et al. Elastic, viscoelastic and fibril-reinforced poroelastic material properties of healthy and osteoarthritic human tibial cartilage. *Ann Biomed Eng*. 2019;47(4):953−966.
27. Ponche A, Bigerelle M, Anselme K. Relative influence of surface topography and surface chemistry on cell response to bone implant materials. Part 1: Physico-chemical effects. *Proc IME H J Eng Med*. 2010;224(12):1471−1486.
28. Sophia Fox AJ, Bedi A, Rodeo SA. The basic science of articular cartilage. *Sports Health*. 2009;1(6):461−468.
29. Warnecke D, Meßemer M, de Roy L, et al. Articular cartilage and meniscus reveal higher friction in swing phase than in stance phase under dynamic gait conditions. *Sci Rep*. 2019;9(1):1−9.
30. Makris EA, Hadidi P, Athanasiou KA. The knee meniscus: structure−function, pathophysiology, current repair techniques, and prospects for regeneration. *Biomaterials*. 2011;32(30):7411−7431.
31. Buehler MJ, Yung YC. How protein materials balance strength, robustness and adaptability. *HFSP J*. 2010;4(1):26−40.
32. Zhu J, Hoop CL, Case DA, Baum J. Cryptic binding sites become accessible through surface reconstruction of the type I collagen fibril. *Sci Rep*. 2018;8(1):1−12.
33. Gottardi R, Hansen U, Raiteri R, et al. Supramolecular organization of collagen fibrils in healthy and osteoarthritic human knee and hip joint cartilage. *PloS One*. 2016;11(10): 1−13.
34. Murakami Y, Suzuki R, Yanuma H, et al. Synthesis and LC-MS/MS analysis of desmosine-CH2, a potential internal standard for the degraded elastin biomarker desmosine. *Org Biomol Chem*. 2014;12:9887−9894.
35. Trowbridge JM, Gallo RL. Dermatan sulfate: new functions from an old glycosaminoglycan. *Glycobiology*. 2002;12(9):117R−125R.
36. U. G. K. Wegst, H. Bai, E. Saiz, A. P. Tomsia, R. O. Ritchie, Bioinspired structural materials. 2015;4(1):23−36.
37. Elgazzar AH. *Orthopedic Nuclear Medicine*. 2nd ed. Cham, Switzerland: Springer Nature; 2017.
38. Hertz K, Santy-Tomlinson J, Santy-Tomlinson Editors, J, Falaschi P. *Fragility Fracture Nursing Holistic Care and Management of the Orthogeriatric Patient Perspectives in Nursing Management and Care for Older Adults Series Editors*. Cham, Switzerland: Springer International Publishing; 2018.
39. Lasanianos NG, Kanakaris NK, Giannoudis PV. Singh Index for osteoporosis. In: Lasanianos NG, Kanakaris NK, Giannoudis PV, eds. *In Trauma and Orthopaedic Classifications: A Comprehensive Overview*. London: Springer; 2015:1−547.
40. Wang JH-C. Mechanobiology of tendon. *J Biomech*. 2006;39(9):1563−1582.
41. Docheva D, Müller SA, Majewski M, Evans CH. Biologics for tendon repair. *Adv Drug Deliv Rev*. 2015;84:222−239.
42. Camarero-Espinosa S, Weder C. Articular cartilage: from formation to tissue engineering. *Biomater Sci*. 2016;4(5):734−767.
43. Park J, Lakes RS. *Biomaterials: An Introduction*. New York, NY: Springer Science Business Media, LLC; 2010.

Composite biomaterials

Learning objectives

At the end of this chapter the readers will:

- Become familiar with the concepts related to the design and use of biomaterials that are composed of two or more materials from different classes.
- Gain knowledge of the different types of materials that are used together as biomaterial composites in biomedical devices, their design criteria, and types of applications.
- Understand the principles of effective application of composites in biomedical devices.
- Learn what are the main advantages, challenges, and drawbacks that each of the materials described present to their applications in medicine.

Introduction

Composite materials are made of two or more different individual materials or two or more distinct phases of the same material. In composites, the scale of separation between the dissimilar phases needs to be larger than the atomic level. What matters when designing composite materials is to choose components in such a way that their combination results in significantly different properties than those of the individual materials. For example, bone is a composite material.[1] It contains more than two different phases. Chapter 6 provided a comprehensive view of bone structure and the corresponding components. Briefly, the main mineral phase is hydroxyapatite (HA) and the main organic phase, aside from cellular components, is collagen protein. Without collagen (see Chapter 6 for information on collagen structure), bone will be only composed of HA and cellular components and it would be brittle. Collagen fibers reinforce HA and increase the tensile strength, and also lower Young's modulus.[2] Other examples of composite materials found in nature are cartilage, tendon, wood, and dentine. The composition and structure of most of these materials are also detailed in Chapter 6. Grafts for repair of injury of orthopedic tissues can therefore be designed by using composite biomaterials with matching structure and properties.[3–6] Often such materials also have structural features such as pores or

Introductory Biomaterials. https://doi.org/10.1016/B978-0-12-809263-7.00007-X

fibers, which vary in geometry and size, and each has a unique influence on the final properties of the particular natural material of which they are part.

Biomaterials science can get inspiration from the natural world in designing composite materials that can use the large array of flexibility in geometry, size, porosity, or mechanical properties of dissimilar materials and bring them together in a way that mimics natural tissues and renders these composites useful for biomedical applications such as tissue engineering. By careful design, a scientist or an engineer working in this field can control mechanical properties, such as altering Young's modulus, stiffness, or mechanical strength.

Example 7.1

Derive an equation for the mechanical load on the mineral phase only, if bone is modeled as a composite of collagen and HA mineral and ignoring the cellular components. For detailed information on bone structure, refer to Chapter 6.

Answer When bone is modeled as a two-phase composite, the total load that it will withstand will be calculated by adding the loads on the mineral phase to the load on the collagen phase:

$$F_t = F_m + F_c \tag{7.1}$$

where F_t is the total load on the bone, F_m is the total load on mineral, and F_c is the total load on collagen.

The formula for mechanical stress, σ, is

$$\sigma = \frac{F}{A} = E \cdot \varepsilon \tag{7.2}$$

Thus,

$$F_m = A_m \cdot E \cdot \varepsilon_m \tag{7.3}$$

where A is the area, E is Young's modulus, and ε is the strain. The strain on collagen will be the same as the strain on the mineral part of the bone composite, so $\varepsilon_m = \varepsilon_c$.

Thus,

$$\frac{F_m}{A_m E_m} = \frac{F_c}{A_c E_c} \tag{7.4}$$

From Eqs. (7.1) and (7.4), we calculate:

$$F_m = \frac{F_t A_m E_m}{A_m E_m + A_c E_c} \tag{7.5}$$

Materials selection is also critical for the successful application of composite biomaterials in medicine. Tissue compatibility, nontoxicity, or a noncarcinogenic potential are significant parameters to control for in the design of such materials. Moreover, the effect of physiological environment on the composite material, and in particular on the interface between the two dissimilar phases, needs to be taken into consideration. It is well known that the physiological environment can have aggressive effects on materials, and interfaces between dissimilar materials can be especially sensitive to chemical attack, leading to degradation.

Some examples of the most well-known applications of composite biomaterials include metallic orthopedic implants that have porous HA coatings, bone cements reinforced with polymers, or composites used for dental fillings.[7-14]

Rational design of composites for biomedical applications

Since composite materials are composed of two or more dissimilar materials, their properties can be designed to match the application requirement if we know the properties of the individual components.[15,16] Moreover, the chemistry of each component or phase is not the only determinant of the final properties and applications. The geometry of the individual phases as well as of the composite biomaterial as a whole can have a large impact on the applications.[17,18] For example, composite materials can be fabricated in such a way that particles, fibers, or disks are used to reinforce a main phase. Or they can be designed to have certain porosities to satisfy a tissue engineering requirement (porosity is a requirement for materials used in tissue engineering), or any application that will require cellular attachment. Tissue engineering aims to design and assemble functional systems that either replace tissues or even whole organs that are damaged by injury or disease. A review of Chapter 5 will offer a refresher on key concepts related to tissue engineering and on the requirements that tissue engineering constructs need to meet for successful implementation into functional biomedical systems. The advantage of composite materials is that, as opposed to single-component biomaterials such as metals, polymers, or ceramics, there is a large amount of flexibility of design and use.[19,20] For example, we can make these materials more flexible if we incorporate long fibers into a ceramic or polymeric matrix, we can increase strength by using reinforcing particles or short fibers. One phase in a composite could be considered as the matrix phase, while a second phase could be called an inclusion. Inclusions can be particles, fibers, disks, or even pores.

The main factors ultimately influencing the properties of composite biomaterials are (i) the structure of the individual phases; (ii) the shape of the inclusions; and (iii) the interfacial properties between the two (or more) phases.[21-29]

One example of synthetic composites with biomedical applications is hydrogel composites (Fig. 7.1), in which hydrogels can have their properties altered by

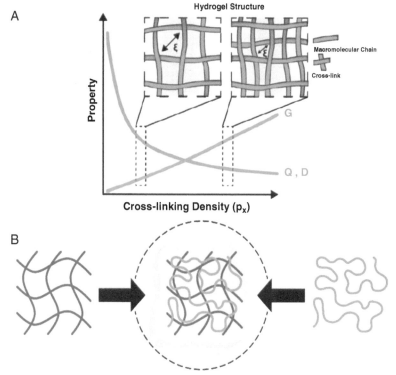

FIGURE 7.1

Structure of hydrogels. Relationship between the cross-linking density and hydrogel properties (A), where two network structures (low and high cross-linking densities) are presented to illustrate the relationship between the cross-linking density and basic hydrogel properties such as shear modulus *(G)*, equilibrium volumetric swelling ratio *(Q)*, and diffusivity *(D)*. Illustration of two polymeric chains cross-linked into a three-dimensional structure, showing their ability to swell and collapse in response to stimuli (B).

Adapted with permission from Chyzy et al.[40]

inclusion of a secondary, nanostructured phase.[30–38] Doing so leads to the design of nanocomposite hydrogels. Hydrogels are polymeric materials that are hydrated and highly cross-linked at a three-dimensional (3D) level and have high elasticity and the ability to swell and collapse depending on the hydration level of their structure (see Chapter 5 for details on the structure and properties of hydrogels). Due to the ability to swell and collapse in aqueous environments in response to stimuli such as changes in temperature, they have applications in medicine, and especially in drug delivery (Chapter 5), tissue engineering (Chapter 5), or sensing (Chapter 11). The type and properties of nanoinclusions (inclusions with sizes under 100 nm) can determine what stimuli lead to structural changes in these hydrogels (see Chapter 5 for details).

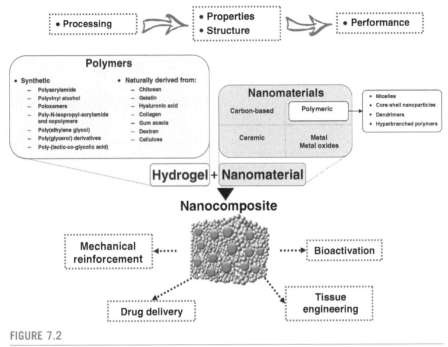

FIGURE 7.2

Chemistry, structure, and properties of materials components of nanohydrogels for biomedical applications.

Reproduced with permission from Biondi et al.[41]

Many types of inclusions can be used to control and design the properties of hydrogels, according to the intended application. Processing methods, or the way these composite hydrogels are synthesized, will determine their structure and properties, and ultimately the performance. Fig. 7.2 shows how the chemistry and processing influence properties and applications and the type of materials most commonly used in creating these composites.

In Fig. 7.2, the inclusion in the hydrogel-based composite is a nanomaterial, which is a material with size at the nanometer level, general under 100 nm. The resulting composite is called a nanohydrogel. Among the nanomaterial inclusions, polymers can have several forms. The can be core-shell nanoparticles, which are particles that have a nanosize particle in the core, which is further covered with a polymeric shell, as the name suggests. Micelles are nanosized polymeric structures and are another type of core-shell particle. These are created by the self-organization of polymers containing both hydrophobic (water-repelling) and hydrophilic (water-attracting) components called amphiphilic polymers. When amphiphilic polymers are added to an aqueous environment, they tend to self-organize with the hydrophobic parts oriented inward toward the center, and the hydrophilic groups outward in structures called micelles. Dendrimers are another type of polymer that are

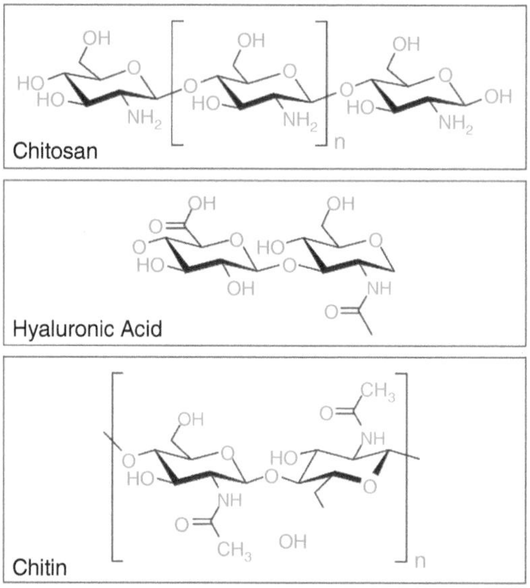

FIGURE 7.3

Chemical structure of chitosan, hyaluronic acid, and chitin.

organized in a radially symmetric fashion and have a tree-like branched structure. The polymeric branches are organized around a central linear polymeric core chain. A type of polymer structure that is very similar to dendrimers are the hyperbranched polymers.

Different applications of hydrogels have requirements for inclusion size to be either as small as in the nanometer range, or as large as in the millimeter range. In some cases, the size of the inclusion tends to have a larger influence on properties, especially Young's modulus, when in the nanometer size than when the inclusions are larger (micron to millimeter).

Polymers that have the ability to organize as hydrogels and then become the matrix material in a hydrogel-based composite include both synthetic and natural polymers, such as chitosan, hyaluronic acid, and collagen (Fig. 7.3).

As shown previously, these natural polymers can become matrix materials, which combined with inclusions such as metals or metal oxides, but also carbon-based materials, other polymers, or ceramics, can form composites. Some reinforcing materials have simply the role of enhancing mechanical properties, while others can have a role in making the material bioactive and usable for biomedical applications.

Indeed polymeric nanoparticles such as those shown in Fig. 7.2 can act as secondary phases when incorporated into a hydrogel. Polymeric inclusions can be polymeric drugs themselves as well. Polymers are rarely used for mechanical reinforcement. Instead, the objective of their utilization is often to improve the drug release properties of the nanocomposite gel. Without additional polymeric phase inclusions it is difficult for simple hydrogels to load hydrophobic materials, which poses a problem when the drugs being released are hydrophobic in nature. However, hydrophobic drugs can be encapsulated into a polymeric micelle or capsule that has a hydrophilic character and then manipulated for drug release (see Chapter 5 for more information on drug delivery) by changing pH, temperature, or ultraviolet (UV) light, which in turn can lead to disassembly of such micelles and the release of the drug that they encapsulate in the center.

Composites for orthopedic tissue repair
Overview

Although we discussed the use of hydrogels for drug delivery applications, it is very important to note that they also play a major role in the fabrication of tissue engineering scaffolds, with orthopedic tissues such as bone or cartilage being the main beneficiaries. To design tissue engineering scaffolds with adequate properties, one first has to have a good understanding of the relationship between structure and function of the natural tissues that are intended to be repaired. Once this understanding is gained, it becomes much more straightforward to design a composite material that will mimic tissue properties as closely as possible. A well-designed tissue engineering scaffold will be able to promote cellular attachment, proliferation, and ultimately the repair of the damaged tissue.

One-phase materials that can fulfill all the requirements for tissue repair do not exist, and most natural tissues themselves are composed of multiple different materials. For tissue reconstruction and repair, we need to design composites where one matrix material is reinforced with other materials with different geometries and properties (mechanical, chemical, biological). By doing so, parameters of the matrix material that do not match the properties of the original tissue of interest for repair can be improved and corrected by using custom synthesis, microstructure, and chemistries.

Most often, for orthopedic tissue repair, the composites that are being designed have an organic/inorganic character. That means they are composed of both inorganic materials, which generally impart adequate strength to the scaffold, and organic components. Organic phases can be either natural materials (proteins such as collagen, silk fibroin, or chitosan) or synthetic polymers (e.g., polyvinyl chloride, polyethylene). Their role is generally to decrease brittleness of the inorganic phase and promote cellular adhesion. Porosity is another critical parameter that needs to be tailored for appropriate cellular migration and subsequent generation of natural tissue. Both inorganic and organic (polymeric) materials can act either as the matrix or the enforcing material, depending on the application. For example, a polymeric matrix can be reinforced with inorganic particles. An example of such is the bone tissue

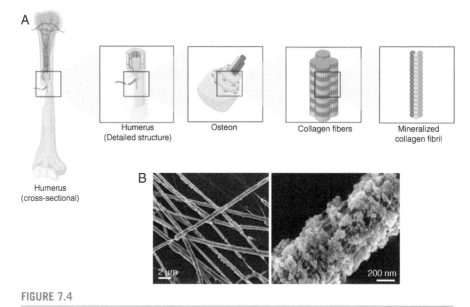

FIGURE 7.4

Hierarchy structure of composite materials in bone tissue (A); scanning electron microscopy images of mineralized collagen fibril (B).

Adapted with permission from Ping et al.[42]

itself, where long collagen fibers are reinforced with short HA needles, as presented in Fig. 7.4. The reinforcing material can improve properties such as stiffness, tissue biocompatibility, and thermal stability.

A uniform dispersion of the filler material is often critical for the successful applicability of the as-designed scaffold. If the reinforcing material aiming to increase the mechanical properties is not uniformly distributed throughout the matrix material, the resulting composite will have zones with weak mechanical properties, which will be the first to initiate failure under normal operating conditions of the final tissue repair material. This is especially true for load-bearing applications. In the case of composites between inorganic materials—among which HA is the most well known and used—and polymers, the polymers do not promote bioactivity. Instead, it is HA which fulfills this role. That is because most synthetic polymers have a hydrophobic character and cannot bind to natural tissues (e.g., bone), whereas HA is bioactive and osteogenic.

On the topic of morphology, nanosized inclusions are preferable because they promote faster degradation, superior bioactivity, and enhanced mechanical strength. The ability of scaffolds to degrade is significant since the goal of such repair materials is to initially support healing, while slowly allowing for cells to produce natural tissue that completely and gradually replaces the synthetic scaffold.

Natural biopolymer-hydroxyapatite composites for orthopedic repair

HA is the main inorganic material used for bone and cartilage repair scaffolds. When nanosized HA is used, the clinical results are superior than when the material has micron scale morphology, due to the reasons mentioned previously. Collagen accounts for 95% of the organic component of the bone and thus is often used in combination with HA to fabricate bone tissue engineering scaffolds. Collagen's structure was discussed in a previous chapter. Being an extracellular matrix protein, it is clearly tissue compatible, as it is also biodegradable and nontoxic, making it ideal for such tissue engineering scaffolds. Moreover, collagen promotes excellent cellular compatibility and both collagen I and collagen II have been proven to be osteogenic, thus being a good choice for bone healing when used in combination with HA in bone defect repair. Osteoblasts (mature bone cells) have excellent adherence to HA, especially when the HA is nanosized (nHA). They can also grow, divide, and produce extracellular matrix within the HA matrix material. Osteoblasts can normally adhere to nHA and both grow and divide in the material. Lin et al.[39] designed nano-HA-collagen I biocomposites and tested them for fixing defects in rabbit femurs. A 4-week examination revealed that these bone replacement composites indeed led to enhanced bone healing.

Another natural material that has been used in combination with nano-HA is chitosan. Chitosan is not a protein but a polysaccharide that is hydrophilic. Hydrophilicity is important because this property is beneficial for the promotion of cell adhesion, proliferation, and differentiation. Chitosan is biocompatible, biodegradable, and can be made into porous structures through freeze-drying. Porosity, in addition to the other characteristics mentioned previously, makes chitosan an interesting candidate for bone tissue engineering application. As with all the other organic components of such scaffolds, chitosan alone does not have the appropriate material properties and thus is used as part of composites with HA. Such chitosan-HA composites have been tested in the fabrication of porous scaffold that performed well in bone tissue engineering. In animal studies, such scaffolds were osteoconductive, biocompatible, and biodegradable.

Chitosan is not the only natural biomaterial that can be used in combination with HA to produce scaffolds for orthopedic tissue repair. Another fibrillar protein, fibroin, which is synthesized via extraction from silk and has 18 amino acids in its composition, has been also used. Similar to chitosan, this protein is nontoxic and affordable but, as expected, does not meet the mechanical property requirements to be used on its own in orthopedic scaffolds. In composites with HA, fibroin has been used for bone defect repair. Silk fibroin has the ability to form strong bonds with HA, resulting in composites that have adequate porosity for cellular migration and an overall similar structure to that of the natural bone tissue. In addition to bone defect repair, fibroin-HA hybrid composites can be used as materials for artificial cartilage and ligaments, as well as nerve tissue.

Composites for hard-tissue repair

Materials from all classes covered so far in this book can be incorporated in composite biomaterials used for hard-tissue repair, including orthopedics. Fig. 7.5 shows examples of multilayered composites with multiple components—metals, alloys, carbon nanotubes, and HA ceramics—and some of the orthopedic biomedical devices in which they can be incorporated, including devices with drug delivery potential.

Although natural polymers like those mentioned previously in this chapter are good candidates for the creation of composites with applicability in tissue repair, synthetic polymers can also be used for the same purpose. One such polymer class that displays the appropriate combination of chemical and mechanical properties that allow them to form porous composites with HA is polyamide (PA). Polyamide 66 (PA66) can be impregnated with short HA needles, which results in a porous composite with a structure that is very similar to that of the natural bone tissue. The properties of this composite material are also close to those of the natural bone in terms of similar values for the modulus of elasticity, flexural strength, and compressive strength. The pore sizes of this composite are in the 300 to 500 μm range, which is appropriate for cellular (osteoblasts) migration and vascularization.

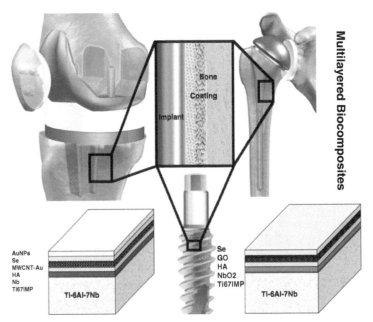

FIGURE 7.5

Illustration of multilayer biofilm-Ti67IMP systems that may contribute to facilitate low-risk bone regeneration with antibacterial and drug delivery potential and long-term mechanical satisfaction.

Reproduced with permission from Rafieerad et al.[43]

HA/PA composites were used by Lin et al.[39] to prepare artificial bone that was applied to the repair of bone defects in rabbit tibia. This artificial bone composite was implanted in vivo and led to its replacement with natural bone after a period of 26 weeks. During this time period the composite resorbed, while new bone formed to replace it and this resulted in healed bone tissue, free of any artificial implants. Other studies used growth factors, such as the basic fibroblast growth factor (bFGF), in conjunction with HA/PA66 porous composites and reported positive outcomes when used as artificial bone in animal studies. Low-load-bearing areas (e.g., the jaw) can have a superior outcome when such composites are used to repair bone defects, when compared with high-load-bearing areas.

Polyesters, such as poly-lactic acid (PLA), polyglycolic acid (PGA), poly(ε-caprolactone) (PCL), and polyhydroxybutyrate (PHB), are fully biodegradable synthetic polymers that can be used in conjugation with HA and calcium phosphates (CaPs) to prepare composites for orthopedic applications. PLA has not only been used in composites, but is also processed into resorbable orthopedic screws, pins, and plates, or in drug delivery applications. The degradation rates of PLA vary between 10 months and 4 years. This slow degradation rate is due to the presence of a hydrophobic methyl group in its chemical structure. The degradation reaction occurs via hydrolysis of the ester bond and its speed depends on the molecular weight and polymerization number, as well as the crystallinity level. On the other hand, PGA degrades faster, between 6 and 12 weeks. Thus copolymers of PLA and PGA, PLGA, can often be used to modulate the degradation rate for various United States Food and Drug Administration (FDA)—approved biomedical device or drug delivery applications. These polymers can be used in composites with HA ($Ca_{10}(PO_4)_6(OH)_2$) or tricalcium phosphate and used as artificial bone in orthopedic applications. One disadvantage of PLA, PGA, and PLGA is that these polyesters release acidic degradation products in the surrounding environment, which changes the local tissue pH and often triggers a foreign body response and inflammatory reactions. When used in composites with CaP materials (HA and tricalcium phosphates), this effect can be mitigated, although not completely eliminated.

Another example of fibers reinforcing a matrix in a composite material is those of ultra-high-molecular-weight polyethylene or UHMWPE. UHMWPE itself is used in joint replacement, especially the knee, as the articulating surface. UHMWPE fibers are also included in composites to improve wear properties (beyond 10 years of UHMWPE alone) and reduce creep (tendency of a material to deform slowly under the influence of continuous mechanical forces). Creep occurrence allows indentation of the UHMWPE component, which can in turn influence the behavior of the joint, as illustrated in Fig. 7.6.

When polyesters are used in composites with CaP materials, the matrix (also called the continuous phase) is considered to be the PLA/PLGA polymer, while the CaP phase is referred to as the inclusion or dispersed phase. The matrix usually serves as support for the inclusions and helps to keep their positions, while the inclusions act like reinforcers of the mechanical properties. As discussed previously, neither the polymers alone nor the CaP materials can meet the mechanical

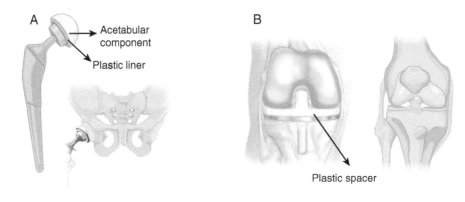

FIGURE 7.6

Illustration of composite materials used in (A) femur and (B) knee replacement.

performance criteria for orthopedic use, which highlights the value of composite design for the successful use of biomaterials in orthopedics.

Bone cement is another composite material that is typically used to anchor joint prostheses and various fracture fixation devices. The main component of the most well-known bone cement is polymethyl methacrylate (PMMA). It acts like grout on tiles and creates a tight space that will anchor an implant to the bone. Bone cements are not adhesives. Instead, they function by using the surface roughness of the bone and fill out the empty spaces to interlock the bone mechanically with the implant. Besides PMMA-based bone cements, there are also calcium phosphate cements (CPCs) and glass polyalkenoate cements (GPCs). The latter have low mechanical strength and so are mostly used for small bone defects and in areas where the mechanical loads are small, such as maxillofacial applications. The GPCs are tissue compatible (do not elicit negative tissue reactions) and bioresorbable, which provides an advantage in that new bone can replace the cement upon healing of the initial injury.

Often, bone cements contain reinforcing carbon fibers when used in joint replacement. Carbon fiber inclusion in such composites improves mechanical properties, but at the same time increases the viscosity of unpolymerized cement, making it hard to use. Bone cements typically start as unpolymerized materials, and the viscosity needs to be controlled by easy manipulation. Bone cements have also been reinforced with metallic wire for spinal stabilization surgery, which successfully increased tensile strength.

Dental composites are resins consisting of a polymer matrix and stiff inorganic inclusions. Stiff, angular inclusions (particulate) provide increased strength and wear resistance. These are typically barium glass or silica (quartz). They have similar appearance to dental enamel, making this a cosmetically acceptable material for anterior teeth. The material is mixed and then placed in the prepared cavity, where it polymerizes. For the preparation to be successful, low viscosity and controlled polymerization are necessary. Use of additional chemicals to lower the matrix viscosity and allow controlled triggering of the polymerization process is common. Polymerization is initiated by thermochemical treatment or photochemical treatment (UV light). Resins are made with varying type and amount of filler (particle) to give varied properties. Some composites contain colloidal silica as a filler, which allows the composite to be polished to reduce both wear and plaque accumulation.

Mechanical properties of composites

The most important advantage of using composite biomaterials rather than their single-phase counterparts is the ability to tailor the mechanical properties to those specifically required for applications. When designing composite biomaterials, natural composites such as tissue materials serve as inspiration and provide the mechanical property targets in each category. For example, bone, which is one of the most often used tissues for inspiration, contains collagen, HA, water, and cells. Collagen properties are dominated by its ductility and high Young's modulus, whereas the HA crystals are brittle and stiff, properties that allow HA crystals to be the reinforcers of the collagen matrix. More specifically, the elastic modulus of collagen is 1.5 GPa and that of HA is 100 GPa. Composites between these two components, however, can be formed with a rather large variation of properties, depending on the geometry and distribution of the components. For example, in trabecular bone, the high porosity leads to lower Young's moduli (between approximately 10 and 20 GPa) compared with the more dense cortical bone or individual trabeculae.

At a larger scale, the mechanical properties of composites are strongly influenced by their microstructure. In other words, these will change not only depending on the intrinsic mechanical properties of the individual phases, but also depending on whether the inclusion phase is a fiber, particle, or disk (e.g., platelets, laminae). The models that are most used to characterize the properties of composite biomaterials are the idealized Voigt and Reuss models (Fig. 7.7).

The Voigt model is represented by a Newtonian damper and a Hookean elastic spring in parallel and represents a solid under reversible viscoelastic strain. We can calculate Young's modulus of a composite viewed under the idealized Voigt model as:

$$E = E_i V_i + E_m (1 - V_i) \tag{7.6}$$

where E_i is Young's modulus of the inclusion, V_i is the volume fraction of the inclusion, and E_m is Young's modulus of the matrix. Notice that the Young's modulus formula under this model follows the rule of mixtures for the inclusion and matrix materials.

According to the Reuss model, the composite stiffness is given by:

$$E = \left[\frac{V_i}{E_i} + \frac{1 - V_i}{E_m} \right]^{-1} \tag{7.7}$$

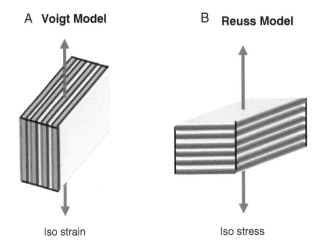

FIGURE 7.7

Illustration of (A) Voigt and (B) Reuss models: Young's modulus prediction.

The E value in the Reuss model is lower than that in the Voigt model. They each provide the upper (Voigt) and lower (Reuss) limits of a composite's stiffness, where the composite is considered to have a random phase geometry. When the moduli of the individual phases are very different, the upper and lower bounds of the composite stiffness are also far off from each other, which is often the case, since most composites are anisotropic which means that their properties are dependent on direction. Accordingly the stiffness of a composite is anisotropic and their stress-strain relationships follow the general trend for anisotropic materials:

$$\sigma_{ij} = \sum_{k=1}^{3} \sum_{k=1}^{3} C_{ijkl}\varepsilon_{kl} \tag{7.8}$$

where σ_{ij} is the stress, ε_{kl} is the strain, and C_{ijkl} is the elastic modulus tensor. This equation is the tensorial form of Hooke's law.

The advantage in mechanical properties that composites bring stems mostly from their anisotropic nature, which leads to higher stiffness and strength than those of isotropic, or single-phase, materials. There are gains in mechanical properties in a certain direction, which is possible at the expense of a reduction in properties in another direction. Therefore it is necessary to know the direct application of a certain composite biomaterial to understand where the major directions of stress are expected to lie.

Example 7.2

Calculate the Reuss and Voigt model Young's modulus for an artificial bone composite with 60 vol% collagen and 40 vol% HA, assuming only collagen and HA as components. Assume that Young's modulus for collagen is 1 GPa and for HA is 117 GPa.

Answer The composite Young's modulus for the Voigt model is:

$$E_{bone} = V_{collagen} E_{collagen} + V_{hydroxyapatite} E_{hydroxyapatite} \tag{7.9}$$

After inputting values given in the text, we calculate $E_{bone} = 47.4$ GPa.

The composite Young's modulus for the Reuss model is given by solving the equation:

$$\frac{1}{E_{bone}} = \frac{V_{collagen}}{E_{collagen}} + \frac{V_{hydroxyapatite}}{E_{hydroxyapatite}} \tag{7.10}$$

After calculations: $E_{bone} = 1.6$ GPa

Composite materials-tissue interactions

As previously discussed, the term biocompatibility is not a material property but encompasses the sum of properties of a biomedical system that enables its safe and adequate functioning in the context of being used as intended in a biomedical device. Thus it is more accurate to discuss the composite-tissue interactions rather than referring to their biocompatibility, as was often the case in older textbooks and scientific works.

In the context of a composite being used in orthopedics, often its performance as far as tissue compatibility refers to whether the material forms a direct bond with the bone after implantation. Such a composite is deemed to be bioactive. Both PLA and PLGA have been shown to be able to form a direct bond with the bone in studies in vivo. Also, some polymers, including PLA and PLGA, which biodegrade via a hydrolysis reaction, form −COOH groups that can further bind calcium ions and lead to the nucleation and growth of apatite and increase their bioactivity. Surface modification strategies that increase the number of acetate groups on the polymers' surface can be used in order to further enhance apatite formation. Another strategy is the use of CaP in composites with these polymers, which has been shown to further enhance this effect.

Aside from bioactivity, the mechanical integrity of composites used for biomedical applications, such as bone grafts, is to be examined before establishing the adequacy for use of a given composite biomaterial in a biomedical implantable device. For bone graft composites in particular, the necessity for biodegradability poses questions related to how the composite architecture and mechanical integrity for the required period of time can be affected by the precursor materials, geometry of the components, and composite fabrication methods. It is, for example, difficult to disperse CaP particles in a polymer matrix, and particulate aggregation can lead to issues related to uneven distribution, further affecting both composite mechanical performance and bioactivity. Solutions include ultrasonication with the goal of dispersing particulate agglomerates.

Polymer-ceramic composites where the polymer is the matrix material can swell when placed in a water-rich physiological environment. The swelling can result in modified mechanical properties, such as Young's modulus (stiffness). This can be the result of a shift of the glass transition temperature of the polymer toward lower values, which in turn can result in stiffness reduction. However, swelling can also be used as a positive effect in certain applications, such as in dentistry, where it can counteract shrinkage during dental composite curing and polymerization.

Other biocompatibility challenges that can result from polymer matrix composites in particular include the creation of debris when used in orthopedic implants, for example, fracture fixation devices. Polymers tend to create particulate debris, which is then likely to elicit a foreign body response and implant failure.

Summary

The use of composite materials is a more recent practice in the biomedical field when compared with the use of other classes of materials such as metals or ceramics. When rationally designed, composite biomaterials can improve success in the use of biomedical implants, by offering the ability to customize mechanical properties as well as, in some cases, bioactivity. However, as it is the case with all biomaterials classes, attention must be paid to how the properties of individual materials as well as their microstructure, orientation, and fabrication methods affect the in vivo performance of composite-containing biomedical implants. Corrosion, swelling, debonding, or toxic degradation byproducts are all potential factors that can lead to implant failure and need to be carefully considered and investigated before use.

Problems

1. The elastic moduli of collagen and HA are 0.1 GPa and 100 GPa, respectively; Young's moduli are 40 MPa for collagen and 6 GPa for HA. Calculate the load of the mineral phase on the bone only, when a compressive stress of 64 N is applied to a 1-mm^3 bone sample where HA has a surface area of 200 mm^2 and collagen has a surface area of 500 mm^2. In this scenario, bone is modeled as a two-phase composite of mineral and collagen.

2. If the values for the elastic moduli and HA in a bone sample are 40 MPa for collagen and 6 GPa for HA, calculate the load on the collagen phase only in a bone sample with a 1-mm^3 volume, when a compressive stress of 150 N is applied. Assume HA has a surface of 200 mm^2 and collagen has a surface area of 500 mm^2. In this scenario, bone is modeled as a two-phase composite of mineral and collagen.

3. Bone is modeled as a two-phase composite of collagen and HA. Calculate the composite stiffness under the Voigt model in a scenario where the collagen matrix is reinforced with 25% volume fraction of HA crystals. The values for Young's moduli are 50 MPa for collagen and 5 GPa for HA.

4. A poly(ε-caprolactone) (PCL)-based hydrogel is combined with conductive polyaniline (PANI) conductive fibers with a high length-to-diameter aspect ratio, in an attempt to create a bone scaffold that is responsive to electrical stimuli. How would you expect the composite hydrogel scaffold to compare to the pristine PCL hydrogel in terms of elastic modulus?

5. A polymeric hydrogel scaffold is tested for the fabrication of artificial cartilage. Cationic latex particles with dimensions of 0.5 μm and 50 nm diameter, respectively, are used as cross-linking agents in two variations of the scaffold fabrication process. Discuss your expectations of the effect of the latex particle size on the maximum fracture stress of the composite hydrogel.

6. Macromolecular microspheres are used as cross-linkers in composite hydrogels and impart superior compressive strength compared with pristine hydrogels but in many cases their tensile properties are not seeing a significant improvement. Discuss reasons behind this behavior and propose a design solution for a composite hydrogel with improved tensile strength.

7. Chemically cross-linked hydrogels have generally superior mechanical properties compared with their physically cross-linked counterparts. What disadvantages could you envision when using chemically cross-linked hydrogels in implantable tissue engineering constructs?

8. Various inclusion shapes are used in composite materials. For a composite biomaterial that is to be used in an application where tensile strength is an important factor, what shape of an inclusion would you propose as being ideal?

9. A fiber-reinforced resin is subjected to a compressive stress applied in the direction parallel to fiber. In what location do you expect to find initial cracks before failure?

10. Are fiber-reinforced composite materials likely to be capable of withstanding higher compressive stress or higher tensile stresses? Explain your answer.

11. Ligamentum nuchae is made of elastin and collagen (others do not contribute to the mechanical properties). The relative amounts excluding water are 70, 25, and 5 wt% for elastin, collagen, and others, respectively. Assume they are homogeneously distributed. Calculate the percentage contribution of the elastin toward the total strength assuming elastic behavior. Assume $E_E = 0.6$ MPa for elastin and $E_c = 1000$ MPa for collagen; $E_0 = 0$ for others and that the densities of elastin and collagen are 1 g/cm^3.

12. Calculate the density of the HA phase in a dried bone sample with a density $= 2.5$ g/cm^3 if the density of water and organic components both are assumed to be 1 g/cm^3 and the mineral and organic phases are each 50 vol% in the total sample volume.

13. For a wet bone sample with phase distributions of 10, 68, and 22 wt% for water, mineral, and organic phase, respectively, calculate the volume percentage for each of these phases. Consider the density values for the mineral and organic phases to be 3.2 and 1.05 g/cm^3, respectively.

References

1. Weiner S, Wagner HD. The material bone: structure mechanical function relations. *Annu Rev Mater Sci*. 1998;28:271–298.
2. Weiner S, Traub W, Wagner HD. Lamellar bone: structure-function relations. *J Struct Biol*. 1999;126(3):241–255.
3. Olkhov A, Gur'ev V, Akatov V, Mastalygina E, Iordanskii A, Sevastyanov VI. Composite tendon implant based on nanofibrillar polyhydroxybutyrate and polyamide filaments. *J Biomed Mater Res A*. 2018;106(10):2708–2713.
4. Parsons JR, Rosario A, Weiss AB, Alexander H. Achilles-tendon repair with an absorbable polymer-carbon fiber composite. *Foot Ankle*. 1984;5(2):49–53.
5. Purbrick M, Ambrosio L, Ventre M, Netti P. Natural composites: structure-property relationships in bone, cartilage, ligament and tendons. In: Ambrosio L, ed. *Biomedical Composites*. Cambridge, UK: Woodhead Publishing Limited, CRC Press; 2009:3–24.
6. Tong SY, Wang ZY, Lim PN, Wang W, Thian ES. Uniformly-dispersed nanohydroxapatite-reinforced poly(epsilon-caprolactone) composite films for tendon tissue engineering application. *Mat Sci Eng C-Mater*. 2017;70:1149–1155.

7. Sivaraj D, Vijayalakshmi K. Novel synthesis of bioactive hydroxyapatite/f-multiwalled carbon nanotube composite coating on 316L SS implant for substantial corrosion resistance and antibacterial activity. *J Alloy Compd*. 2019;777:1340−1346.

8. Gnedenkov SV, Sinebryukhov SL, Zavidnaya AG, Egorkin VS, Puz AV, Mashtalyar DV, Sergienko VI, Yerokhin AL, Matthews A. Composite hydroxyapatite-PTFE coatings on Mg-Mn-Ce alloy for resorbable implant applications via a plasma electrolytic oxidation-based route. *J Taiwan Inst Chem E*. 2014;45(6):3104−3109.

9. Kim SM, Kang MH, Kim HE, Lim HK, Byun SH, Lee JH, Lee SM. Innovative micro-textured hydroxyapatite and poly(L-lactic)-acid polymer composite film as a flexible, corrosion resistant, biocompatible, and bioactive coating for Mg implants. *Mat Sci Eng C-Mater*. 2017;81:97−103.

10. Maistrelli GL, Mahomed N, Garbuz D, Fornasier V, Harrington IJ, Binnington A. Hydroxyapatite coating on carbon composite hip implants in dogs. *J Bone Joint Surg Br*. 1992;74(3):452−456.

11. Irbe Z, Loca D, Bistrova I, Berzina-Cimdina L. Calcium phosphate bone cements reinforced with biodegradable polymer fibres for drug delivery. *Eng Mater Tribol*. 2014;604: 184−187, Xxii.

12. Mirjalili A, Zamanian A, Hadavi SMM. The effect of TiO_2 nanotubes reinforcement on the mechanical properties and wear resistance of silica micro-filled dental composites. *J Compos Mater*. 2019;53(23):3217−3228.

13. Dionysopoulos D, Tolidis K, Gerasimou P. Polymerization efficiency of bulk-fill dental resin composites with different curing modes. *J Appl Polym Sci*. 2016;133(18):43392.

14. Eke C, Er K, Segebade C, Boztosun I. Study of filling material of dental composites: an analytical approach using radio-activation. *Radiochim Acta*. 2018;106(1):69−77.

15. Alberti KA, Sun JY, Illeperuma WR, Suo ZG, Xu QB. Laminar tendon composites with enhanced mechanical properties. *J Mater Sci*. 2015;50(6):2616−2625.

16. Dehestani M, Adolfsson E, Stanciu LA. Mechanical properties and corrosion behavior of powder metallurgy iron-hydroxyapatite composites for biodegradable implant applications. *Mater Design*. 2016;109:556−569.

17. Yu XM, Gu BQ, Zhang B. Effects of short fiber tip geometry and inhomogeneous interphase on the stress distribution of rubber matrix sealing composites. *J Appl Polym Sci*. 2015;132(16): 41638.

18. Ghosh R, Gupta S. Bone remodelling around cementless composite acetabular components: the effects of implant geometry and implant-bone interfacial conditions. *J Mech Behav Biomed*. 2014;32:257−269.

19. Ravanbakhsh H, Bao G, Latifi N, Mongeau LG. Carbon nanotube composite hydrogels for vocal fold tissue engineering: biocompatibility, rheology, and porosity. *Mat Sci Eng C-Mater*. 2019;103:109861.

20. Yang Y, Zhu XL, Cui WG, Li XH, Jin Y. Electrospun composite mats of poly[(D,L-lactide)-co-glycolide] and collagen with high porosity as potential scaffolds for skin tissue engineering. *Macromol Mater Eng*. 2009;294(9):611−619.

21. Wang J, Lu YY, Xu YC, Ruan YJ, Li LY, An LJ. Effects of block copolymer compatibilizers on phase behavior and interfacial properties of incompatible homopolymer composites. *Acta Polym Sin*. 2016;3:271−287.

22. Garcia-Martinez, J. M.; Collar, E. P., In-phase and out-of-phase tensile properties of polypropylene/mica composites modified by a novel industrial waste based interfacial agent. Responses at the alpha and beta transitions of the polymer phase. *VIII International Conference on Times of Polymers and Composites: From Aerospace to Nanotechnology*. 2016;1736:020021.

23. Tian B, Cheng ZG, Tong YX, Li L, Zheng YF, Li QZ. Effect of enhanced interfacial reaction on the microstructure, phase transformation and mechanical property of Ni-Mn-Ga particles/Mg composites. *Mater Design*. 2015;82:77–83.

24. Sung SJ, Jung EA, Sim K, Kim DH, Cho KY. Phase separation structure and interfacial properties of lattice-patterned liquid crystal-polymer composites prepared from multicomponent prepolymers. *Polym Int*. 2014;63(2):214–220.

25. Wilson KS, Antonucci JM. Structure-property relationships of thermoset methacrylate composites for dental materials: study of the interfacial phase of silica nanoparticle-filled composites. *Abstr Pap Am Chem S*. 2004;228:U354–U354.

26. Sun JR, Shen BG, Yeung HW, Wong HK. Formation of interfacial phase and its effects on the magnetic and transport properties of the $La_{0.82}Ca_{0.18}MnO_3/La_{0.18}Ca_{0.82}MnO_3$ composite. *J Phys D Appl Phys*. 2002;35(3):173–176.

27. Vidal-Setif MH, Lancin M, Marhic C, Valle R, Raviart JL, Daux JC, Rabinovitch M. On the role of brittle interfacial phases on the mechanical properties of carbon fibre reinforced Al-based matrix composites. *Mat Sci Eng A-Struct*. 1999;272(2):321–333.

28. Inem B. Crystallography of the 2nd-phase/Sic particles interface, nucleation of the 2nd-phase at beta-Sic and its effect on interfacial bonding, elastic properties and ductility of magnesium matrix composites. *J Mater Sci*. 1995;30(22):5763–5769.

29. Janowski GM, Pletka BJ. The influence of interfacial structure on the mechanical-properties of liquid-phase-sintered aluminum-ceramic composites. *Mat Sci Eng A-Struct*. 1990;129(1):65–76.

30. Hu XC, Qu SX. Inclusion size effect on mechanical properties of particle hydrogel composite. *Acta Mech Solida Sin*. 2019;32(5):643–651.

31. Desfrancois C, Auzely R, Texier I. Lipid nanoparticles and their hydrogel composites for drug delivery: a review. *Pharmaceuticals (Basel)*. 2018;11(4):118.

32. Villalba-Rodriguez AM, Parra-Saldivar R, Ahmed I, Karthik K, Malik YS, Dhama K, Iqbal HMN. Bio-inspired biomaterials and their drug delivery perspectives—a review. *Curr Drug Metab*. 2017;18(10):893–904.

33. Mohammadi A. Oscillatory response of charged droplets in hydrogels. *J Non-Newton Fluid*. 2016;234:215–235.

34. Pafiti K, Cui ZX, Adlam D, Hoyland J, Freemont AJ, Saunders BR. Hydrogel composites containing sacrificial collapsed hollow particles as dual action pH-responsive biomaterials. *Biomacromolecules*. 2016;17(7):2448–2458.

35. Pafiti K, Cui ZX, Carney L, Freemont AJ, Saunders BR. Composite hydrogels of polyacrylamide and crosslinked pH-responsive micrometer-sized hollow particles. *Soft Matter*. 2016;12(4):1116–1126.

36. Shah K, Vasileva D, Karadaghy A, Zustiak SP. Development and characterization of polyethylene glycol-carbon nanotube hydrogel composite. *J Mater Chem B*. 2015; 3(40):7950–7962.

37. Tardani F, La Mesa C. Effects of single-walled carbon nanotubes on lysozyme gelation. *Colloid Surface B*. 2014;121:165–170.

38. Lin GY, Cosimbescu L, Karin NJ, Tarasevich BJ. Injectable and thermosensitive PLGA-g-PEG hydrogels containing hydroxyapatite: preparation, characterization and in vitro release behavior. *Biomed Mater*. 2012;7(2):24208–24117.

39. Lin BN, Whu SW, Chen CH, Hsu FY, Chen JC, Liu HW, Chen CH, Liou HM. Bone marrow mesenchymal stem cells, platelet-rich plasma and nanohydroxyapatite-type I collagen beads were integral parts of biomimetic bone substitutes for bone regeneration. *J Tissue Eng Regen M*. 2013;7(11):841–854.

40. Chyzy A, Tomczykowa M, Plonska-Brzezinska ME. Hydrogels as potential nano-, micro- and macro-scale systems for controlled drug delivery. *Materials*. 2020; 13(1):188.
41. Biondi M, Borzacchiello A, Mayol L, Ambrosio L. Nanoparticle-integrated hydrogels as multifunctional composite materials for biomedical applications. *Gels-Basel*. 2015;1(2): 162−178.
42. Ping H, Xie H, Wan YM, Zhang ZX, Zhang J, Xiang MY, Xie JJ, Wang H, Wang WM, Fu ZY. Confinement controlled mineralization of calcium carbonate within collagen fibrils. *J Mater Chem B*. 2016;4(5):880−886.
43. Rafieerad AR, Bushroa AR, Banihashemian SM, Amiri A. Not-yet-designed multilayer Nb/HA/MWCNT-Au/Se/AuNPs and NbO_2/HA/GO/Se biocomposites coated Ti6Al7Nb implant. *Mater Today Commun*. 2018;15:294−308.

Tissue-biomaterials interactions

Learning objectives

This chapter will focus on understanding the effects of tissue-biomaterials interactions. At the end of the chapter, the students will:

- Be able to understand the main concepts related to biomaterials-tissue interactions.
- Become familiar with concepts such as inflammation, protein absorption, and immune response.
- Learn the mechanisms of tissue healing and repair after injury.
- Understand the effects of the biological environment on biomaterials properties.
- Gain a general understanding of the main biomaterials testing methods, along with concepts related to carcinogenicity and mutagenicity as related to biomedical implants.

Introduction

So far, we discussed the main classes of biomaterials and how they can be used in various biomedical implants in the body. The term biocompatibility was defined and discussed in some detail in Chapter 1. Briefly, biocompatibility[1−4] is a complex problem and the term is inherently linked to the success of the implant. An implant is biocompatible when not only is it not toxic or carcinogenic but also when it functions as intended over the entire lifetime of the device. In other words, when it is able to perform with appropriate host response for the intended application. Specific guidance for biocompatibility testing of a biomaterial can be found in ISO 10993, which can be consulted for detailed information on standardized testing methods.

Today, with longer life spans and an aging population, understanding not only the mechanical, structural, and functional properties of the implants but also the short- and long-term effects these have on host tissues becomes critical. In this chapter, we will discuss the main challenges related to the biomaterials-tissue interactions, tissue repair, and the effects of the biological environment on the properties of the biomaterial.

Introductory Biomaterials. https://doi.org/10.1016/B978-0-12-809263-7.00008-1

The biological environment is rich in a variety of chemical elements present in large ranges of concentration. For example, oxygen and carbon are part of the human body in high concentrations and are the basic elements that sustain life. On the other hand, low or trace elements are also vital for the functioning of tissues and organs. Calcium is a major component in bone and is involved in other physiological functions, such as a role in cardiac action potentials and in smooth muscle contractions, blood clotting, and cellular injury. Chlorine is present in lower concentrations than oxygen or carbon but has significant roles in the maintenance of the acid-base balance in the body, muscle physiology, as well as in the production of stomach acid, which enables digestion.

Trace elements include metallic ions such as magnesium, that are present in minimal concentrations but also play a significant role in biological functions when present in a normal physiological range. However, when metal ions exceed the physiological range, they can become toxic to the body and can lead to a multitude of diseases such as neurological diseases (e.g., Parkinson and Alzheimer), or cancers. Polymeric materials that are part of biomedical implants have the potential to degrade after implantation and release free radicals, or reactive oxygen species (ROS), which are also harmful and could be the source of diseases such as neurodegenerative diseases or cancers. To summarize, although many elements are normally present in the body in concentrations within the physiological range, this does not ensure that implants can be constructed out of these elements without negative effects on tissues and organs. Moreover, a material can be nontoxic by itself, but it can react with the biological environment and result in harmful reaction products. Mechanical loading conditions, in addition, can also have an effect on the kinetics of the reactions between the biomaterial and the tissues. A young athletic individual will likely have a greater outcome for a hip implant, for example, than an elderly, inactive patient. Only estimations of such outcomes can be brought forward when evaluating the likelihood of success of an implant, but these considerations should nevertheless not be ignored.

The effect of the biomaterial on the tissue is called the host response.[5] The effect of the physiological environment on the implant is termed material response. They should be considered in tandem when implant design decisions are made. All of these concepts will be covered in this chapter. As part of understanding the host response to a biomedical implant, it will be important to understand what types of cells are involved at all stages of biomaterials-tissue interaction, including wound healing, tissue remodeling, and the immune response. Table 8.1 lists the most common types of cells that this chapter will mention as being involved in different stages of tissue-biomaterials interactions.

Table 8.1 Partial list of cell types involved in tissue-biomaterial interactions.

Cell types	Definition
Erythrocytes	Red blood cells; contain hemoglobin, which carries oxygen from the lungs to the rest of the body
Leukocytes	White blood cells; part of the immune system. There are several types of leukocytes
Platelets	Large blood cells that help form blood clots
Granulocytes	A type of leukocyte that contains small protein-containing granules; help fight infections
Macrophages	A type of white blood cell that surrounds foreign particles, removes them, and activates other immune system cells
Phagocytes	A type of cell that functions to ingest or engulf foreign particles
Foreign body giant cells	A cell with multiple nuclei that forms around foreign particles by fusion of macrophages
Neutrophils	A type of immune cell that arrives first at the site of tissue injury or infection; can ingest pathogens and release enzymes that kill them. Neutrophil is a type of white blood cell, is a granulocyte, and a phagocyte
Fibroblasts	A type of cell found in connective tissues such as skin; it secretes collagen and plays an important role in tissue remodeling
Endothelial cells	Cells that line the blood vessels

Injury and repair of natural tissues and the immune response
Wound healing

A list of common tissue effects induced by the presence of biomaterials in the human body is presented in this subsection and it will be useful in setting an overall picture of the macro-level effects that one can expect on the implantation of different biomaterials, depending on their properties. However, there is one property that is ubiquitous to all biomaterials, before considering other effects, and that is the induction of an inflammatory response upon implantation. All biomedical implantation occurs via major surgery, which is an invasive procedure, leading to tissue injury and thus wound healing afterward. What happens next we call wound response and it will occur regardless of whether the injury is a result of surgery or an accident. Immediately upon implantation of a biomaterial, the same events that usually accompany tissue healing after injury will occur. The first effect in this process will be protein adsorption at the surface and the formation of a temporary protein layer around the implant, as will be detailed later in this chapter. This layer is called the provisional matrix. The most abundant protein in the provisional matrix is fibrin.

Since implantation of a biomaterial results in tissue injury, a profound understanding of processes associated with normal wound healing will also help understand how the induced changes in the local environment may affect the implant (material response).

Wound response, either intentional (e.g., surgery) or unintentional (accident, disease), has two main stages: inflammation and cellular response to repair. The cellular response includes cellular invasion at the site of injury and ultimately the remodeling of the tissues upon healing.

Inflammation is the physiological response to trauma, infection, presence of foreign materials, or localized cell death. Inflammation can also occur in tandem with general immune response effects. The main symptoms of inflammation are redness, swelling, pain, and heat. Depending on how severe each of these symptoms is, the severity of the inflammation can also be determined. An implant inducing initial tissue damage after surgery does not trigger additional inflammation because of its presence, but may affect the severity or the length of time of the response. The inflammatory response is an immune response called innate immunity. Fig. 8.1 shows the stages of wound healing, starting with injury up to tissue remodeling.

As seen in Fig. 8.1, immediately after tissue injury, a soft blood clot (thrombus) will form via coagulation. Thrombus formation starts with a protein called coagulation factor XII contacting collagen, foreign materials, or foreign proteins. Thrombus

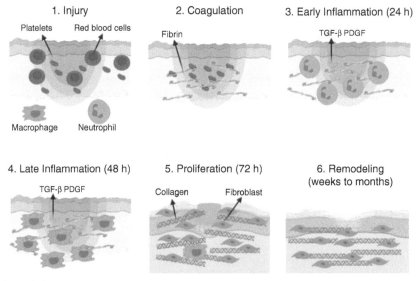

FIGURE 8.1

Stages of wound healing. Transforming growth factor-β *(TGF-β)* is a protein produced by white blood cells, which stimulates extracellular matrix production by fibroblast cells. Another growth factor protein, called platelet-derived growth factor *(PDGF)*, acts to regulate cell growth and division.

is a fibrous clot made from blood platelets and the protein fibrinogen. Thrombus formation helps localize the injury and start the healing process. In the next step after thrombus formation, capillaries constrict to stop leakage of blood, before they dilate again and the cellular activity of the endothelial cells lining their inner surface increases. Thus the capillaries become covered with erythrocytes, leukocytes, and platelets and there is an increased blood flow at the injury site. This increased blood flow is why redness is present. Another effect at this stage is an outflow of plasma to nearby tissues and leakage of plasma from capillaries. This, together with leukocytes and dead tissue, forms the exudate and leads to the swelling effect seen with inflammation. Finally, pain is present due to the pressure applied to the nerve endings, as well as the release of molecules called kinins, that react with nerve endings further leading to pain.

If the injury is very severe and extensive or the wound contains irritants (including implant presence) or bacteria, inflammation may lead to extensive tissue damage. Collagenase, an enzyme capable of digesting collagen, is released after granulocytes are lysed by a lower pH (as low as 5.2) at the site of injury. If destructive inflammation does not subside after 3 to 5 days, a chronic inflammatory process starts. For chronic inflammation, macrophages are activated to remove bacteria or foreign particles, followed by tissue destruction by collagenase.

If severely destructive inflammation persists for more than 3 to 5 days, a chronic inflammatory process begins by activation of macrophages (large phagocytic mononuclear cells). The temporary fibrin tissue is replaced by stronger granular tissue at this phase of the body response process, and this is when macrophages are activated. Macrophages can coalesce and turn into multinuclear giant cells (foreign body giant cells [FBGCs]). Macrophages remove bacteria or foreign materials if they are micron size or below.

The stages of inflammatory response start within minutes to hours from injury with white blood cells (*neutrophils*) moving into the surrounding tissue in the acute phase (initial phase) of inflammation. Macrophages also move to the site of injury in the acute stage of inflammation. As illustrated in Fig. 8.2, the next step is the *phagocytosis* of small fragments of tissue or foreign material. Phagocytosis of larger particles begins later, in the chronic inflammation stage, by *macrophages* and *FBGCs*. Phagocytosis is initiated by the presence of particles (pathogens or simply foreign materials such as biomaterials) with submicron or micron diameter. Particles larger than 50 μm do not initiate a reaction greater than the bulk material.

Receptors on the phagocytic cell surface bind ligand molecules on the target, which can be a pathogen, a dead cell, or a particle of foreign material (Fig. 8.2). As receptors bind more and more ligand molecules, the phagocytic cell membrane progressively engulfs the target. Upon full engulfment, a phagosome (a vacuole in the phagocytic cell that contains the foreign particle) is formed, which fuses with lysosomes, leading to the digestion of the target.

Phagocytosis removes dead tissue and foreign materials from the site of injury. If particles are toxic to the cells, this process may lead to white blood cell death and pus accumulation.

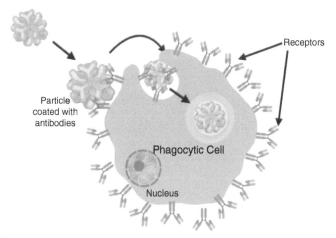

FIGURE 8.2

Graphic illustration of phagocytosis of a target particle.

Macrophages, however, do not just play a role in acute inflammation, but also are involved in the end stages of wound healing, which are cellular proliferation and tissue remodeling. In the initial wound healing stage, inflammation, the macrophages present at the site are what we call M1 macrophages, which are proinflammatory. However, in the proliferation and tissue remodeling stages of healing, they transition to antiinflammatory or M2 macrophages and help tissue remodeling.

Up to this point in the chapter, we discussed the first stages of tissue response to injury. The end stage, tissue remodeling, is the successful completion of the healing process after injury and often results in decreased tissue mass as diseased or injured tissue is removed (Fig. 8.1). As the normal tissue pH is 7.0, a pH of 5.2 at injury sites is acidic and leads to the release of collagenase and tissue damage. As for inflammation, tissue remodeling also goes through different phases. Tissue remodeling starts with cellular division, followed by budding off of small vasculature and penetration into the injured area. This improves blood circulation and in the next steps, proliferation, fibroblasts, keratinocytes, and endothelial cells propagate to the site and produce collagen and polysaccharides, which form scar tissue, as well as grow vascularization (endothelial cells are responsible for vascular growth). Scar tissue acts as a scaffold for continuous tissue remodeling, with cells producing additional tissues until remodeling is complete. Fibroblasts overproduce extracellular matrix (ECM), which results in both scar tissue formation and fibrous encapsulation of biomedical implants. Not all tissues remodel completely, however. The only organs that completely remodel after injury (regenerate) are the bone and the liver. Because of the tendency of some cells such as fibroblasts to overproduce ECM, other tissues can have nearly complete remodeling, but the remodeled tissue will not have the same properties as the original tissue. For example, in the case of skin, the ECM overproduction will result in a residual scar that contains more collagen than healthy skin and there is possible absence of hair follicles. Articular cartilage never completely remodels and is replaced by fibrocartilage, which deteriorates under repeated loading. While in acute wounds, macrophages transition from proinflammatory to antiinflammatory (M2), in chronic wounds, normally associated with disease (e.g., diabetic ulcers), they remain in the proinflammatory state M1.

Example 8.1

From a physiological and cellular perspective, explain why redness, pain, and heat are present during the inflammatory stage of tissue healing.

Answer After injury, capillaries will dilate and there will be an increase in the activity of the endothelial cells lining the capillaries. Capillaries will become covered with leukocytes, erythrocytes, and platelets. This will result in increased blood flow to the injured area resulting in redness due to the presence of more blood cells. Then plasma will outflow to surrounding tissues and leak from the capillaries. Leaked plasma, leukocytes, and dead tissue will accumulate leading to swelling. This will also lead to pain due to the extra pressure on deep pain receptors. Kinins are also produced in this process, and react with nerve endings, causing pain.

The complement cascade

After tissue injury, the inflammatory response triggers thrombus formation as a result of the activation of the coagulation cascade and of what we call the *complement cascade*. Briefly, the complement recruits macrophages, which are usually not found in normal healthy tissues but activate after injury and act to kill pathogens, remove foreign particles and dead tissue, and secrete growth factors that stimulate the ECM and fibroblast production for the tissue remodeling phase of wound healing (Fig. 8.2).

The complement is a series of serum and membrane proteins that are deployed by the immune system as part of its function of removal of pathogens and antigens from the human body. The complement system consists of approximately 30 proteins that are synthesized by the liver cells and by the cells that are involved in the inflammatory response. The more precise functions of the complement system are lysis (breaking down of the cellular membrane) of bacterial cells, opsonization (coating of a particle with antibodies or complement proteins) of viruses, and immune clearance (removal of pathogens by the immune system). As part of the immune clearance, a foreign particle, when it is under 50 μm in size, is engulfed by macrophages and removed. This process in which certain immune response cells such as macrophages engulf any foreign particles of material (pathogens or simply particles of a biomaterial) is called phagocytosis, as discussed earlier in this subsection. Cells involved in this process are macrophages, which are a type of white blood cells that surround and destroy pathogens and other foreign materials that enter the body.

The complement cascade is a process where, once an antibody recognizes a foreign particle and binds to it, the component proteins become activated by each other in complex mechanisms, leading to the removal of the pathogen or foreign particle by phagocytosis at the end of the process. For example, when the immune system identifies the presence of a virus, first antibodies bind to the virus (Fig. 8.1), followed by a cascade of complement proteins that activate each other through a complex mechanism, which ends up with the virus being coated and inactivated (opsonization), thus interfering with the ability of the virus to interact with the membrane of the cells it is targeting and blocking its ability to enter into the cell. For bacteria, the proteins that are part of the complement system activate each other upon recognition of the bacteria by antibodies, to end up with the formation of what we call the membrane attack complex (MAC). The MAC is a ring formed of complement proteins that polymerize together on the bacterial membrane and then undergo a structural change that allows them to bind to the phospholipid membrane of the bacterium and gain access into its cell, thus creating a small pore or channel in it. The pore disrupts the osmotic equilibrium of the cell, allowing ions to pass through this channel, water to enter the cells, and ultimately the cell components to escape from the cell, leading to its destruction (cell lysis).

When any foreign particle, including biomaterials that are implanted, pieces of the dead bacterial cells, or opsonized viruses, is recognized by the immune system, antibody binding to the foreign particle forms a so-called immune complex. Further attachment of complement proteins to the immune complex leads to its breaking down into smaller pieces that can be engulfed and cleared by macrophages via phagocytosis (Fig. 8.2).

Tissue response to permanent implants
Introduction

As discussed, implants cause tissue injury, at least immediately at implantation and after implantation, and some may cause tissue necrosis (death). This can happen due to chemical, mechanical, or thermal injury. Ultimately, after implantation of a prosthesis, the goal is to reestablish homeostasis and a new status quo at the implant site, although this would be different than the original, preimplant tissue environment. The response of the tissue to implantation of biomaterials triggers a series of events at the biomaterial-tissue interface that is summarized by the term foreign body reaction. This is a very complex sum of processes, which is affected by the type of material, its stability (e.g., no corrosion, particles or free radicals leaching in polymers), mechanical properties of the biomedical implant and how closely it matches the tissue with which it interfaces, as well as the implant shape.

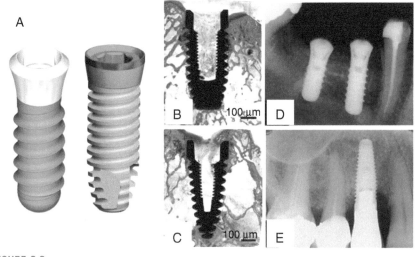

FIGURE 8.3

Effect of geometry and surface composition of implants on bone response. Illustration of threaded cylindrical-shaped and conical (root)-shaped implants (A). Histological image of a threaded implant provided with hydroxyapatite coating; the implant is covered with bone tissue growing into the screw threads (B). Histological image of tapered titanium implant showing its position on trabecular bone: there is no resorption of cortical bone (C). Radiographic images of clinical cases showing both geometries implanted in the jaw bone (D and E).

Reproduced with permission from Alghamdi[11] and Aldosari et al.[12]

Among these factors, tissue response to an implant is influenced by the choice of biomaterial, as well as its geometry (Fig. 8.3). Correct implant geometry ensures a good tissue-biomaterial integration. For example, the thread geometry in orthopedic screws can influence the strength of the tissue-implant bond and ultimately the success of its functional implantation, with no micromotion after healing. By contrast, incorrect geometry can lead to negative effects such as bone resorption, due to micromotion and poor integration.

The first effect that occurs immediately upon the implantation of a biomaterial (metal, ceramic, polymer, or composite part of an implantable biomedical device) in the human body is protein adsorption.

When a biomedical implant is placed into the body, and after tissue healing from the initial injury is complete, there are several types of behavior that could be expected. One of these is implant extrusion, which means there is a pocket that forms around the implant and the implant effectively comes through the tissue and can even become exposed. This happens mostly with implants that are placed under the skin (subcutaneous) and indicates a failure of the implantation procedure. Some reasons include issues related to tissue healing after the surgery, chronic diseases in the patient (e.g., diabetes), or patient overexertion too soon after surgery.

Implant resorption occurs when there is a collapse into scar tissue in the case of soft tissue implants, or when the implant resorbs and completely disappears and leaves behind the newly formed tissue, such as in the case of bone defects being corrected with bone substitute materials. Another implant response is integration, which means that a permanent implant adheres very well to the tissue, without any intervening fibrous layer or capsule. This is a positive response that is met in very few cases, such as integration between titanium implants or glass ceramics and bone.

The most common response, however, is fibrous encapsulation, which is in fact the final stage of tissue healing in response to biomedical implants. In this case, the implant is blocked off from the rest of the tissue by a fibrous capsule. The capsule is thicker at irregular, sharp edges, around chemically active implants, and if there is leaching of corrosion products. The way this capsule forms is similar to what was described previously for regular tissue healing and scar tissue formation, ending up with the deposition of high levels of collagen. When there is micromotion between the implant and the surrounding tissue, the thickness of the fibrous capsule is also larger than normal. Micromotion is not desirable, and if it is extreme, a painful liquid-filled, bursa can form, and the implant needs to be revised.

The next subsections will detail the main effects that one could expect when a biomaterial is implanted into the human body, including protein adsorption, as well as the effect of the biomaterial's chemistry on the type of tissue interactions that are to be expected.

Protein adsorption on the surface of biomedical implants

Protein adsorption refers to the almost instant deposition of proteins to the biomedical implant surface, and it is influenced by protein diffusion to the implant surface as well as by whether the protein has a high affinity for the biomaterial surface. It is critical for the success of a biomedical implant because it helps the implant to be recognized as self. This concept means that because protein adsorption occurs very rapidly, after implantation at the millisecond level, by the time cells arrive at and make contact with the surface of the biomaterial, a protein layer has already been formed and coats the implant, thus causing the cells to not recognize the implant as a foreign object, but as "self." Cells are much larger in size compared with protein and thus move slower and arrive at the implanted surfaces later. The chemistry, surface geometry, and roughness of an implant will have a strong influence on the formation, composition, stability, and overall properties of the protein adsorption layer coating it. Controlling the cellular and therefore tissue response to an implanted biomaterial is necessary to elicit the proper operation of the implant in vivo. Chapter 10 will cover the topic of protein adsorption on surfaces in more detail, present examples on protein adsorption on specific implants, and discuss how implant surfaces can be engineered to elicit a specific desired response (e.g., prevent coagulation or encourage strong integration between the outer surface of an implant, such as a cardiovascular stent or hip implant, and the surrounding tissues) by using intentional protein adsorption, functionalization for imparting certain surface properties (e.g., hydrophobicity), and stimuli-responsive polymer coatings.

Proteins are macromolecules, more specifically polypeptides, that contain a multitude of functional groups, and typically have in their primary backbone structure of the form [-NH-CHR-CO-]. The functional group R is a side group whose sequence imparts the protein its specific properties. Depending on the chemical structure of the side group R, proteins can be either nonpolar (hydrophobic), polar (hydrophilic), or ± charged. The primary structure is then assembled into the secondary structure into α-helix, β-sheet, and loops connecting helix and sheet structures. We discussed the general structure of proteins in Chapter 6 and the reader is also urged to consult additional chemistry and biochemistry undergraduate textbooks for more in-depth details on the topic. Briefly, the hydrophilic or hydrophobic character of the primary protein structure, combined with the type of secondary structure and spatial distribution, leads to the creation of specific bioactive domains of proteins. The bioactive domains, in turn, are necessary for a specific protein to be functional and perform as expected for its biological function. In general, proteins tend to fold in such a way that the hydrophobic groups are located deep inside the protein, while the hydrophilic and charged groups are exposed to the surrounding environment, in an attempt for the protein to minimize its free energy and decrease the surface area available to the surrounding solvent. However, it is commonly the case that proteins have both hydrophilic and hydrophobic residues at their surface, which impart it an amphiphilic character. When a protein contacts a biomaterial surface, its adsorption will be dependent on the surrounding pH as well, which changes the native charge of the protein surface by changing the protonation level. Moreover, the kinetics of protein adsorption at a biomaterial surface results in both reversible and irreversible protein adsorption and results in water release in the surrounding environment (Fig. 8.4).

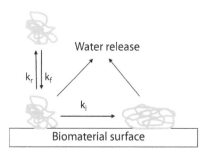

FIGURE 8.4

Schematic of protein adsorption at the surface of a biomaterial. Once a biomaterial is implanted in the human body, proteins are adsorbed at a surface either reversibly or irreversibly.

The kinetics of irreversible protein adsorption is governed by the reaction rate constants k_f (for forward reaction), k_r (for reverse reaction), and k_i (for the irreversible adsorption). Thus depending on the protein concentration in solution, the available surface area for adsorption, the surrounding pH, and the surface chemistry, some proteins will be reversibly adsorbed and will tend to leach out from the biomaterials' surface over time, while others will be irreversibly adsorbed to the surface and will not leach out. Water is always complexed to the native protein and tends to be released upon adsorption at the biomaterials' surface. Adsorption of proteins at a biomaterial's surface generally occurs in an irreversible manner because of the much faster adsorption rate compared with the desorption rate. As a general rule, a hydrophobic character of the surface will lead to a higher tendency for protein adsorption than a hydrophilic surface.

In addition to interactions between proteins and the surface of the biomaterial the final properties of the protein layer adsorbed at the surface of an implant will also be influenced by protein-protein interactions. Once a biomaterial's surface is saturated with a monolayer of adsorbed protein, the proteins spread out and expose their core and further protein adsorption is not likely. The biological environment contains a heterogeneous composition of different proteins and this multicomponent character will influence the in vivo protein adsorption behavior on the surface of biomedical implants. The rule of thumb in this scenario is that the adsorption rate of a protein is directly proportional to its concentration and inversely proportional to its molecular weight. Thus the small proteins tend to adsorb first on the blood-interfacing implants but can be later displaced by larger proteins arriving at the same surface. This phenomenon is called the Vroman effect and is illustrated by the fibrinogen exchange from materials' surfaces. It has been shown that fibrinogen that is adsorbed at the implant's surface gets easily displaced by other plasma proteins at neutral pH. However, this effect is not exclusive to fibrinogen and has been proven for other proteins such as albumin, antibodies, or fibronectin.

Biomaterials degradation

One of the other factors that influence tissue response to implants is composition and surface chemistry (physical & chemical phenomena occuring at the interface) of biomaterials.

Pure metals, for instance, are generally very chemically active and can react with tissue and blood components, and it is their tendency to corrode under the physiological environment that can elicit a severe tissue reaction. There are some exceptions to this rule: metals such as tantalum and titanium that can spontaneously form passivating oxide layers that are adherent and prevent further direct contact between the pure metal and tissues. However, while the formation of passivating oxide layers is the basis of allowability of metallic alloys to be used in implantation, these layers can become disrupted under use and leave the biomaterial vulnerable to corrosion by various mechanisms. Refer back to Chapter 3 for details on various mechanisms of corrosion in metallic biomaterials. Corrosion of biomedical alloys can lead to degradation of their mechanical properties and ultimately mechanical failure of the implant under the mechanical loading of normal use. It is also important to mention that corrosion of metallic biomaterials, when it occurs in permanent implants, can lead to significant negative tissue effects such as allergenic reactions, inflammation, discoloration, or erosion from the corrosion products diffusing into such tissues. As we learned in earlier chapters, the main metallic biomaterials used in permanent implants are Ti alloys, Co-Cr alloys, and stainless steels. These alloys are typically protected from corrosion by the formation of stable passivation oxide layers; however, they could still be affected by corrosion if this layer becomes disrupted. Chapter 3 discusses the rate at which each type of alloy can reform the passivating layer after its disruption. When corrosion occurs, corrosion species in these materials could include ions of Ni, Co, Cr, V, and Al, the release of which can lead to the negative tissue reactions mentioned previously, or can diffuse away from the implant to produce systemic reactions, discussed later in this chapter. Thus degradation of metals and metallic alloys via corrosion can occur even in the case of well-designed materials as far as corrosion protection goes, with negative effects on the via mechanical properties and successful operation of the implant, as well as on the surrounding tissues.

Many, although not all, ceramics (see Chapter 4 on different types of ceramics) are chemically inert and thus produce minimal tissue response stemming from chemical reactions. Instead, there is generally a thin layer of fibrous tissue similar to the encapsulation effect.

Polymers are generally inert if additives are not present; however, if there is leaching of monomers or free radicals, or release of polymer particles with high surface area, both severe tissue reaction and implant property degradation and failure can occur.

Free radicals are groups of atoms with one or more unpaired electrons. Because of the unpaired electrons, they are very chemically reactive and when generated in excess can have negative effects on the structure of macromolecules. For biological systems, including tissues, the most concerning free radicals are the so-called ROS. These are radicals that have oxygen with two unpaired electrons. While ROS are involved in a series of essential biochemical reactions including playing a positive role in defense against pathogens, they can also have negative effects on cells. More precisely, because of their very high chemical activity they can have oxidizing effects that can be harmful to cells by damaging lipid membranes, proteins, or nucleic acids.

It is significant to note that the involvement of ROS in immune protection can be a double-edged sword. ROS are generated by phagocytes with a goal of attacking pathogens but also foreign materials, which implants are. When ROS are generated by phagocytes, these can oxidize implanted polymeric biomaterials. The result can be pitting, embrittlement, and cracking, all leading to the ultimate property degradation and mechanical failure of the implant. Similarly, the ROS can attack metallic implants, disrupting their passivating layers and resulting in corrosion, degradation of mechanical properties, and implant failure. Degradation of certain types of biomedical implants, such as Mg alloys or polylactic acid implants, can be engineered to occur in temporary, bioresorbable devices. Chapter 12 discusses such temporary biomedical devices in detail.

In addition to the potential formation of ROS upon polymer degradation, polymer swelling and leaching are also effects of them being immersed in the biological solution. This means that there is material being transferred from or into the implant material but without a chemical reaction occurring. For example, polymeric materials can absorb blood and blood components and swell, or they can absorb different molecules from the blood, which in turn also leads to swelling because of an increase of the mass of the material. Swelling and leaching of polymeric implants can lead to a change in materials properties, which in turn can lead to a change in implant behavior.

Both absorption and leaching are governed by the diffusion laws. Fick's first law of diffusion describes the rate of solute transfer:

$$F = -D\left(\frac{dC}{dx}\right) \tag{8.1}$$

where F is the rate of transfer per unit cross-sectional area, D is the diffusion constant, and dC/dx is the gradient of concentration normal to surface. The diffusion coefficient D is a material-dependent constant and it also depends on the matrix and the diffusion type.

Another relevant law is Fick's second law of diffusion, which describes the change in concentration with respect to time for one-directional flow in an infinite medium:

$$\frac{dC}{dt} = D\left(\frac{d^2C}{dx^2}\right) \tag{8.2}$$

When a solute is transferred from a fluid to a solid and a stagnation layer is present, there is a reduced concentration of solute, which reduces the rate of transfer compared with the one that would otherwise be theoretically predicted. In the case when together with the solute there is also absorption of the solvent, there is a "solvated" surface layer that forms within the solid matrix. This layer decreases the concentration of the diffused solute within this area of the matrix. This area accepts solute particles more easily than the unsolvated matrix.

There is a situation when there is a transfer of solute from a fluid to a solid and a solvated layer develops in the solid because of the absorption of water molecules along with the solute. This effectively reduces the concentration of solute in the solid and therefore increases the transfer rate of the solute. All materials can absorb molecules from the surrounding environment, but this occurs easily in loosely bonded polymers compared with metals and ceramics.

Example 8.2

Fick's law of diffusion indicates that the rate of transfer of a gas through a sheet of material is proportional to the material area and the difference in gas partial pressure between the two sides and inversely proportional to the material thickness. Use Fick's law to calculate how much oxygen diffuses through a polymeric membrane that has a surface area of 75 m^2 and a thickness of 0.7 µm, which should be included into a membrane oxygenator when the partial pressures on each sides of the membrane are 100 mmHg and 40 mmHg, respectively. The diffusion coefficient is 20 ml·min^{-1}·mmHg^{-1}.

Answer

$$\frac{dV}{dt} = \frac{\text{Material area}}{\text{Material thickness}} \cdot D(P_1 \cdot P_2) = D_{LO_2}(P_1 - P_2)$$

Inputting the values for each parameter results in $\frac{dV}{dt} = 1.66$.

Swelling can cause internal stresses to exceed the threshold creep stress and allows continued deformation of the solid (expansion) without an externally applied stress. As volume increases, the concentration of solute decreases and further absorption is allowed. In this case, no equilibrium is reached in the absorption process. Swelling also reduces the remaining strain available before the elastic limit is reached. Thus microcracks can form in the polymer past this elastic limit, which leads to static fatigue (crazing). Cracks cause a reduction in the strength of the material without necessary cyclic loading, which can lead to mechanical failure.

A type of artificial heart valve consists of a polymeric sphere locked in a metal cage (Fig. 8.5). When adsorption of lipids from blood occurs, the polymeric sphere can increase in diameter and lead to malfunctioning of the valve.

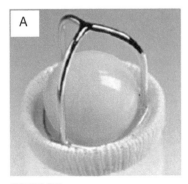
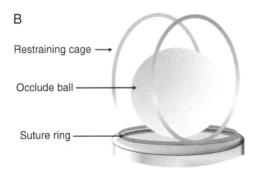

FIGURE 8.5

Prosthetic heart valve. Starr-Edwards caged-ball valve (A) and its components (B).

Adapted with permission from Movafaghi et al.[13]

The possible changes in mechanical effects, besides changes in physical size, include a *decrease* in modulus, change in ductility, change in the friction coefficient (important for joint components), and a decrease in wear resistance.

Leaching is the removal of material from the surface or bulk of a biomaterial, and in an idealized case it occurs at a constant rate. If the fluid receiving the components is in motion but not fully mixed, a rate of transfer must be assumed proportional to the difference between the concentrations in the solid and fluid. The rate of transfer also varies with time. Leaching has a much smaller effect on the implant properties than absorption, but it does result in local and systemic biological reactions. Excess leaching can cause a reduction in mechanical properties. If the defects resulting from leached molecules join with any macroscopic voids, porosity-induced changes will occur. For example, Young's modulus and fracture strength can decrease.

Polymeric implants can also deteriorate due to chemical reactions. For example, linear polymers can be broken down by random molecular bond scission, or depoly-merization. Both lead to a reduced molecular weight and a change in properties. Cross-linking is another effect that can either be induced or damaged by chemical attack. For example, induced cross-linking occurs in low-density polyethylene and increases the stiffness of this material. Change in bond structure can result in different compounds with very different properties. Polymer degradation products can be toxic to the surrounding tissues.

All biomedical devices must be sterilized before implantation.[6] This brings up issues related to the sterilization process that can come with relatively high temperature, reactive chemicals, or gamma radiation.

It is important to note here that implant failure may occur also because of infection after implantation. It is for this reason that all biomedical implants must be sterilized, which means to be subjected to treatments that aim to destroy all microorganisms on their surface and prevent infection and disease. As discussed in earlier chapters, there are a few different sterilization methods that can be used for this purpose, depending on what types of biomaterials are part of a given implant. For example, dry sterilization involves exposing the device to temperatures between 160 and 190°C. While these temperatures are not excessively high for metals and ceramics, and their properties will not be affected, they are higher than the melting temperature of several linear polymers. Thus some polymeric biomaterials cannot be sterilized by dry sterilization. In other cases, even when the melting temperature of the polymer is above the sterilization temperature, oxidation can affect its structure (e.g., nylon). Thus among the polymeric materials used in the biomedical industry, only Teflon and silicon rubber can be dry sterilized.

Steam sterilization, or autoclaving, is another sterilization technique that requires high pressure but lower temperatures than dry heat sterilization (120−135°C). However, because water vapors are used, the method is not useable for polymers that have hydrolyzable bonds. These include polyvinyl chloride, poly-acetals, polyesthers, polyanhydrides, low-density polyethylenes, and polyamides. On the other hand, corrosion will be a predominant mechanism for failure of biomedical implants that have metals and alloys as components. Please refer to Chapter 3 for a description of the various mechanisms through which corrosion can occur and negatively affect implant properties, ultimately leading to failure. It is passivation with stable oxide layers which makes possible the use of metals in biomedical implants, including TiAl6V4 (titanium oxide), 316L stainless steel, and Co-Cr alloys (chromium oxide passivation layer for both).

Sometimes, implant responses, including corrosion, are desirable and often engineered to occur. This is the case for bioresorbable metallic alloys, such as Mg alloys, which are engineered to corrode at a specific rate with few adverse tissue interactions, and can be used in temporary implants such as in resorbable cardiovascular implants, as presented later in Chapter 12. While bioresorbable alloys are generally engineered to not elicit adverse tissue effects, the rate of corrosion and the type of ions being released need to be controlled, as they will have a significant effect on whether or not a negative host response will actually be induced upon implantation.

Porosity effects on implant-tissue interactions

Another good example of an engineered positive host response upon implantation of a biomaterial is tissue ingrowth. Ingrowth has been seen to occur in a wide variety of materials, including metals, ceramics, and polymers. Good tissue ingrowth aids the mechanical stability of the implant, as it allows little micromotion between the implant and the surrounding tissue. Exceptions are situations such as artificial blood vessel grafts and stents, where motion between the blood and the inside surface of the implant is continuous and on a macroscopic level. In these cases, cellular elements must deposit on the outer graft surface for ingrowth and connection with the outer connective tissues to occur. The nature of ingrowing tissue is dependent on the minimum size of the interconnections between pores.

Tissue ingrowth is the desired response for many implants and occurs with all classes of biomaterials. The positive effect of tissue ingrowth results from reducing the micromotion, except for blood-contacting implants (vascular grafts and stents). Examples include porous hydroxyapatite coatings applied on hip implant stems or porous coatings on the back of some types of breast implants. For most biomedical implants, the porosity is limited to coatings that enhance tissue-implant integration. However, for tissue engineering constructs, such as artificial bone scaffolds aiming to repair bone defects, implants must have interconnected porosity with interconnects larger than 1 µm in size for tissue ingrowth to occur for most soft tissues. Tissue engineering is not the objective of this chapter, and the reader is advised to consult textbooks on this topic for more information. However, the size of the pores on the surface coatings, even in the most common biomedical implants, needs to be matched to the size of the specific cells relevant to the tissue the implant is interfacing. For bone, for example, the pore spaces need to be large, with a minimum of around 250 µm.

Blood compatibility

Most implants will be in contact with blood and blood components, and that is not limited to cardiovascular implants such as heart valves or vascular grafts. Thus how compatible a biomaterial is with the blood environment is the most important factor of biocompatibility for blood-interfacing implants.[7−10]

To be functional, implants should not induce blood clotting, damage blood cells, or induce red blood cell rupture (hemolysis). Blood clotting (coagulation) can happen via either intrinsic or extrinsic stimulation. Intrinsic stimulation means that clotting is initiated by blood contact with either a damaged portion of the blood vessel wall or another thrombogenic (clot-causing) surface. It takes between 7 and 12 minutes to form a soft clot. Extrinsic stimulation is the result of the presence of a foreign body or tissue damage (other than blood vessel). In this case, it takes between 5 and 12 seconds to form a soft clot. It is the complement cascade which plays a role in hemocompatibility, as explained in the complement cascade section.

When looking for avenues to affect the blood compatibility of a material, several parameters can be manipulated, one of which is the surface roughness. If an implant material has a rougher surface, it will have a greater surface area and contact surface with blood compared with smooth surfaces, which will lead to faster coagulation. Rough surfaces are sometimes used to promote initial clotting at porous interfaces to minimize leakage of blood, such as for vascular grafts.

Surface charge is another factor that affects blood coagulation. The intima, the interior surface of blood vessels, has a negative charge due to the presence of proteins. On the other hand, red blood cells, white blood cells, and platelets also have a negative charge. Thus imparting a negative charge to a blood-contacting material leads to repulsive forces between its surface and the blood components and minimizes coagulation. This effect has been proven in experiments using negatively charged and noncharged segments of copper tubing as "vessels." Copper is normally thrombogenic (clot causing) but imparting a negative charge delayed the onset of clotting. For artificial biomaterials, surface charge only occurs naturally in polymers, while metals and ceramics generally need to be surface modified to induce it.

To fabricate implants with thromboresistant properties, one of the approaches is to mimic the properties of the natural blood vessels as presented in Fig. 8.6. Imparting negative charges to biomaterials is one route, while another is to induce low surface tension. Low surface tension materials tend to reduce the deposition of blood components. Heparin is a polysaccharide with negative charges that is used as an anticlotting agent in many instances, such as in blood samples and patients with blood clotting problems. For implants such as artificial blood vessels, heparin is being attached to their surface to impart anticoagulation properties. Other molecules used for coating artificial blood vessels include albumin or gelatin.

An alternate approach to the fabrication of coagulation-resistant blood-contacting implants makes use of natural tissues. These could be human, porcine, bovine, or cultured tissues, which are stripped of cells to avoid the immune response.

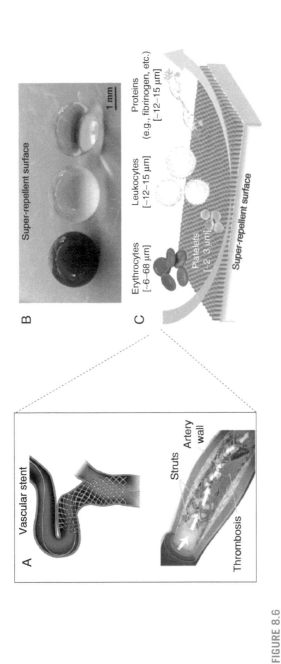

Reproduced with permission from Movafaghi et al.[13]

FIGURE 8.6

Hemocompatibility of super-repellent surfaces. A few examples of blood-contacting medical implants (A). Droplets of blood, plasma, and water on a super-repellent surface (B). Illustration of the performance of these surfaces "repelling" blood components due to the reduced blood-material contact area (C).

These decellularized tissues are highly thromboresistant but have the disadvantage of not having a strong tissue-tissue bond with the nearby tissues. The reason behind this disadvantage is that the same properties that reduce blood clotting also retard the cell adhesion that is necessary for tissue ingrowth and integration. While the inside of a vessel, for example, needs to be thromboresistant, the outer parts need to integrate with neighboring tissues and thus have to have somewhat opposite surface properties than the inside.

Overall, in the efforts to create thromboresistant surfaces, the most successful were materials that could sustain pseudointimal layers, such as some polymers and pyrolytic carbon. When anticoagulant coatings are used on materials, any defects and disruptions of the coatings lead to the initiation of clotting and render the coating ineffective.

Another successful approach to the creation of successful blood-interfacing implants is the use of viable biomaterials. These are resorbable materials that are seeded with cells (e.g., smooth muscle and endothelial cells). The resorbable material, usually a polymer, acts to provide a structure for the cells to produce the ECM. At the same time as matrix resorption, new natural tissue grows and remodels to resemble the host tissue.

For a blood-interfacing implant, there is a continuous flow of blood at motion at a blood/surface interface that may damage red and white blood cells, resulting in cell death. The shear stresses that can lead to hemolysis can be less than 500 dyn/cm^2. Hemolysis can also happen in the absence of blood interaction with surfaces, simply due to the high shear stresses of 1500 to 3000 dyn/cm^2 or more.

Designing materials with high surface tension may reduce hemolysis at a blood/surface interface because layers of platelets and fibrin that are attracted to the surface discourage further adhesion (surface charge becomes negative due to formed cell elements). However, there is no right surface tension for a blood-interfacing implant. Materials with high surface tension attract proteins and form a protective layer, helping with tissue integration, and materials with low surface tension minimize blood coagulation.

Biomaterials testing

When deciding whether a biomaterial is to be chosen to be part of an implantable device, it has to be tested to ensure it will behave as intended. To be relevant, the testing needs to be performed with the biomaterial in the final form that will be used when the biomedical implant will be functional in the body. Individual parts of the biomedical implant need to also be tested to ensure their safety to the surrounding tissues. Most biomaterials testing strategies focus on the host (tissue) response to the implant. The biomaterial response is generally easier to predict based on prior knowledge of material properties.

Most biomaterials testing goes through the same general steps. First, a biomaterial is selected based on engineering data and previous information on host response information. Once such a biomaterial is proposed, experiments are designed that will verify that indeed the host response is adequate for the functioning of the biomedical implant, and data are collected to support the decision being made on whether or not the biomaterial is appropriate to be used for implantation under the functional conditions that are intended. The biomaterials testing can be performed in vitro, ex vivo, and in vivo. The ultimate proof for a biomaterial being acceptable for a certain application is the in vivo test, that is, testing of the biomaterial when it is part of the actual device and is implanted and functioning as intended in a clinical setting.

In vitro testing refers to tests that are done in the laboratory and outside of the physiological environment under conditions that aim to simulate the biological response after the biomaterial implantation. Such testing is generally performed in cell culture, in contact with tissues, or even with entire organs. Here, the biomaterial (bulk material, particles, or a solution containing the material of interest) is put in contact with cells, tissues, or organs, and parameters such as pH, temperature, or cell type are being varied and their effect understood. The effect of the biomaterial on the cells, tissues, or organs is then measured and understood. Information related to the chemical composition of the biomaterial is necessary for the identification of the expected risk resulting from exposure to these chemicals. As shown in the previous section, metals, polymers, and ceramics can all degrade and their degradation products can be potentially harmful. The methods related to characterization of biomaterials and, most importantly, their potential for degradation in the context of being used in biomedical implants are described in the ISO 10933 standards. Based on the information, it can become clear what are the potentially toxic materials so they can be considered as part of a risk-assessment process. Briefly, the testing of the biomaterials needs to be performed based on expected use. For example, where will the biomedical device that the tested biomaterial is part of be implanted will provide some guidance on the type of testing. This can be, for example, information on the contact site (blood interfacing, tissue interfacing), and the expected contact time will be relevant. The contact time can be limited (less than 24 hours), prolonged (less than 30 days), or permanent (more than 30 days). Based on the collected information, assays will be designed differently and not all of the assays recommended in standards will be applicable to every biomedical device. At the minimum, however tests need to be performed on cytotoxicity, sensitization, and intracutaneous reactivity. For blood-contacting implants, tests determining acute and subchronic toxicity and hemocompatibility may be necessary. Other tests may be necessary, depending on the functional needs of the biomaterial. Table 8.2 shows the types of assays that are necessary for biomedical devices based on their intended applications and the in vivo conditions that need to be considered. The table is not comprehensive and other tests may be required, depending on the type of implant.

When in vitro testing is performed by putting the material in contact with tissue culture, it can serve as an initial screening method. However, this testing cannot replace in vivo functional testing because of problems related to the lack of systemic pathways within culture, no systemic feedback loops providing physiological

Table 8.2 Selected biomaterials testing assays for different biomedical devices and for different contact times.

Device type	Type of contact	Duration contact	Sensitization	Irritation	Cytotoxicity	Acute system toxicity	Subchronic toxicity	Hemocompatibility
Surface interfacing device	Skin	<24 hours	Y	Y	Y	N	N	N
		24 hours to 30 days	Y	Y	Y	N	N	N
		>30 days	Y	Y	Y	N	N	N
	Mucosal membrane	<24 hours	Y	Y	Y	N	N	N
		24 hours to 30 days	Y	Y	Y	N	Y	N
		>30 days	Y	Y	Y	N	N	N
External communicating device	Blood interface indirect	<24 hours	Y	Y	Y	Y	N	Y
		24 hours to 30 days	Y	Y	Y	Y	N	Y
		>30 days	Y	Y	Y	Y	Y	Y
	Tissue/bone/ dentine	<24 hours	Y	Y	N	N	N	N
		24 hours to 30 days	Y	Y	Y	Y	Y	N
		>30 days	Y	Y	Y	Y	Y	N
	Circulating blood	<24 hours	Y	Y	Y	Y	N	Y
		24 hours to 30 days	Y	Y	Y	Y	Y	Y
		>30 days	Y	Y	Y	Y	Y	Y
Implanted device	Tissue/bone	<24 hours	Y	Y	Y	N	N	N
		24 hours to 30 days	Y	Y	Y	Y	Y	N
		>30 days	Y	Y	Y	Y	Y	N
	Blood	<24 hours	Y	Y	Y	Y	Y	Y
		24 hours to 30 days	Y	Y	Y	Y	Y	Y
		>30 days	Y	Y	Y	Y	Y	Y

Y = yes; N = no.

response, or no local circulation. For example, because of the lack of local circulation, cell death may occur because of the lack of nutrients reaching the cells that are located too distant from the nutrient source. Also, some physiological processes that occur in vivo do not occur in vitro, such as fracture healing in bone.

Cytotoxicity testing is required for almost all biomedical implants and their biomaterials components and refers to a material having a toxic effect on cells, by affecting their growth, reproduction, or metabolism, in a way that results in cell death.

Sensitization refers to the fact that exposure to degradation products of biomaterials after implantation, such as leached ion metals or radicals, can result in local or systemic hypersensitivity reactions. The body's immune response is at the root of implant-induced sensitization, via complex mechanisms that involve the recognition of degradation products (e.g., ion metals)—protein complexes as antigens by T cells. T cells are involved in the response to foreign particles such as allergens and will bind to the degradation products of biomaterials, eliciting an inflammatory response.

Hemocompatibility of a biomedical device such as catheters, endovascular grafts, or stents, which come in contact with circulating blood, refers to the evaluation of risks for hemolysis (breaking down of blood cells) or thrombosis (formation of blood clots). Complement activation, hemolysis, and thrombogenicity potential are tested for implants that are in direct contact with circulating blood. On the other hand, devices that are not in direct contact with circulating blood can be tested only for the hemolysis potential.

For blood contact studies it is difficult to standardize response because blood is not a standard material. Blood properties can vary depending on individual health, diet, medication use, age, as well as on whether the blood contact is in vivo or has been drawn. The blood contact studies that are performed on biomaterials are separated into in vitro static tests, ex vivo dynamic tests, and in vivo dynamic tests, which are performed sequentially. It is especially important that blood contact studies in biomaterials are performed in the correct environment because of the great differences in the results, for example, in response to a high-flow artery compared with a static blood contact location. However, if a biomaterial performs poorly in a static blood contact test, it will not perform well upon implantation because dynamic processes such as blood flow increase both hemolysis (breaking up of blood cells) and blood coagulation. If in a static blood contact test, the biomaterial proves to induce blood clotting, it is only expected that the coagulant effect will be enhanced in a dynamic environment when blood flow is present.

In a typical static blood contact test, the biomaterial is tested in comparison with a control silicone-coated soft glass material. The silicon-coated soft glass is chosen as control sample due to its low and reproducible host response in the presence of blood.

When dynamic ex vivo tests are set up, there is a connection being established with the circulation of a living test subject and the behavior of biomaterial as far as its hemocompatibility is tested. The problem with this approach is that thrombus formation outside of the in vivo environment can change the flow dynamics and may threaten the donor's safety. Another challenge is in the difficulties in reproducing in vivo flow dynamics due to the large effect of vascular intima (the inside surface of blood vessels) on flow.

Histology and immunohistochemistry are some of the tools used to evaluate tissues after they have been exposed to a biomedical implant and/or its biomaterial components. In this context, histology refers to microscopic examination of tissues to evaluate any changes to the tissue structure or cells induced by the presence of a biomedical implant. Immunohistochemistry uses antibodies and enzymes (usually horseradish peroxidase) that together result in the formation of colorful compounds that can be visible by microscopy in a tissue sample. The antibody helps target a certain area of interest of a tissue, and the enzyme labeling gives color to the area, which then becomes easier to visualize with microscopy for evaluation of any changes induced by the implant presence.

In vivo animal studies of biomaterials are conducted over an extended period of time using a living animal and are the next step after the in vitro testing. Higher-level, mature animals, such as horses or sheep, are preferred to smaller animals due to increased life span and mechanical stresses that are closer to those present in humans. Although in vivo animal testing provides a better idea of how the biomaterial is expected to perform upon implantation and involves exposure to systemic physiological processes, there are still some challenges in translating the results to human use. In one version, nonfunctional testing, the materials are typically implanted in a soft-tissue site and effects on tissues are observed (see Table 8.2). The results, however, are fairly general. The soft tissues are chosen because implantation involves generally minor surgery, with the exception of cardiovascular implants and hard tissue (bone, teeth) implants. Typical sites for implantation are subcutaneous (under the skin), intramuscular, intraperitoneal (fluid-filled space under the abdominal wall), transcortical, or intramedullary. Implants are of an arbitrary shape, usually determined by later testing that is to be conducted on the biomaterial. For example, for biomaterials that are implanted and later tested for mechanical properties, the specimens implanted are cylindrical. These tests are normally conducted for periods that range between a few weeks and up to 24 months. Both the implant itself and the site of implantation can be evaluated for effects induced by implantation.

Although not mentioned in Table 8.2, assessing the potential of a device material that is implanted in the human body to induce the growth of malignant cells is one of the tests that are considered for implants that will be in permanent contact with tissues and will be covered in the next section.

Carcinogenicity and mutagenicity

Before we delve deeper into the issues related to the testing of biomaterials for carcinogenicity, we must first define a few terms that we will use throughout this discussion. Another note that is of importance is that while we have to discuss carcinogenicity here as one of the concerns whenever a foreign material, including a biomedical implant, is present in the body, implant-induced carcinogenesis is rare in humans. However, due to the potential catastrophic effects on the patient's health, this test is necessary for any new permanent implant. Carcinogenesis means cancer production, and a carcinogen is a material that can produce cancer. Cancer is a disease characterized by uncontrolled multiplication and growth of abnormal forms of the body's own cells. Mutagenesis, on the other hand, is different from carcinogenesis, and it means the production of genetic changes in cells. A neoplasm is a mass of tissue that contains abnormal cells, and a primary neoplasm is a neoplasm that arises at that location, not via spreading from a different location. Tumor is the term used for the swelling associated with neoplasms. Metastasis is a neoplasm that forms via spreading of abnormal cells from a primary neoplasm at a different site in the body. A benign tumor is a tumor that does not have the ability to metastasize and has a limited ability to grow. A malignant tumor, on the other hand, is one that has uncontrolled growth, is invasive, and can metastasize.

Different cancers have different names depending on the primary affected site. For example, leukemia or lymphoma refers to the cancer of the cells of the blood circulation system. Sarcoma is the cancer arising from cells of connective tissue. A carcinoma is a malignant neoplasm formed of epithelial cells.

Materials can be carcinogenic and carcinogenesis can arise via different mechanisms. All carcinogenic materials are mutagenic (produce genetic mutations in cells), but not all mutagenic agents are carcinogenic. The Ames test is a test that verifies mutagenicity of a material. A culture of bacterial cells that require histidine to grow is placed in contact with the material being tested. If in the presence of this material being tested, the cells mutate back to the histidine-free form and are able to proliferate, the material is mutagenic (Fig. 8.5).

One of these mechanisms that are relevant to biomaterials is chemical carcinogenesis. As the name suggests, for this mechanism the cancer is produced due to chemical modifications in the biological environment that arise due to chemical reactions. Some chemical interactions can lead to abnormal cell growth by inducing an alteration in the cellular metabolic processes, cellular replication, or mutagenesis.

In the Ames test (Fig. 8.7), a strain of the bacterium *Salmonella typhimurium* is used that requires histidine to grow but cannot synthesize it. Rat liver extract is added to both the negative control sample, which does not have the material being tested for mutagenicity, and the test sample containing the suspected mutagenic material, because it contains certain enzymes. These help better simulate the metabolic processes in mammalian organisms. In early versions, the rat liver extract was not used.

For biomedical implants that can produce cancer via a chemical route, the biomaterials they are composed of can be divided into complete carcinogens, pro-carcinogens, and co-carcinogens. An important must be made, as it relates to the safety of biomedical implants, is that biomaterials-induced carcinogenicity has a very low incidence in humans.

Chemical carcinogenesis can be induced in the immediate vicinity of implants when direct contact with chemical carcinogens leads to diffusion into tissues and

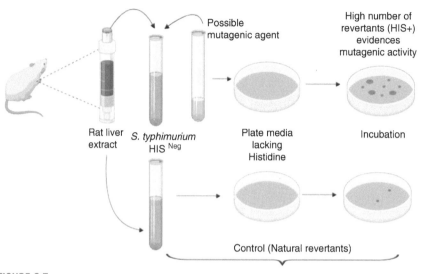

FIGURE 8.7

Schematic illustration of the Ames test for mutagenicity assessment. *HIS*, Histidine.

alteration of cellular processes. However, chemical carcinogens from an implant can also diffuse away from implants and produce cancer at a remote location. Of course, chemical carcinogenesis does not have to be connected to the presence of an implant and can be due to the body's contact with these chemicals in a different way, such as ingestion, inhalation, or adsorption through the skin.

Materials that are complete chemical carcinogens can interact with the cellular environment and produce neoplastic transformation directly. On the other hand, procarcinogens are not initially able to produce neoplastic transformation but can react with other chemicals in the biological environment and thus become complete carcinogens. Co-carcinogens do not have any ability to produce neoplastic transformation; however, they can act to enhance the carcinogenicity of complete or pro-carcinogens.

Example 8.3
A new composite material has been developed, carbon fiber–reinforced hydrogel. The following test results were obtained:

Tissue culture
- Cell growth normal in culture of muscle cells
- Positive Ames test (mutagenic)

In vivo (animal, nonfunctional)
- Thick fibrous capsule forms around implant
- Local tissue discoloration

What conclusions do you draw from each of these tests?

Answer The tissue culture test suggests that while the material may be mutagenic, it does not necessarily mean that it is carcinogenic, since the cell growth in culture is normal. The thick fibrous capsule suggests a reactive implant, while tissue discoloration suggests that there may be polymer leaching going on at the implant site.

The human body contains an array of elements that could be deemed carcinogenic under certain conditions but do not prove harmful in the low concentrations of the physiological range. For example, heavy metals in high concentrations are harmful, but not in the physiological level range. Although metal-induced carcinogenesis has been studied by many research groups, there is not a common view as to what is its exact mechanism. However, most evidence points to ROS produced by the oxidative action of metallic ions as being one very likely factor in carcinogenesis, as well as other diseases, such as neurodegenerative diseases.

Example 8.4

As a bioengineer, you designed a new material to use in the development of a new artificial bone material. What is the initial test that you should perform to test for carcinogenic potential? Describe the test in detail and explain what question about your material it answers.

Answer The Ames test. The Ames test is a test that verifies the mutagenicity of a material. A culture of bacterial cells that require histidine to grow is incubated with the artificial bone material. If in the presence of the material being tested, the cells mutate back to the histidine-free form and are able to proliferate, the material is mutagenic. A material that is mutagenic is not necessarily carcinogenic, but all carcinogens are mutagens.

Despite the danger induced by metallic cations, many metallic implants are allowed. The question that comes to mind is whether there is a certain threshold of concentration over which using transition metals in biomedical devices is safe. In fact there is no common minimum concentration; however, the higher the concentration of free metallic ions that are released in the body, and the higher the exposure time, the higher the probability of carcinogenesis. Thus each implantable device containing potentially carcinogenic materials has to be thoroughly tested and decisions should be taken on a case-by-case basis. Long-term clinical studies are the most useful source of reliable information in this respect. However, most implants have been in place for less than 15 to 20 years and therefore not even clinical observations may completely paint an accurate picture of the true carcinogenicity of a material. The best course of action is to understand the effects of each biomaterial and make decisions based on the benefits of the implant. It is safe to state that if corrosion is avoided in metallic implants and leaching of polymeric degradation products occurs at low rates, the material can be used in implants.

Summary

In this chapter, we discussed concepts related to tissue repair, the immune response, and the effects of the biological environment on the properties of the biomaterial. The physiological environment is very chemically active and aggressive toward the biomaterial components of implanted biomedical devices. At the same time, the immune system reacts to the inevitable trauma of the surgery that is associated with the implantation of the biomaterials components of implants. Implants, on the other hand, are foreign to the physiological environment and are recognized as such by the immune system. Understanding the intricate mechanisms that are at play immediately after a biomaterial is implanted and interfacing with a particular tissue and at a certain anatomical location is key to the long-term success of implantation. Biomaterials testing, both in vitro and in vivo, brings to the table a set of tools that can predict a biomaterial's performance and is indispensable for the design and selection of biomaterials.

Problems

1. One of the most important challenges with vascular grafts is hemocompatibility and vessel occlusion. Heparin is an anticoagulant. Would the use of heparin on hydrophilic surfaces be an ideal solution to prevent these negative effects? Explain in detail why or why not.
2. Explain why pain is one of the signs of tissue inflammation.
3. What blood components are involved in the initial stages of tissue inflammatory response? Explain your answer.
4. You are working as a biomedical engineer for a company that designs hip implants. There have been reports of sensitization testing giving unsatisfactory results for one of the products that included polyethylene-on-metal components. What could be the sources of this negative outcome? Explain.
5. What are the characteristics that differentiate the acute phase of inflammation from the chronic phase?
6. The complement system acts to attack pathogens. How is the complement involved into the inflammatory response?
7. Use Fick's law to calculate how much oxygen diffuses through a sheet of tissue with a surface area of 50 m^2 and a thickness of 0.5 μm, when the partial pressures on each side of the tissue are 90 mmHg and 50 mmHg, respectively. The diffusion coefficient is 22 $ml \cdot min^{-1} \cdot mmHg^{-1}$.
8. A new polymeric material is considered for use in vascular grafts. List three requirements that the bioengineer should consider when evaluating the hemocompatibility of the new material. Discuss why these factors are important.

9. Explain what are the requirements that need to be met for the interior vs. the exterior surface of a vascular graft. What materials and strategies would you propose, from the point of view of hemocompatibility and tissue integration, to ensure the success of this prosthesis?
10. Describe what experimental approach would you take as a bioengineer for tasks with testing the tissue reaction of a device intended to be implanted in brain tissue?
11. What does the term sterilization mean? Describe two of the most commonly used sterilization techniques and for what types of biomaterials would they be most appropriate to be used?
12. Explain why some polyanhydrides are susceptible to undergoing degradation in the physiological environment.
13. Why is fibrinogen often displaced from a biomedical implant's surface?
14. Is size the most important factor governing protein adsorption on surfaces? What other parameters may influence this process? Is desorption of proteins likely to happen after implantation?

References

1. Boutrand J-P. *Biocompatibility and Performance of Medical Devices.* Oxford; Philadelphia: Woodhead Publishing; 2012.
2. Smith DC, Williams DF. *Biocompatibility of Dental Materials.* Boca Raton, FL: CRC Press; 1982:1−3.
3. Smith DC, Williams DF. *Biocompatibility of Prosthodontic Materials.* Boca Raton, FL: CRC Press; 1982:274.
4. Williams DF. *Biocompatibility of Tissue Analogs.* Boca Raton, FL: CRC Press; 1985.
5. Williams DF. *Biocompatibility of Clinical Implant Materials.* Boca Raton, FL: CRC Press; 1981.
6. Monje A, Insua A, Pikos MA, Miron RJ, Wang HL. The use of allografts for bone and periodontal regeneration. *Next-Generation Biomaterials for Bone & Periodontal Regeneration.* 2019:35−57.
7. Weidenbacher L, Muller E, Guex AG, et al. In vitro endothelialization of surface-integrated nanofiber networks for stretchable blood interfaces. *Acs Appl Mater Inter.* 2019;11(6):5740−5751.
8. Chernonosova VS, Gostev AA, Chesalov YA, Karpenko AA, Karaskov AM, Laktionov PP. Study of hemocompatibility and endothelial cell interaction of tecoflex-based electrospun vascular grafts. *Int J Polym Mater Po.* 2019;68(1−3):34−43.
9. Nakielski P, Pierini F. Blood interactions with nano- and microfibers: recent advances, challenges and applications in nano- and microfibrous hemostatic agents. *Acta Biomater.* 2019;84:63−76.
10. Recek N. Biocompatibility of plasma-treated polymeric implants. *Materials.* 2019;12(2):240.

11. Alghamdi HS. Methods to improve osseointegration of dental implants in low quality (type-IV) bone: an overview. *J Funct Biomater.* 2018;9(1):7.
12. Aldosari AA, Anil S, Alasqah M, Al Wazzan KA, Al Jetaily SA, Jansen JA. The influence of implant geometry and surface composition on bone response. *Clin Oral Implan Res.* 2014;25(4):500−505.
13. Movafaghi S, Wang W, Bark DL, Dasi LP, Popat KC, Kota AK. Hemocompatibility of super-repellent surfaces: current and future. *Mater Horizons.* 2019;6(8):1596−1610.

Orthopedic and dental biomedical devices

Learning objectives

In the previous chapters, we have discussed both the main classes of materials used in biomedical devices and the properties of biological tissues. This chapter presents an overview of the main applications of biomaterials in specific biomedical devices that interface with bone and teeth, with a focus on implantable devices. At the end of the chapter, the students will:

* Become familiar with the biomedical alloys, bioceramics, and biopolymers most commonly used in implants that are replacing or help in the healing of hard tissues.
* Understand the differences between requirements imposed by load-bearing and non–load-bearing anatomical locations on biomaterial selection.
* Gain a general understanding of concepts related to biomedical implants that interface with bone, such as fracture fixation devices, joint replacement prostheses, and prostheses for spinal repair.
* Understand the standards that biomaterials need to meet in terms of mechanical and tissue compatibility for their application in such devices, and what are the most common modes of implant failure.

Introduction

Many of the hard-tissue replacement and repair devices focus on orthopedic implants, while others are designed for dentistry.

As discussed extensively in previous chapters, bone is a living tissue, which has the ability to model and remodel in response to the mechanical forces it withstands. When trauma or disease affects bone health and leads to injury, fracture of skeletal system defects will ensue. The same is true for teeth and dental-related injuries or disease. Therefore the effective design of orthopedic and dental restoration devices to repair such trauma is a significant area of research and development in the biomedical industry. Orthopedic implants aim to restore the skeletal function, and in general include devices for fracture fixation and repair, or devices that repair

Introductory Biomaterials. https://doi.org/10.1016/B978-0-12-809263-7.00009-3

bone defects, both of which will be covered in the following subsections of this chapter. Design of both orthopedic and dental implants requires an understanding of the tissue compatibility, mechanical properties, and modes of failure of the biomaterials. The ultimate biomaterial to be chosen for such devices needs to display microstructural and mechanical properties that are as close as possible to those of the hard tissue of interest, while having the ability to be tissue compatible. The implants should not have a negative effect on the healing and regeneration process of the hard tissue (bone, teeth). Bone tissue remodeling after intentional or nonintentional injury follows the same steps described in Chapter 8 on the topic of tissue healing. It is relevant to mention that in a situation when an orthopedic implant is present, bone growth happens both in the direction of the bone edges and from the edges of the bone toward the implant. In an ideal situation, the bone-implant bond should be secure with no micromotion. However, when stress shielding or disease affects new bone growth in the remodeling stage of tissue healing, this bond will not stay secure due to bone degradation, which will ultimately result in implant failure. Implant failure may also occur due to problems related to how the biomaterial is affected by the in vivo environment. Issues include corrosion of metals and alloys (see Chapter 3 for details), degradation of ceramics and polymers (Chapters 4 and 5), microstructural and structural defects, or adverse immune or tissue reactions (Chapter 8).

This chapter will cover some, although not all, common types of orthopedic and dental devices, along with the types of biomaterials used in their design and fabrication and their properties, as related to the ability to achieve the ultimate goal of successful clinical use of these devices. Although many more types of devices may be used clinically today, the information in this chapter aims to provide a general understanding of what are the main characteristics of the biomaterials that interface with hard tissues, which could then be used when working with other orthopedic and dental implants that may not have been discussed herein.

Biomaterials properties and tissue interactions for hard-tissue implants

Bone is a tissue that is able to model and remodel continuously for the entire lifetime of a patient. That means most bone fractures heal without major intervention in the form of bone grafts or other bone substitutes. Dental and orthopedic devices aim to reconstruct or repair dental or skeletal tissue. With orthopedic fracture fixation devices, joint replacement devices, as well as materials used in dentistry, metallic biomaterials, such as Ti6Al4V, SS 316L stainless steel and Co-Cr alloys, are most often the materials of choice. When clinical outcome is poor it is often associated with poor integration with either bone tissue or soft tissue (e.g., gums in dental implants). It is thus of interest to investigate the interfacial interactions between metallic biomaterials and bone tissue.

The concept of integration between a biomaterial and bone tissue that is observed clinically and experimentally is termed osseointegration. It can be definitely stated that osseointegration between an implant and bone tissue has been achieved successfully when there is no intervening fibrous soft tissues between the implant and the bone.[1] This can be verified via histological studies that show that the integration is complete and there are no patches of fibrous tissues along the interface. Mechanical stability of the implant does not verify osseointegration, and neither does the absence of radiolucent lines obtained by performing X-rays at the interface. However, the presence of radiolucent lines (distance intervals, measured in millimeters, between the cement used to anchor orthopedic implants and the bone) in radiographs proves with no uncertainty that osseointegration has not been achieved. Testing of implants in animal studies or of removed implants in clinical settings via immunohistochemistry is the gold standard to verify bone-implant integration. Good integration will reveal the absence of soft-tissue cells at the metal-bone interface and only bone tissue being in direct contact with the metal.

The implant's surface properties play a major role in osseointegration. A high surface energy is favorable for tissue integration. The microstructure of the biomaterials incorporated in the implant has to be carefully controlled to achieve the right balance between the mechanical properties required for successful functioning and the success of tissue integration. In general, high porosity, along with higher surface roughness, is favorable for osseointegration. In turn, the presence of surface impurities due to inferior cleaning has the opposite effect and prevents tissue integration. To achieve the balance between high porosity required for tissue integration and high mechanical strength required for proper functioning under load bearing during implant service, in many situations porous coatings (such as hydroxyapatite [HA] coatings) are deposited on the surface of metallic implants (e.g., SS or Ti alloys). Thus the surface coating offers the porosity and roughness, whereas the metallic alloys provide the strength required to withstand the mechanical forces during use.

Cleaning and sterilization are mandatory in all biomedical implants. Once an implant is introduced in the body, a conditioning protein film immediately forms, followed by colonization of cells. Implant surfaces that have properties preventing protein film formation performed poorly in terms of integration with the surrounding tissue. The bonding between bone tissue and implants is generally of the Van der Waals type and these bonds form with the passivating oxide layers rather than with the metallic elements themselves.

Aside from osseointegration, metal ion leaching from an implant can have an adverse reaction on tissues. For example, Ti ions can leach out of Ti6Al4V alloys and lead to blackened regions around the bone. Metallic ions such as Ti or aluminum can also travel through the body sometimes reaching organs as far as the lungs.

As mentioned previously, coatings on metallic implants tend to improve biofunctionality and osseointegration capacity. Cells react well to rough surfaces such as pores, grooves, kinks, or wells of micron-scale dimensions. There are studies that show that cell behavior responds to certain controlled topography by affecting the differentiation of mesenchymal stem cells to produce bone tissue. HA coatings, as well as reinforcements with HA-collagen composites, also improve bioactivity and bone formation at the bone-implant interface. Biofunctionality can be positively influenced by incorporation of peptide motifs that are recognized by the body as self. These include, for example, arginine-glycine-aspartic acid motifs or heparin, and thus can be incorporated into various implant configurations to increase tissue compatibility.

Biomaterials for orthopedics and dentistry
Overview

As discussed previously, orthopedics and dentistry are the main areas where the replacement and repair of hard tissues is an ongoing therapeutic necessity. The main hard tissues that are being replaced or repaired are bone and dental hard tissue and associated dental components. For orthopedics, the main types of biomedical implants are fracture fixation devices and prostheses that replace missing bones or joints. Both categories aim to restore mobility and reduce pain.

As mentioned, many repairs have to do with fracture, in which case fracture fixation devices are employed.[2–5] These are generally the metallic materials we discussed in the previous chapters, Ti alloys[6,7] and 316 SS.[8–12] These materials have good mechanical resistance, are useful for load-bearing applications, as well as have adequate tissue biocompatibility. For prostheses, joint prostheses are the most well known. Examples are knee and hip replacement devices, as well as devices for elbow, shoulder, ankle, small joints of the foot, wrist, and hand arthroplasty (the term arthroplasty means the surgical replacement or reconstruction of a joint).

Tissue ingrowth is an issue that needs attention in all cases of hard-tissue reconstruction or replacement, to avoid micromotion at the bone-implant interface and implant failure.[13,14] Thus porosity is a parameter that is engineered to maximize function. Porous ceramic coatings, such as HA or fluorapatite, on metallic hard-tissue implants are sometimes used to promote bone ingrowth.[15–21] Fluorine, which is present in fluorapatite, has been shown to be beneficial in dental restorative work due to its ability to prevent dental caries and promote biomineralization of calcium phosphate (CaP).

There are also fracture fixation devices that do not contain metals, and instead incorporate ceramics, polymers, or mixtures of the two.[22–28] In all cases, to ensure proper functioning, tissue integration and mechanical properties are the parameters that are given maximum attention by biomedical professionals.

Most of these fracture fixation devices are permanent and nondegradable or must be removed via a second surgery. There have been more recent developments in the field of biomaterials that introduced biodegradable poly(lactic) acid (PLA) as bone replacement.[29] However, the degradation products of PLA, which are acidic and induce tissue reactions, are still a matter of concern in the long term, although PLA is generally tissue compatible in the short term. Inflammation and premature mechanical failure are also problems reported with PLA. Often, these issues are mitigated by fabricating ceramic-polymer composites with calcium-based ceramics such as HA, calcium carbonate, or tricalcium phosphate.

Besides bone fracture and dental restoration, joint replacement, such as the hip and spine, is a large area of application for metallic, polymeric, and ceramic biomaterials. In most cases these implants are constructed in a ball and socket design (Fig. 9.1). There is constant friction between the joint components while these implants are in use, which creates concerns related to wear and particulate debris migration into the surrounding tissues. The interaction of wear particles with tissue can create complications such as inflammation, osteolysis (bone loss), toxicity, and implant loosening and failure.[30-36] Artificial lubricants, such as saline solutions or polyvinyl alcohol (PVA), are used to mitigate the production of wear debris. Proper material selection for the interfacing socket and ball is also critical. For example, a polymer being interfaced with a ceramic can lead to more wear debris than ceramic-ceramic interfacing components (Fig. 9.1).

Major applications of 316L SS for hard-tissue replacement or repair include cranial plates, orthopedic fracture plates, screws or pins, spinal rods, joint replacement prostheses, and dental implants. Co-Cr alloys are used for orbit reconstruction, orthopedic fracture plates, spinal rods, as well as joint replacement prostheses.

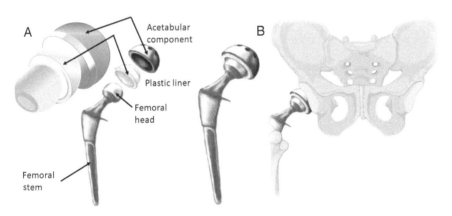

FIGURE 9.1

Ball and socket joint. The acetabular component often contains a polymer (e.g., Poly (L-lactide) [PLLA]) and metal in close contact (A). The femoral head can be made of ceramics (e.g., alumina) or metals (e.g., stainless steel or Ti alloys), while the femoral stem could be made of Ti alloys or stainless steel, and often coated with a porous hydroxyapatite or fluorapatite coating for enhanced tissue integration (B).

Ti alloys are used for cranial plates, orbit reconstruction, maxillofacial reconstruction, dental implants, or dental and orthopedic wires. For all the biomedical systems that will be mentioned in this chapter, metallic components are always susceptible to corrosion-induced failure. Refer to Chapter 3 for a detailed discussion of corrosion and its mechanisms in biomedical alloys.

Fracture fixation devices

Bone fracture is a very common occurrence in all age groups and thus fracture fixation devices are one of the most commonly used. An effective fracture fixation device offers adequate mechanical stability, proper bone alignment, and allows the patient to continue using the limb in a functional manner. Other goals for fracture fixation devices include short healing/recovery time, good tissue biocompatibility, and ease of use.

Casts are one of the most common, cheap, and noninvasive fracture fixation devices.[37–39] Casts allow immobilization of the broken bone and proper alignment while healing. The two main types of casts are plaster and fiberglass. Plaster casts can align and hold the bones in the correct position for healing and are effective at supporting healing. However, they are heavier than fiberglass casts and must not get wet. Fiberglass casts, on the other hand, are used when bone alignment is not necessary, and the bone fragments are in the proper position already or after the healing already commenced. They are lighter, sturdier, and more breathable than plaster casts. Casts, in general, are cost effective and offer adequate protection from external trauma, do not require invasive surgical procedures, and are easy to use. However, they cannot provide direct mechanical fixation for complicated fractures and do not allow facile repositioning of the bone, due to their external character.

By contrast with casts, which are external fracture fixation devices, internal fixation devices can better restore bone position and allow the patient to become mobile sooner. These devices are implanted in the affected limb and thus eliminate the discomfort of wearing the cast and the poor aesthetics. When the internal fixation is successful, rehabilitation and regaining limb function are both achieved early.

Metal wires are the simplest fracture fixation devices. These can be used to keep bone fragments together, which becomes important when shattered bones are present, in repairing facial fractures, or spiral and oblique fractures. Wires are usually made of SS or Ti alloys and are flexible, which allows deformation to join bone pieces.[40] Because they are thinner than screws, plates, or rods, wires produce less tissue damage, are cheap, and easy to manipulate. Because of the bending that is common with wires, they can encounter fatigue due to cyclic loading or stress corrosion. At the same time, wires are not very stable and thus cannot be used in any areas subjected to any significant mechanical loading.

Metallic pins[41] are used to connect bone segments and are implanted percutaneously. Bone pins are connected with an external fixator, which is a disadvantage.

However, when it is desirable to implant lower amounts of metal implanted are desired they can be useful to hold bone in place and are relatively easy to use, when other means of fracture fixation are not possible. The percutaneous fixation, however, can lead to infection and cannot hold any mechanical load without using external fixation, which is visually unattractive.

Screws are some of the most widely used devices for fixation of bone fragments to each other or in conjunction with fracture plates.[42,43] We know that there are two different microstructures in natural bone, cortical and cancellous. The cortical bone is denser and less porous than the cancellous bone. There are thus two types of bone screws, one for each type of bone. Cortical bone screws are usually fully threaded and are thinner and have a smaller pitch than cancellous bone screws. They are used in general in the diaphysis of the bone. Cancellous screws have to connect long segments of porous bone and keep the bone mechanically stable in the process. Thus they have a larger thread diameter and a greater pitch. They are also only partially threaded. The pullout strength of screws is directly proportional to their diameter, thus thicker screws have a greater pullout strength than thinner screws. Other screw parameters have less influence on pullout strength.

Regardless of the type of bone that they are used for, bone screws can also be self-tapping or non–self-tapping. The self-tapping screws are first screwed in and in the process create their own tapping. The non–self-tapping screws need the hole to be pretapped. Both designs, however, have the same pullout strength.

Screws can be used alone or together with bone plates[44] to ensure greater stability and minimize stress concentration at the interface between the screw shaft and head. However, since two different metallic components are in contact with each other in the screw-plate designs, these designs are more susceptible to corrosion.

Regardless of their design or whether they are used in conjunction with plates or on their own, the screws can either be removed or left in place after bone healing is completed. The initial trauma of screw insertion can result in nearby tissue necrosis and resorption initially. However, in time, bone tissue is expected to ingrow into the screw threads and lead to firm fixation. Micromotion between the screw and the bone should always be avoided, as is the case with other fixation devices or implants. Micromotion at the implant-tissue interface will lead to the formation of a fibrous capsule between the implant and the tissue, which prevents tissue ingrowth and leads to implant loosening, patient pain, and implant failure.

Bone screws provide good fixation and can be used in conjunction with plates when high loading is expected. When bone plates are not being used, the presence of smaller pieces of metal than plates or bone rods induces less stress shielding. Since these are internal fixation devices, however, the procedure of inserting them is invasive and results in both soft- and bone-tissue damage initially. The thread design can result in crevice corrosion, and the friction on insertion can produce metallic particulate debris. If screws are removed after healing, stress concentrations in the bone can be present due to a bone defect being left behind.

Example 9.1

In a fracture fixation device design, a plate is used with two screws in a femur fracture. Considering that the force during movement is completely transferred from below the patient's waist to the device and all the force is absorbed by the device, calculate the shear stress on the screws. In this scenario, the screw diameter is 4 mm and the patient's weight is 800 N.

Answer The shear stress, τ, is:

$$\tau = \frac{1}{2} \cdot \frac{800 \, (N)}{\pi \, (0.004)^2 / 4} = \frac{63.69 \, \text{MPa}}{2} = 31.84 \, \text{MPa}$$

To avoid the metallic screws either being left in place or removed, researchers have been putting forward resorbable screws, which can support bone healing and then disappear in time (average 24 months), leaving behind healthy bone tissue. Researchers developed a moldable composite made of PLA and HA, which promotes bone growth and can be processed into resorbable screw designs.

Bone plates have been mentioned previously and they are also internal fixation devices, which use bone screws placed at various positions along the plate. Plate placement has a significant influence on the load-bearing properties of the system, since fractured bones can withstand compressive loading but not tensile. Thus bone plates should provide tensile support to the bone during normal loading. Bone plates are removed after fracture fixation in the great majority of cases.

One type of bone plates is the dynamic compression plate (DCP), which induces compression of bone segments during fixation.[45,46] The DCPs can be either self-compressing, with compression being induced by screw position, or can use a "jack" to induce compression. The DCP functions in different modes: compression, neutralization, tension band, or as a buttress. In the DCP, the areas around the plate holes are less stiff than the areas between them, and during bending the plate tends to bend only at the hole sites. Newer DCP designs are low contact and reduce the area of contact between the plate and bone for improved cortical perfusion and reduction of stress concentration at one of the screw holes.

DCPs offer good fixation and mechanical support in tension and torsion and can induce compressive loading at the fracture site. Thus this type of plate can be used for multiple fracture fragments. However, due to their relatively large size, they can cause stress shielding. Plate insertion requires major surgery, plus an additional removal surgery is usually performed, adding to the costs and tissue damage. The large size of the plates and their use in conjunction with screws also lead to a higher possibility of corrosion than for bone pins or screws alone.

Trabecular bone fractures, due to their high porosity, are usually repaired with a combination of plates, screws, bolts, and nuts, which means that a large amount of material is used. Fixation of trabecular bone fracture with nails alone may be possible if the trabecular bone is sufficiently dense.

Intramedullary devices, or rods, are used for fractures of long bones and are inserted into the medullary canal. These devices are more resistant to bending than cortical bone plates, but more susceptible to torsional loading. Their disadvantage is that they may destroy intramedullary blood supply, but they do not affect periosteal supply. Rods can be inserted through a small incision near the proximal or distal end of the bone. The cross-sectional shape of rods can vary from diamond to cross to clover-leaf. Rods provide good fixation and stabilization and require only a small incision. However, because of their relatively large size, these device can damage blood supply and lead to stress shielding and higher susceptibility to corrosion.

Producing artificial bone[47–49] from mixtures of mono- and tricalcium phosphates, calcium carbonate, and sodium phosphate has been attempted to hold bone fragments in place or fix bone defects. Artificial bone mineral can be useful for fractures of the trabecular bone and it is resorbable. The mineral paste hardens to provide fixation in 10 minutes and the compressive strength of this initial material is approximately 10 MPa. It typically takes 12 hours for the bone mineral to be cured and the compressive strength increases to 55 MPa, while tensile strength is approximately 2.1 MPa. The main advantage of the artificial bone mineral is that a second removal surgery is not necessary. It also provides quick fixation, minimizes stress shielding, and does not corrode. The hardening of the paste within minutes requires short surgery time.

Bone grafts are also used to repair bone defects due to injury or disease. These are pieces of trabecular or cortical bone. Autografts, allografts, and bone graft substitutes are all used for this purpose.

It bears repeating that all the devices mentioned in this section need to adhere to the same requirements mentioned previously in the Overview of the Biomaterials for orthopedics and dentistry section, and provide a good balance between tissue integration and mechanical properties.

Biomaterials for joint arthroplasty and prostheses

In the Overview subsection under the Biomaterials for orthopedics and dentistry section, we mentioned that biomaterials are used in joint arthroplasty and prostheses such as knee and hip replacement devices, as well as devices for elbow, shoulder, ankle, small joints of the foot, wrist, and hand arthroplasty. This section provides an overview of such devices. They typically have a finite lifetime that is determined by whether there are complications related to their fixation, biomaterial mechanical failure, wear, infection, corrosion, or immunological related issues.

Fig. 9.1 shows a general image of a device used for total hip replacement. Total hip replacement is necessary when the patient suffers from various afflictions such as osteoarthritis, hip fractures, or tumors located around the hip area, to mention only a few. More than 100,000 total hip arthroplasties are performed in the United States every year. The main components of such a prosthesis include the femoral stem with a femoral head at its end and an acetabular component. The acetabular component is composed of a metal shell and an ultra-high-molecular-weight polyethylene (UHMWP) lining. The acetabular component snaps together with the femoral head. The femoral head can be made of ceramics (e.g., alumina) or metals (e.g., SS or Ti alloys), while the femoral stem could be made of Ti alloys or SS, and is often coated with a porous HA or fluorapatite coating for enhanced tissue integration. Motion occurs at the interface between the internal part of the acetabular component and the femoral stem, as well as at the interface between the external surface of the acetabular component and the native bone. Such movement involves frictions, which could lead to wear debris. Wear debris involves particles of either metal or ceramic, which can become a problem as they can induce inflammation due to attack by the immune system. At the same time, friction can lead to material failure, dislocation, wear of the acetabular articular cartilage, and ultimate device failure. Early hip replacement devices, as well as most of the other joint replacement devices, were fixated with bone cement, which is composed of polymethyl methacrylate (PMMA) polymer. This acts to fill the space between the implant and the bone via a mechanical interlocking mechanism. There is no special adhesive property of PMMA, and instead the mechanical interlocking is made possible by the irregular bone surface because of its ability to blend well with bone morphology. After materials science advances were made in the 1980s, successful cementless joint replacement devices have become more common. The key factor that contributed to this success was an understanding that fixation through bone ingrowth into a porous implant is possible if porosity levels and pore size are correctly chosen. Controlling surface roughness is another way to achieve fixation. The maximum bone ingrowth was found to occur for pore size between 100 and 200 μm. Ti fiber meshes were used to induce porosity in Harris-Galante total hip replacements.[50] Other methods for inducing fixation were using Co-Cr beads in the AML® total hip replacement and plasma-sprayed Ti porous coatings for the Mallory-Head total hip replacement device[50] (Fig. 9.2).

In some cases, cementless prostheses are provided with additional fixation with screws. Cementless joint replacement devices have the advantage of being easier to remove than the cemented counterparts, with less bone loss after surgery. Removal is necessary when revision surgery is required because of a failure, such as aseptic loosening of the prosthetic device. The expectation is for total joint replacements, as well as for other types of joint replacement devices such as knee replacement devices, to function well enough to have a survival rate of 95% at 15 years after implantation.[51]

Knee replacement devices (Fig. 9.3) are necessary in case of injury or disease of the knee joint, such as degenerative and inflammatory arthritis. The knee is a rather complicated joint that connects the tibia and fibula to the femur, and also includes

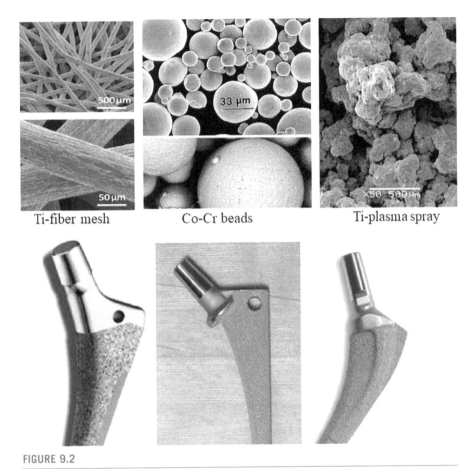

Ti-fiber mesh   Co-Cr beads   Ti-plasma spray

FIGURE 9.2

Porous coating types used in three commercially available total hip replacement devices.

Reproduced with permission from Yamada et al.[70]

the kneecap patella. The knee is protected by a joint capsule, ligaments, and a synovial fluid—containing membrane that lubricates the joint and reduces friction.

When injury or disease affects damaged cartilage and bones in the knee, the damaged parts are replaced by a prosthesis (Fig. 9.3). Sometimes not all the components need to be replaced, but we will discuss here the total knee replacement for a more comprehensive understanding of these devices. In this case the prosthesis connects the upper part of the tibia to the lower part of the femur. The biomaterials used are a metallic femoral component and a tibial component made of a metal-backed polyethylene or of polyethylene alone. The kneecap patella is made of high-density polyethylene, which can also be metal backed in some versions of the device. As with the hip replacement, most failures are due to aseptic loosening mainly caused by particulate debris due to polyethylene wear and osteolysis (bone resorption).

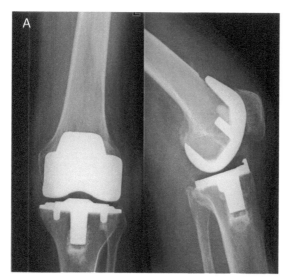
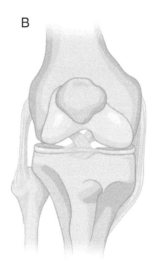

FIGURE 9.3

Total knee replacement. Radiograph showing a total knee prosthesis after 3 years of implantation (A); illustration of a healthy knee joint (B).

Adapted from Saouti et al.[71]

Similar to other joint implants, total ankle replacement prostheses are necessary when there is injury, osteoarthritis, or inflammatory arthropathy. Examples of such prostheses that are in use are the Scandinavian total ankle replacement (STAR), the Buechel-Pappas total ankle prosthesis (Endotec, South Orange, NJ, USA), the TNK ankle prosthesis (Nara, Japan), and the Agility total ankle replacement device (DePuy, Warsaw, IN, USA). With the exception of the Agility device, which is cemented, the rest are cementless designs. There is a high level of rotation and translation because of the three different articulations that are needed in these prostheses: tibial and superior talar trochlea, fibula and lateral talus, and medial malleolus and talus. Total ankle replacement is not performed as often as other joint replacement surgeries due to the complicated structure and also because alternative surgeries are able to solve most of the problems associated with ankle function. As with the other joint prostheses, the components of such systems, as far as biomaterials are concerned, are UHMWP, alloys (Co-Cr is typically used rather than Ti alloys or SS), and sometimes HA coatings.

Problems related to upper-extremity joint functioning can also be resolved by joint replacement procedures and prostheses. The prosthesis systems include shoulder, elbow, as well as wrist and hand arthroplasties, performed in cases of fracture or arthritic disease. As with lower-extremity prostheses, these implants can be either cemented or cementless designs. Metals (e.g., Ti alloys) and UHMWP are the biomaterials used in these prostheses as well, and surfaces can be coated with HA or rendered porous in another fashion, with a goal of tissue integration and ultimately implant stability. Complications associated with lower-extremity prostheses include loosening, particulate debris, osteolysis, infection, or mechanical failure of the component biomaterials.

Biomaterials for dental implants

Loss of teeth due to trauma, disease, or age can have a profoundly negative influence on one's psyche. Everyone's smile is symbolic of their personality and can contribute to social integration and developing healthy relationships. Traditionally when loss of teeth occurs, removable partial dentures, fixed partial dentures (bridges), or even complete removable dentures have been used as a solution. However, dentures are difficult to accept and suffer from disadvantages of not being permanent, difficulties in being worn, lack of stability, poor aesthetical appearance, and can lead to jaw bone resorption. An ideal emerging alternative is the use of whole teeth implants, which look closest to the natural teeth (Fig. 9.4).

Dental (endosseous) implants are composed of a metallic artificial root that is inserted surgically into the jawbone[52] and is covered by a crown for a single tooth, or by a maxillofacial prosthesis. After being covered by soft tissue, the bone and implant are allowed to heal for about 14 months to ensure good stability. Materials used for endosseous implants are metals, ceramics, and polymers, with good success rate. Ti and tantalum, as well as zirconia, are examples of materials used in dental implants.[53–56]

Good osseointegration is critical for the success of the procedure, and often HA coatings are used to encourage this effect, due to its ability to induce bone formation. In addition, implant positioning will have a major effect on the function and aesthetics of the final implant. Dental implants are subjected to significant cyclic loading and thus fatigue properties are critical for success, as well as strength in compression, shear, and torsion.

Biomaterials for spinal implants
Overview

The spine is an organ composed of bone material that connects the head to the pelvis and protects the spinal cord. The spinal cord is a critical organ supplying nerves from the brain to the body. The spine, along with spinal cord protection, has to support the entire body mass and at the same time provide mobility, flexibility, and impact protection. The cervical spine has seven vertebrae (C1–C7) and six intervertebral discs and connects the skull to the top of the trunk. Next, the rib cage and thoracic vertebrae support the trunk and connect the top part of the trunk to the pelvis. If the spinal cord suffers damage, this can lead to severe impairment and even paralysis. The intervertebral discs in the spine can be rather easily affected by traumatic injury, due to their exposure to high stresses and strains during normal body movements.

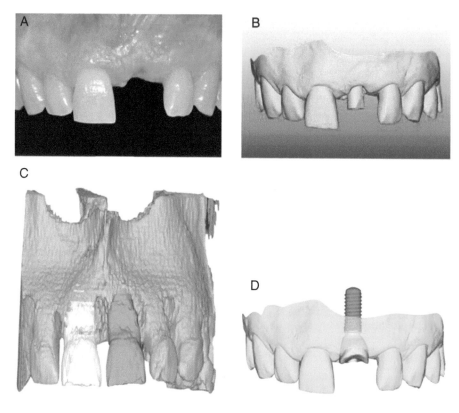

FIGURE 9.4

Soft-tissue healing after implantation. Clinical situation with single tooth gap in position 21 (A), image of reconstruction of final implant location (B); maxillary DICOM data with segmented natural tooth 11 *(white)* for visualization of implant reconstruction *(pink)* (C); prefabricated bonding base in position 21 (D).

Reproduced with permission from Joda et al.[72]

The intervertebral disc has three main components: a core structure, nucleus pulposus; an outer cartilage tissue ring, called annulus fibrosus; and the cartilage vertebral endplates. The annulus fibrosus is composed of concentric lamellae with a composition of water, collagen type I and II, proteoglycans, and minor amounts of other proteins. The composition and angles of the collagen fibers vary along the radial distance from the nucleus to the outer edge of the annulus fibrosus.

The nucleus pulposus is composed of disorganized collagen II fibers, along with elastin arranged radially and a proteoglycan hydrogel with chondrocyte-like cells sparsely populating it. However, the great majority of the nucleus pulposus (80% −90%) is water.

The vertebral endplates are composed of both bone and cartilage tissue. There are both hyaline and articular cartilage in its composition, depending on the location.

The bone component of the endplates is composed of porous layers of trabecular bone of 0.1- to 0.2-mm thickness. The intervertebral disc has very few blood vessels in a mature skeleton and relies on diffusion to supply nutrients to the chondrocytes.

Surgery to improve or correct spinal stability, such as for scoliosis or Pott's disease, has been attempted since the first half of the 19th century. The first spinal fixation devices for the C6–C7 vertebrae were invented by Dr. Hadra in the second half of the 19th century. Since then the field has expanded to using various devices for spinal surgery. These included wires, lateral mass screws, rods, and plates. Since the early stages the biomaterials used in such devices continued to evolve, along with the technologies for the design of such biomedical devices.

The main requirements that spinal implants need to meet are biostability, which includes resistance against microbial attack, and good tissue compatibility, along with appropriate mechanical properties for the site of application. For the latter, Young's modulus is the parameter that gained most attention since, as with any other orthopedic device, this needs to be as close as possible to that of the bone. Most commonly used materials include Ti alloys, SS, Co-Cr alloys, nitinol, tantalum, and polyetherketone (PEEK). Additional materials properties that need to be taken into account include a high tensile and fatigue strength and minimal imaging artifacts. For spinal fixation and repair, implantable devices include cages, screws, plates, and rods, and the biomaterials used for each of these will be discussed here.

Spinal cages

Spinal cages are used in cases of intervertebral disc failure and act as a scaffold for re-fusing vertebrae.[57–59] Their role is to stabilize the force distribution between vertebrae and repair the height of the foramina and intervertebral space. One option of biomaterial to use in spinal cages is autologous bone graft. Autologous materials, however, proved to have high failure rates mainly due to mechanical failure or resorption of the graft, which along with donor site morbidity means these are not considered an ideal option for spinal repair. More success has been achieved with metallic, metal-ceramic composites, or polymeric cages.[60]

The most common metallic biomaterials used for spinal cages are Ti6Al4V alloys[61] owing to their lower Young's modulus compared with both Co-Cr alloys and SS 316L, coupled with high resistance to fracture and good tissue compatibility. However, Ti alloys do not easily form a direct bond with the bone tissue and require bioactive coatings.

The other material that is often used for spinal cages is PEEK.[62,63] This material has a Young's modulus that is close to that of the bone, good biocompatibility, and minimal imaging interference. However, as it is often the case with polymeric biomaterials for implantation, its mechanical properties, such as compressive strength, are lower than those of Ti alloys. At the same time, its hydrophobic character prevents good bonding with the bone. The advantage of PEEK over Ti alloys is that when used for spinal cages there is a comparatively lower rate of subsidence (34.5% for Ti6Al4V vs. 5.4% for PEEK) (Fig. 9.5).

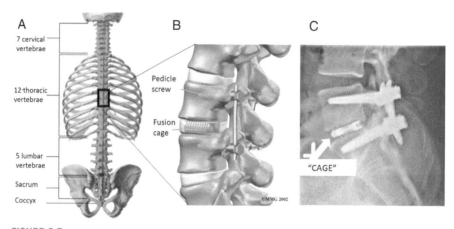

FIGURE 9.5

Spinal stabilization with cage (A), rod (B), and pedicle screws (C).

Reproduced with permission from Frost et al.[73]

In addition to Ti alloys and PEEK, experimental materials that have been tested to improve bone adhesion include silicon nitride (Si_3N_4), which showed very similar performance in a study comparing it with PEEK. An apatite-wollastonite (A/W) ceramic spinal cage also showed no improved performance. For all of these experimental materials, however, there have not yet been any clinical studies released at the time this book chapter was written.

Spinal rods

The function of spinal rods is to improve the stability of a spinal implant. Their design is usually tailored for each patient. The first spinal rod was introduced by Harrington in 1962 and called the "Harrington rod." This was an SS rod and was used initially for the treatment of scoliosis. The fatigue strength of this rod is impaired by rod contouring, which introduces defects in the material structure.

Although early rods were initially composed of SS, modern rods are, like spinal cages, made of Ti6Al4V or PEEK.[64,65] As one can guess, this shift occurred due to the lower stiffness of these materials, that is their Young's modulus is closer to that of bone, thus improving the effects of stress shielding. This is particularly important for rods, which are large enough and sustain enough load that stress shielding is a significant concern.

Other metallic materials have been tried for spinal rods, such as Fe-Cr-Ni, SS 316L, and Co-Cr alloys. Although all of these metals form stable passivating oxide layers, SS is the least corrosion resistant. In addition, all of these alloys do not match Ti's favorable Young's modulus and biocompatibility and are thus less favorable to be used in rods. Another alloy that has been tested in rods is nitinol. Nitinol showed no inflammatory response and has shape recovery that was used toward scoliosis correction. However, this alloy has higher costs and thus is not widely used.

PEEK rods have also been used. These are semirigid rods and impart similar stability to that provided by Ti spinal rods. The advantage of PEEK rods compared with Ti rods is an improved contact with bone due to PEEK having better elasticity and biocompatibility compared with Ti. It also produces fewer imaging artifacts. Problems with using PEEK rods include difficulty in forming a strong interface with screws and a higher failure rate than that of Ti rods.

Other materials that have been tested but are not yet in use include polycarbonate-urethane rods and β-type titanium-molybdenum alloys.

Pedicle screws

Pedicle screws are frequently used in conjunction with spinal rods for spinal stabilization surgery.[66–68] They are often used for scoliosis correction and generally withstand the large forces that are necessary for the redirection of force through vertebral bodies. It is expected that pedicle screws will most often be made of similar materials to those used for spinal rods and cages. Thus the most popular material for pedicle screws is Ti6Al4V, due to its biocompatibility and favorable mechanical properties, such as good mechanical strength and Young's modulus being closest to that of the bone among metallic materials used in orthopedic and spinal surgeries.

Because of the difficulties encountered by Ti alloys in forming a direct bond with bone, failure may occur due to screw pullout or loosening, and coatings are being used to improve bone compatibility. Examples of coating materials include HA, CaP, PMMA bone cement (PMMA-BC), tantalum, or biological coatings made of extracellular matrix components.

Studies have shown that the screws coated with bone cement have a higher pullout strength than uncoated screws. Bone cement–coated screws, however, have a higher rigidity and come with more difficulty in handling after surgery.

Tantalum-coated screws have been shown to improve cellular proliferation by in vitro studies, as well as have a higher pullout strength than uncoated Ti screws. Similar improvements in the pullout strength of screws coated with extracellular matrix components have been noted in other studies.

Causes of pedicle screw failure include bending, breakage, fatigue failure, dural tears, nerve root injury, cerebral spinal fluid leaks, loss of fixation, or defective placement. Besides the biomaterial composition, the pullout strength is influenced by factors such as osteoporosis, pedicle morphology, bone density, screw thread area, or screw orientation.

Spinal plates

As with fracture fixation devices described in the previous section, spinal plates are extremely valuable in spine stabilization and alignment. Again, porous Ti or Ti6Al4V are the preferred biomaterials for spinal plate design, with the same advantages and disadvantages described previously. In addition, there are some plate designs that include polyethylene rings for screw locking. Newer proposed materials for spinal plates include biodegradable composite plates that have been tested in vitro with good results in terms of no imaging artifacts and similar fixation performance to that of Ti plates.

Three-dimensional printing for spinal implants

Three-dimensional (3D) printing is the most preferred method currently used for the design of spinal implants that can be customized to each patient. First, polymer templates are printed to match the desired geometry for the patient. Materials for such design include acrylonitrile butadiene styrene, acrylate resin, PLA, and polyamide photosensitive resin. 3D printing of metals such as Ti has also been used. Printed pedicle screws have been tested successfully in cervical spine surgeries. Printed models have been shown to be beneficial for spinal fusion surgeries, where reducing blood loss and fluoroscopy time led to superior results in terms of lowering the rate of screw misplacements and other failures. However, 3D-printed models come with associated additional costs to the patient, as well as increased time leading to surgery to allow for their creation. The most positive aspect of using 3D-printed models is the ability to customize implant design to every patient. The most printed scaffolds were done for Ti implants, but more recently biodegradable 3D-printed polyurethane intervertebral disc implants have been put forward.[68,69]

Summary

Biomedical implants that interface with bone and teeth are typically designed using the most common biomedical alloys, that is, TiAl6V4, SS 316L, and Co-Cr alloys, along with ceramic (e.g., alumina, zirconia) or polymer components (UHMWPE, PMMA). Devices interfacing with hard tissues include fracture fixation devices, joint prostheses, spinal repair devices, and dental implants.

Each of these devices aims to restore function that has been limited or lost due to trauma or disease. Although the combination of biomaterials typically used in each of these devices is generally common to all of them, the individual design and specific material choice will vary based on the function that the system needs to restore. For example, a fracture fixation device may not contain any polymers or ceramics, whereas a hip or shoulder replacement prosthesis may include all three types of materials. Yet other devices (e.g., wrist or ankle replacement) will only include metals and polymers (typically polyethylene).

The ultimate goal when designing such implants is to provide a long lifetime of the device that will restore function with no pain or loosening. Typical success rates for most devices are at 95% after 15 years in use. Friction between polymers and metals, two metals, or ceramics and metals can result in particulate debris, or corrosion and corrosion products. These events in turn can activate the immune system, which will result in inflammation and failure, or will degrade the mechanical properties of the implant components, also resulting in premature failure. Thus frequent radiological checkups while these implants are in use are necessary to verify any loosening or bone loss.

Problems

1. You work as an engineer in a biomedical device company. A new design for a fracture fixation device with plate and screw components is proposed. What would you select as the best materials to be used in this configuration?

2. In the scenario proposed in problem 1, your boss recommends a Ti alloy screw and a Co-Cr alloy plate. You believe this is not an appropriate design because of potential galvanic corrosion. Write balanced electro-chemical equations for the galvanic cell formed, considering only Ti and Co ions.

3. The most used SS for biomedical applications is the 316L. What does L stand for? Explain how this specific composition affects the material properties and what is the relevance for biomedical applications.

4. Fracture fixation is achieved with a bone plate with a 55-mm^2 cross-sectional area, made out of Ti6Al4V, with a modulus of elasticity of 110 MPa. If the bone and the plate share the same 1000 N load under isostrain condition, what is the cross-sectional area of the bone? Assume the Young's modulus of the bone is 20 MPa.

5. Compare the advantages and disadvantages of a PEEK vs. a Ti6Al4V spinal cage for use in spinal stabilization.

6. A composite with a volume of 3.2 cm^3 is made of a polyethylene matrix rein-forced with 30 vol% Ti fibers and 15% vol% C fibers. This composite was designed to replace a Ti implant. The weight of the composite was measured to be 65% less than that of a conventional Ti implant. Given that the matrix polymer has a Young's modulus of 3.2 GPa and a density of 1.3 g/cm^3, the Ti fibers have a density of 2.8 g/cm^3 and a Young's modulus of 135 GPa, and the C fibers have a density of 1.95 g/cm^3 and a Young's modulus of 180 GPa.
   a) What is the weight percent of the Ti fibers in the composite material?
   b) Calculate the density of the composite.

7. A spinal stabilization system with cage, rod, and screws was removed after failure. Upon examination, fibroblasts were observed in some, but not all, areas of the metallic interface. What is the likely reason for failure of this device?

8. Pedicle screws were used for spinal stabilization. The pullout force required to remove the screws from the bone tissue was measured for $n = 15$ samples with results showing a mean pullout force of 300 N and a standard deviation of 54.3 N. Construct a 95% confidence interval for the true average pullout force for the pedicle screw under the specified conditions.

9. Describe three strategies for improving osseointegration and explain your choice.

10. Why would a Ti6Al4V rod be used for fracture fixation of a femur over an SS one? Explain your choice.

References

1. Coelho PG, Jimbo R. Osseointegration of metallic devices: current trends based on implant hardware design. *Arch Biochem Biophys*. 2014;561:99–108.
2. Johnson J, Deren M, Chambers A, Cassidy D, Koruprolu S, Born C. Biomechanical analysis of fixation devices for basicervical femoral neck fractures. *J Am Acad Orthop Sur*. 2019;27(1):E41–E48.
3. Chapman T, Zmistowski B, Krieg J, Stake S, Jones CM, Levicoff E. Helical blade versus screw fixation in the treatment of hip fractures with cephalomedullary devices: incidence of failure and atypical "medial cutout". *J Orthop Trauma*. 2018;32(8):397–402.
4. Dudko OG, Dudko GY. 30-Year experience of open reduction internal fixation of limb fractures using biodegradable polymeric devices. *Zaporozhye Med J*. 2018;2018(4): 562–567.
5. MacLeod AR, Pankaj P. Pre-operative planning for fracture fixation using locking plates: device configuration and other considerations. *Injury*. 2018;49(suppl 1):S12–S18.
6. Disegi JA. Titanium alloys for fracture fixation implants. *Injury*. 2000;31(suppl 4): S14–S17.
7. Zou YZ, Wang G, Xu YM, Bai YH. Comparative study of the proliferative ability of skeletal muscle satellite cells under microwave irradiation in fractures with titanium alloy internal fixation in rabbits. *Exp Ther Med*. 2018;16(6):4357–4366.
8. Uhthoff HK, Bardos DI, Liskovakiar M. The advantages of titanium-alloy over stainless-steel plates for the internal-fixation of fractures—an experimental-study in dogs. *J Bone Joint Surg Br*. 1981;63:427–434.
9. Vanderelst M, Dijkema ARA, Klein CPAT, Patka P, Haarman HJTM. Tissue reaction on PLLA versus stainless-steel interlocking nails for fracture fixation—an animal study. *Biomaterials*. 1995;16:103–106.
10. Mani US, Sabatino CT, Sabharwal S, Svach DJ, Suslak A, Behrens FF. Biomechanical comparison of flexible stainless steel and titanium nails with external fixation using a femur fracture model. *J Pediatr Orthoped*. 2006;26(2):182–187.
11. Hayes JS, Richards RG. The use of titanium and stainless steel in fracture fixation. *Expert Rev Med Devic*. 2010;7(6):843–853.
12. Couzens GB, Peters SE, Cutbush K, et al. Stainless steel versus titanium volar multiaxial locking plates for fixation of distal radius fractures: a randomised clinical trial. *BMC Musculoskel Dis*. 2014;15:74.
13. Collinge C, Merk B, Lautenschlager EP. Mechanical evaluation of fracture fixation augmented with tricalcium phosphate bone cement in a porous osteoporotic cancellous bone model. *J Orthop Trauma*. 2007;21(2):124–128.
14. Pilliar RM, Cameron HU, Binnington AG, Szivek J, Macnab I. Bone ingrowth and stress shielding with a porous surface coated fracture fixation plate. *J Biomed Mater Res*. 1979; 13(5):799–810.
15. Gwinn DE, Keeling JJ, Andersen RC, McGuigan FX. Hydroxyapatite-coated external fixation pins in severe wartime fractures: risk factors for loosening. *Curr Orthop Pract*. 2010;21(1):54–59.
16. Ishii S, Matsusue Y, Furukawa T, et al. Long-term study of high-strength hydroxyapatite/poly(L-lactide) composite rods for internal fixation of bone fractures: histological examination. *Key Eng Mat*. 2002;218–2:673–676.

17. Kanno T, Tatsumi H, Karino M, et al. Applicability of an unsintered hydroxyapatite particles/poly-L-lactide composite sheet with tack fixation for orbital fracture reconstruction. *J Hard Tissue Biol*. 2016;25:329−334.

18. Landes C, Ballon A, Ghanaati S, Tran A, Sader R. Treatment of malar and midfacial fractures with osteoconductive forged unsintered hydroxyapatite and poly-L-lactide composite internal fixation devices. *J Oral Maxil Surg*. 2014;72(7):1328−1338.

19. Moroni A, Faldini C, Marchetti S, Manca M, Consoli V, Giannini S. Improving fixation strength in osteoporotic wrist fracture patients with hydroxyapatite-coated external fixation pins. *Osteoporosis Int*. 2002;13:S13−S13.

20. Pegreffi F, Maltarello MC, Mosca M, Romagnoli M, Moroni A, Giannini S. Morphological analysis of hydroxyapatite-coated external fixation pins removed from osteoporotic trochanteric fracture patients. *Osteoporosis Int*. 2002;13:S25−S25.

21. Sukegawa S, Kanno T, Katase N, Shibata A, Takahashi Y, Furuki Y. Clinical evaluation of an unsintered hydroxyapatite/poly-L-lactide osteoconductive composite device for the internal fixation of maxillofacial fractures. *J Craniofac Surg*. 2016;27(6):1391−1397.

22. Wu CC, Tsai YF, Hsu LH, Chen JP, Sumi S, Yang KC. A self-reinforcing biodegradable implant made of poly(epsilon-caprolactone)/calcium phosphate ceramic composite for craniomaxillofacial fracture fixation. *J Cranio Maxill Surg*. 2016;44(9):1333−1341.

23. Mitchell PM, Lee AK, Collinge CA, Ziran BH, Hartley KG, Jahangir AA. Early comparative outcomes of carbon fiber-reinforced polymer plate in the fixation of distal femur fractures. *J Orthop Trauma*. 2018;32(8):386−390.

24. Ferretti C. A prospective trial of poly-L-lactic/polyglycolic acid co-polymer plates and screws for internal fixation of mandibular fractures. *Int J Oral Max Surg*. 2008;37:242−248.

25. Tucci V, Gibson B, Dibiasio D, Shivkumar S, Borah G. Sharp memory polymers for external fracture fixation. *Antec 95—the Plastics Challenger: A Revolution in Education, Conference Proceedings, Vols I−III. Boston, MA. May 7−11*. 1995:2102−2104.

26. van der Elst M, Klein CPAT, de Blieck-Hogervorst JM, Patka P, Haarman HJTM. Bone tissue response to biodegradable polymers used for intramedullary fracture fixation: a long-term in vivo study in sheep femora. *Biomaterials*. 1999;20:121−128.

27. Mckenna GB, Bradley GW, Dunn HK, Statton WO. Mechanical-properties of some fiber reinforced polymer composites after implantation as fracture fixation plates. *Biomaterials*. 1980;1(4):189−192.

28. Lewis D, Lutton C, Wilson LJ, Crawford RW, Goss B. Low cost polymer intramedullary nails for fracture fixation: a biomechanical study in a porcine femur model. *Arch Orthop Traum Surg*. 2009;129(6):817−822.

29. Butt MS, Bai J, Wan XF, et al. Mg alloy rod reinforced biodegradable poly-lactic acid composite for load bearing bone replacement. *Surf Coat Tech*. 2017;309:471−479.

30. Benitez C, Perez-Jara J, Garcia-Paino L, Montenegro T, Martin E, Vilches-Moraga A. Particulate wear debris osteolysis, acute urinary retention and delirium as late complications of hip arthroplasty. *Age Ageing*. 2018;47(5):756−757.

31. Rose DM, Guryel E, Acton KJ, Clark DW. Focal femoral osteolysis after revision hip replacement with a cannulated, hydroxyapatite-coated long-stemmed femoral component—a new route for particulate wear debris. *J Bone Joint Surg Br*. 2008;90b(4):500−501.

32. Goodman S. Wear particulate and osteolysis. *Orthop Clin N Am*. 2005;36:41−48.

33. MacQuarrie RA, Chen YF, Coles C, Anderson GI. Wear-particle-induced osteoclast osteolysis: the role of particulates and mechanical strain. *J Biomed Mater Res B*. 2004;69b(1):104−112.

34. Maloney WJ, Smith RL. Periprosthetic osteolysis in total hip arthroplasty: the role of particulate wear debris. *Aaos Instr Cours Lec*. 1996;45:171−182.

35. Maloney WJ, Smith RL. Periprosthetic osteolysis in total hip-arthroplasty—the role of particulate wear debris. *J Bone Joint Surg Am*. 1995;77:1448−1461.

36. Clarke IC, Campbell P, Kossovsky N. Debris-mediated osteolysis—a cascade phenomenon involving motion, wear, particulates, macrophage induction, and bone lysis. In: St KJ, John KR, eds. *STP1144-EB Particulate Debris from Medical Implants: Mechanisms of Formation and Biological Consequences*. West Conshohocken, PA: ASTM International; 1992:7−26.

37. Craig JB, Rogan IM. A comparative-study of the incidence of infection in open fractures of the tibia using an external fixation device or plaster cast immobilization. *J Bone Joint Surg Br*. 1984;66:459−460.

38. Harley BJ, Scharfenberger A, Beaupre LA, Jomha N, Weber DW. Augmented external fixation versus percutaneous pinning and casting for unstable fractures of the distal radius—a prospective randomized trial. *J Hand Surg Am*. 2004;29a(5):815−824.

39. Koenig KM, Davis GC, Grove MR, Tosteson ANA, Koval KJ. Is early internal fixation preferred to cast treatment for well-reduced unstable distal radial fractures? *J Bone Joint Surg Am*. 2009;91a(9):2086−2093.

40. Juutilainen T, Patiala H, Rokkanen P, Tormala P. Biodegradable wire fixation in olecranon and patella fractures combined with biodegradable screws or plugs and compared with metallic fixation. *Arch Orthop Traum Surg*. 1995;114(6):319−323.

41. Nourisa J, Rouhi G. Biomechanical evaluation of intramedullary nail and bone plate for the fixation of distal metaphyseal fractures. *J Mech Behav Biomed*. 2016;56:34−44.

42. Roehsig C, Rocha LB, Junior DB, et al. Canine iliac fracture fixation with screws, orthopedic wire and polymethylmethacrylate bone cement. *Cienc Rural*. 2008;38(6):1675−1681.

43. Paul JP. Optimizing the biomechanical compatibility of orthopedic screws for bone fracture fixation. *Med Eng Phys*. 2003;25(5):435−435.

44. Li JL, Qin L, Yang K, et al. Materials evolution of bone plates for internal fixation of bone fractures: a review. *J Mater Sci Technol*. 2020;36:190−208.

45. Aziz S, Ahmed F, Aslam MU. Comparison of fixation between dynamic compression plate vs locking compression plate in lower limb diaphyseal fractures by bridge plating technique. *Pak J Med Health Sci*. 2019;13(4):1011−1014.

46. Asif N, Ahmad S, Qureshi OA, Jilani LZ, Hamesh T, Jameel T. Unstable intertrochanteric fracture fixation—is proximal femoral locked compression plate better than dynamic hip screw. *J Clin Diagn Res*. 2016;10(1):Rc9−Rc13.

47. Gladine K, Wales J, Silvola J, et al. Evaluation of artificial fixation of the incus and malleus with minimally invasive intraoperative laser vibrometry (MIVIB) in a temporal bone model. *Otol Neurotol*. 2020;41(1):45−51.

48. Ghate NS, Cui H. Mineralized collagen artificial bone repair material products used for fusing the podarthral joints with internal fixation—a case report. *Regen Biomater*. 2017;4(5):295−298.

49. Pohlemann T, Gueorguiev B, Agarwal Y, et al. Dynamic locking screw improves fixation strength in osteoporotic bone: an in vitro study on an artificial bone model. *Int Orthop*. 2015;39:761−768.

50. Hallstrom BR, Golladay GJ, Vittetoe A, Harris H. Cementless acetabular revision with the harris-galante porous prosthesis. *J Bone Jt Surg*. 2004;86(5):1007−1011.

51. Taljanovic MS, Jones MD, Hunter TB, et al. Joint arthroplasties and prostheses. *Radiographics*. 2003;23(5):1295−1314.

52. Katsuta Y, Watanabe F. Abutment screw loosening of endosseous dental implant body/abutment joint by cyclic torsional loading test at the initial stage. *Dent Mater J*. 2015; 34(6):896−902.

53. Salantiu AM, Fekete C, Muresan L, Pascuta P, Popa F, Popa C. Anodic oxidation of PM porous titanium for increasing the corrosion resistance of endosseous implants. *Mater Chem Phys*. 2015;149:453−459.

54. Park SJ, Kim BS, Gupta KC, Lee DY, Kang IK. Hydroxyapatite nanorod-modified sand blasted titanium disk for endosseous dental implant applications. *Tissue Eng Regen Med*. 2018;15(5):601−614.

55. Petrovic I, Ahmed ZU, Matros E, Huryn JM, Shah JP, Rosen EB. Endosseous (dental) implants in an oncologic population: a primer for treatment considerations. *Quintessence Int*. 2019;50(1):40−48.

56. Qian HX, Zhang FQ, Jiao T. Mandibule rehabilitation after embolization of hemangioma with implant overdenture using existing endosseous implants: a clinical report. *J Prosthodont Res*. 2016;60(4):332−336.

57. Metcalfe S, Gbejuade H, Patel NR. The posterior transpedicular approach for circumferential decompression and instrumented stabilization with titanium cage vertebrectomy reconstruction for spinal tumors consecutive case series of 50 patients. *Spine*. 2012; 37(16):1375−1383.

58. Anderson W, Chapman C, Karbaschi Z, Elahinia M. A minimally invasive cage for spinal fusion surgery utilizing superelastic hinges. In: *Proceedings of the ASME 2012 Conference on Smart Materials, Adaptive Structures and Intelligent Systems. Volume 1: Development and Characterization of Multifunctional Materials; Modeling, Simulation and Control of Adaptive Systems; Structural Health Monitoring*. Stone Mountain, GA; 2012:363−372. September 19−21.

59. Dimeglio A, Canavese F. The growing spine: how spinal deformities influence normal spine and thoracic cage growth. *Eur Spine J*. 2012;21(1):64−70.

60. van Hooy-Corstjens CSJ, Aldenhoffa YBJ, Knetsch MLW, et al. Radiopaque polymeric spinal cages: a prototype study. *J Mater Chem*. 2004;14:3008−3013.

61. Assem Y, Mobbs RJ, Pelletier MH, Phan K, Walsh WR. Radiological and clinical outcomes of novel Ti/PEEK combined spinal fusion cages: a systematic review and preclinical evaluation. *Eur Spine J*. 2017;26(3):593−605.

62. Wan ZM, Dai M, Miao J, Li GA, Wood KB. Radiographic analysis of PEEK cage and FRA in adult spinal deformity fused to sacrum. *J Spinal Disord Tech*. 2014;27(6): 327−335.

63. Raslan F, Koehler S, Berg F, et al. Vertebral body replacement with PEEK-cages after anterior corpectomy in multilevel cervical spinal stenosis: a clinical and radiological evaluation. *Arch Orthop Traum Surg*. 2014;134(5):611−618.

64. Foltz MH, Freeman AL, Loughran G, et al. Mechanical performance of posterior spinal instrumentation and growing rod implants: experimental and computational study. *Spine*. 2019;44(18):1270−1278.

65. Studer D, Heidt C, Buchler P, Hasler CC. Treatment of early onset spinal deformities with magnetically controlled growing rods: a single centre experience of 30 cases. *J Child Orthop*. 2019;13(2):196−205.

66. Buttermann G, Hollmann S, Arpino JM, Ferko N. A cost-effectiveness analysis of posterior facet versus pedicle screw fixation for lumbar spinal fusion. *Value Health*. 2019;22: S41−S42.
67. Barzilai O, McLaughlin L, Lis E, Reiner AS, Bilsky MH, Laufer I. Utility of cement augmentation via percutaneous fenestrated pedicle screws for stabilization of cancer-related spinal instability. *Oper Neurosurg*. 2019;16(5):593−599.
68. Zhang JN, Fan Y, Hao DJ. Risk factors for robot-assisted spinal pedicle screw malposition. *Sci Rep*. 2019;9:3025.
69. Garg B, Gupta M, Singh M, Kalyanasundaram D. Outcome and safety analysis of 3D-printed patient-specific pedicle screw jigs for complex spinal deformities: a comparative study. *Spine J*. 2019;19(1):56−64.
70. Yamada H, Yoshihara Y, Henmi O, et al. Cementless total hip replacement: past, present, and future. *J Orthop Sci*. 2009;14(2):228−241.
71. Saouti R, van Royen BJ, Fortanier CM. An impinging remnant meniscus causing early polyethylene failure in total knee arthroplasty: a case report. *J Med Case Rep*. 2007;1: 1−3.
72. Joda T, Ferrari M, Braegger U. A digital approach for one-step formation of the supra-implant emergence profile with an individualized CAD/CAM healing abutment. *J Prosthodont Res*. 2016;60(3):220−223.
73. Frost BA, Camarero-Espinosa S, Foster EJ. Materials for the spine: anatomy, problems, and solutions. *Materials (Basel)*. 2019;12(2):253.

Soft tissue replacement and repair

10

Learning objectives

This chapter will cover the main concepts related to implantable devices/materials for organic soft tissue replacement. At the end of the chapter, the reader will:

- Review the chemical and biomolecular composition of soft tissues.
- Become familiar with the types of biomaterials used for soft tissue replacement and repair.
- Understand the structure and mechanical properties of soft tissues and of biomaterials used for soft tissue replacement and repair.
- Review the types of soft tissue replacement and repair devices.
- Understand the mechanisms of protein interaction with the surface of the biomaterials for each type of device.

Introduction

Soft connective biological tissues are singular in terms of composition, intricate organization, and microstructure, therefore providing unique mechanical properties. Their function as a protective barrier for fragile structures such as internal vital organs demands high flexibility and soft mechanical properties. Soft connective tissue includes tendons, ligaments, fascia, skin, nerves, muscle, fat, and lymph/blood vessels. In one way or another, these tissue are continuous within the body, transforming their composition and structure according to specific functions that might either be related to intercellular exchange (e.g., oxygen, nutrients, and gas circulation) or mechanical support (e.g., transmission of muscle force, flexible linkage, smooth articulation of surfaces), as illustrated in Fig. 10.1.

Ordinary soft tissues are complex reinforced composite structures that are built up from three main components: elastin fibers, collagen fibers, and ground substance (which refers to a hydrated matrix regularly composed of proteoglycans); the main differences among the performance of soft tissues will be governed by the proportions and arrangement of their structural components.[1]

Introductory Biomaterials. https://doi.org/10.1016/B978-0-12-809263-7.00010-X

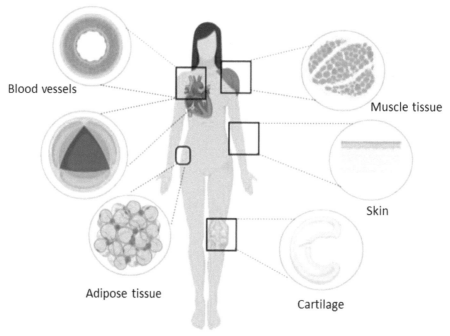

FIGURE 10.1

Soft connective tissue: examples and distribution.

Collagen

Collagen is a protein with a rigid structural motif in which three parallel left-handed polypeptide strands are arranged in a polyproline II-type (PPII) helical conformation coil. The packing of the PPII helices requires glycvine (Gly) present in every three residues, creating a repeating XaaYaa**Gly** sequence (where Xaa and Yaa represent any existing amino acid); this repetition is present in all types of collagen-like proteins.[2] Type I collagen is the most abundant animal protein and the main constituent of soft connective tissue. Due to its fibrillar nature, collagenous tissue generally presents a very organized hierarchical structure, as illustrated in Fig. 10.2A. The main role of collagen proteins within the body is to provide tensile strength to connective tissues (structure, support, shape), following a viscoelastic mechanical behavior at tissue level.

Type I collagen can be found as an homotrimer (composed of three identical polypeptide chains) or a heterotrimer molecule, constituted by two $\alpha 1$ chains and one $\alpha 2$ chain; most collagen types are themselves heterotrimers[3] and their quaternary structure is represented in Fig. 10.2B. As of the time of writing, there are at least 28 types of collagen identified. Collagen type I is the most abundant form of collagen in soft tissue throughout the body (over 85% of the composition of

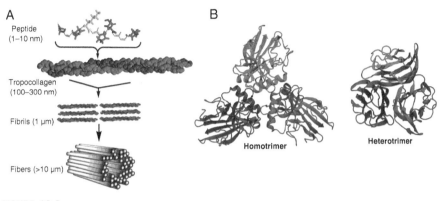

FIGURE 10.2

Collagen fibers. Hierarchical structure (A); examples of homotrimeric and heterotrimeric collagen-like proteins (each monomer is represented by a different color) (B).

Reproduced with permission from Domene et al.[30]

tendons), while type II is typically associated with tissue experiencing compressive loads (such as cartilage, with a composition of nearly 70% collagen type II); other types of collagen are less predominant, but play important roles in certain tissues, as in the case of skin, blood vessels, and other tissues where collagen type III is present, along with large proportions of elastin.[4]

Elastin

Elastin is a key extracellular matrix (ECM) protein and functions to provide elasticity to the tissue fibers; its presence is dominant extensively in tissues and exerts recoil in tissues subjected to repeated stretch (e.g., as for blood vessels). The functional structure of this protein is of a highly cross-linked amorphous polymer (the cross-linking of elastin monomers takes place during formation of desmosine molecules) that forms sheets or fibers in the ECM, as illustrated in Fig. 10.3A. Due to the amorphous nature and complex entanglement of elastin fibers, their mechanical behavior follows an entropic elastic model (as for rubber materials), where the randomness of conformation and therefore the entropy varies with deformation.[5]

Ground substance

The nonfibrous fraction of soft connective tissue is a viscous gel-like matrix, composed of large carbohydrate chains and carbohydrate-protein complexes in aqueous suspension (Fig. 10.4). The function of the ground substance is to not only facilitate intercellular exchange, but also provide some mechanical support

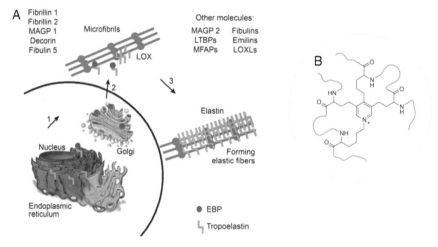

FIGURE 10.3

Schematic diagram of cellular elastogenesis: protein formation, storage, and transport to the extracellular matrix where the elastic fibers are formed (A); chemical structure of elastin interchain cross-links to form desmosine residues (B). *EBP*, Elastin-binding protein; *LOXLs*, lysyl oxidase-like proteins; *LTBPs*, latent transforming growth factor β binding proteins; *MAGP*, microfibril-associated glycoproteins; *MFAPs*, microfibril-associated proteins.

Reproduced with permission from Weihermann et al.[31]

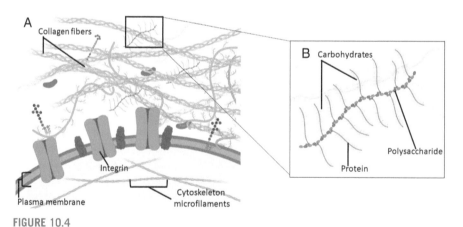

FIGURE 10.4

Illustration of the extracellular matrix (ECM), showing some embedded proteins as key connectors that bridge the inner membrane with ECM proteins such as collagen fibers (A). Zoom-in of a classic proteoglycan structure, typical component of the ground substance (B).

(intercellular joint within main tissues).[6] This fluid matrix pressurizes at high loading rates, thus the ground substance pressurization carries a significant fraction of the total load, serving as an efficient mechanism for mechanical support.[7]

Mechanical properties of soft tissues

As mentioned earlier in this chapter, the mechanical properties of soft tissues are modulated by their composition and conformation (degree of hierarchy and organization), thus providing unique properties according to their physiological function.

In general, soft tissues behave anisotropically regionally and directionally in a three-dimensional space. Now, what does anisotropic mean? In terms of mechanical properties, this definition is applied to those materials that exhibit different properties (tensile strength, tensile strain, Young's modulus) when measured along different axes. In soft tissues, this characteristic behavior is mainly explained by the variation in the distribution of collagen fibers within the assembled tissue, resulting in nonhomogeneous materials.[8] Therefore the tensile (stretching) response of soft tissue is nonlinear and depends on the strain rate, exhibiting steady deformation and recovery when subjected to loading and unloading; this time-dependent material response is called viscoelastic behavior.[9] For viscoelastic materials the relationship between stress and strain can be expressed as:

$$\sigma = \sigma(\varepsilon, \dot{\varepsilon}) \tag{10.1}$$

$$\sigma = \mu \, \dot{\varepsilon} \tag{10.2}$$

Thus:

$$\dot{\varepsilon} = d\varepsilon / dt \tag{10.3}$$

where

σ is stress, ε is strain, $\dot{\varepsilon}$ is strain rate, μ is viscosity, t is time.

There are two characteristic properties of viscoelastic materials. The first one is creep, which refers to the increasing deformation under a constant load, as illustrated in Fig. 10.5A; the second significant behavior is stress relaxation, which refers to a decrease of the stress under constant deformation (strain), as illustrated in Fig. 10.5B. Viscoelastic materials such as soft tissues exhibit hysteresis, known as energy dissipation, showing a hysteresis or difference in between the loading and unloading curves as a result of the energy lost to heat during loading, as illustrated in Fig. 10.5C.

Tissues hold different magnitudes of load in vivo, which relates to the range of collagen structures; Table 10.1 summarizes the mechanical properties of different tissues and correlates to their composition (wt%). A good example is provided by tendons, which are among the strongest soft tissues, with an average ultimate tensile strength (U_{TS}) of 50 MPa and an average composition of collagen of 80%; on the other hand, aortic tissue has an average composition of collagen of 30% and an average U_{TS} of 0.55 MPa.

Fig. 10.6A shows a typical mechanical characterization (stress-strain curve), indicating region of low stiffness modulus (1), toe region (2), high stiffness modulus (3), and maximum tensile strength (4). The plotted curves present mechanical

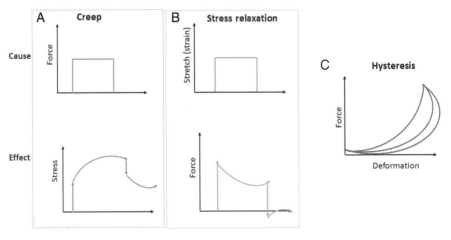

FIGURE 10.5

Viscoelastic behavior of soft tissue. Creep: response under a constant force applied (A); stress relaxation: effect on tissue under a constant strain (B); hysteresis due to energy dissipation (C).

Table 10.1 Mechanical properties of soft tissue and correlation with composition.

Material	Ultimate tensile strength (MPa)	Ultimate tensile strain (%)	Collagen (% dry weight)	Elastin (% dry weight)
Tendon	50–100	10–15	75–85	<3
Ligament	50–100	10–15	70–80	10–15
Aorta	0.3–0.8	50–100	25–35	40–50
Skin	1–20	30–70	60–80	5–10
Articular cartilage	9–40	60–120	40–70	—

Data from Holzapfel[1] and Muiznieks and Keeley.[33]

properties for aortic and pulmonary tissue. The aortic tissue curve shows the combined effect of collagen and elastin, while the pulmonary tissue evidences a lower composition of collagen. As noticed in Fig. 10.6A, the curve is nonlinear since the slope in regions (2) and (3) is different. From a physiological perspective, region (2) will be governed by elastin composition, while region (3) is a response to collagenic tissue. Evidence of this claim is presented in Fig. 10.6B—E, showing nonhomogeneous materials with different compositions of collagen, which relates to their physiological function.

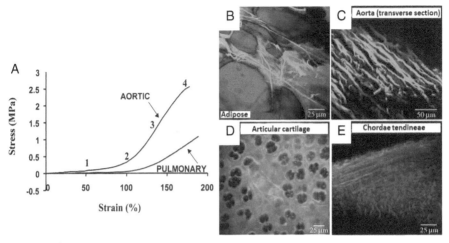

FIGURE 10.6

Mechanical behavior of soft tissue. Typical stress-strain curve for aortic and pulmonary tissue (A). Distribution of elastin *(green)* and collagen *(blue)* fibers and lipid components *(red)* among different soft tissues: adipose tissue (B), aortic tissue (C), cartilage (D), and tendon (E).

-Reproduced with permission from Green et al.[32]

Mechanical properties can be analyzed by different testing approaches. If we neglect the viscoelastic behavior of soft tissues, a typical stress-strain curve is collected by calculating the extension (strain) of a desired material upon application of force (stress). As presented in Fig. 10.7A, the direction of loading/unloading is shown with arrowheads (black and blue, respectively). The load (stress) is calculated by accounting for the cross-sectional area of the testing material as follows:

$$\sigma = F/A \qquad (10.4)$$

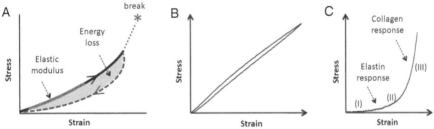

FIGURE 10.7

Measuring mechanical properties. Schematic stress-strain curve (A); typical stress-strain curve for elastin, showing a linear elastic extension (B); typical J-shaped stress-strain curve of tissues containing both collagen and elastin (C).

Reproduced with permission from Muiznieks and Keeley.[33]

where F is the applied force (N) and A is the average cross-sectional area of the samples (mm^2).

Fig. 10.7A indicates the stretch to break with a dotted line; failure of the tissue is represented by an asterisk. When the stress is removed before achieving plastic deformation and failure, information on relaxation can be collected (represented by a dashed blue line). This relaxation modulus is defined as:

$$E(t, \varepsilon) = \sigma(\tau)/\varepsilon_0 \qquad (10.5)$$

where ε_0 is the strain input and $\sigma(\tau)$ is the measured stress, which can be calculated by:

$$\sigma(t) = F(t)/\sigma_0 \qquad (10.6)$$

where $F(t)$ is the recorded force and σ_0 is the undeformed area.

The elastic modulus represented in Fig. 10.7A by a red line is the slope of the curve in the linear region and provides information on the stiffness of the tested material. This property is calculated by:

$$E = \sigma/\varepsilon \qquad (10.7)$$

Therefore,

$$E = \frac{F/A}{\Delta L/L} \qquad (10.8)$$

where $\Delta L/L$ is the material extension per unit length.

The illustration in Fig. 10.7A represents the percentage energy lost to heat (hysteresis) as a shaded area between the stretch and the relaxation curve.

A classic stress-strain curve for elastin shows a linear elastic extension with minimal hysteresis between the stretch and the relaxation curve (Fig. 10.7B), while Fig. 10.7C features a typical J-shaped curve representative of tissues containing both collagen and elastin. The initial response is due to elastic deformation (I), followed by a stiffer response due to molecular deformation (II), ending with plastic deformation and later failure (III).

Example 10.1

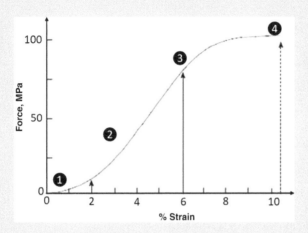

a) Identify the phases in a typical stress-strain curve of soft tissue (tendon), and how they correlate to tendon composition.
b) Calculate Young's modulus of elasticity and yield stress of the tested tissue.

Answer
a) Phase (1) is known as the toe phase; at this stage, collagen fibers behave isotropically, as initial low stress is required to achieve large deformations of the individual collagen fibers without requiring stretch, and the elastin fibers are mainly responsible for the stretching mechanism with a low elastic modulus. During phase (2) as load increases, collagen fibers tend to line up, gradually elongating and increasing the stiffness of the tendon. At this point, the crimp pattern disappears, and the collagen fibers are aligned. At strain levels between 4% and 6%, the tendon becomes easier to extend; however, by releasing the load, the length will return to its initial value. At phase (3), strain levels from 8% to 10% will trigger intrafibrillar damage leading to reduction of stiffness, this phase is known as partial failure. Finally, phase (4) represents complete rupture (failure) of the tendon.
b) Young's modulus is the slope of the curve within the elastic response.

$$E = \frac{\sigma}{\varepsilon}$$

$$E = \frac{(75-25)}{(0.06-0.03)} = 1667\ \text{MPa} = 1.67\ \text{GPa}$$

Yield stress is the material property that defines the stress where tissue starts to deform plastically. From the tendon stress-strain curve, the yield stress is 87.5 MPa.

Vascular soft tissue

The vascular system, also known as the circulatory system, is a network of blood vessels that distributes blood from and back to the heart across all other organs and tissues in the body. This system is characterized by a branching pattern, starting with arteries taking blood from the heart and progressively branching into narrower vessels until reaching microscopic structures known as arterioles, and capillary beads, as illustrated in Fig. 10.8A.

Vascular architecture

Vascular tissue is composed of functional segments, and each of these segments has very specific physiological functions:

Heart: This organ is composed of several layers of muscular tissue. It generates the pressure required to actuate the flow of blood through the entire body.

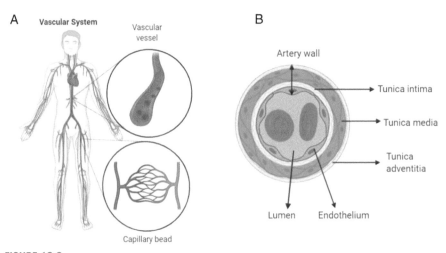

FIGURE 10.8

Graphic illustration of the human vascular/circulatory system shown in a male standing (A), and the functional segments of a vessel (B). Note the representation in *red* for oxygen-rich blood (arterial) and *blue* for deoxygenated venous blood.

Blood vessels: These structures transport oxygen and nutrients to tissues (arteries), and mobilize wastes from the tissues (veins) back to the heart. Typically, arteries and veins share the same architectural organization with a wall composed of three layers (Fig. 10.8B).

- *Tunica adventitia*: This is the most external and strongest layer; it is composed of collagen and elastic fibers and provides a limiting barrier to prevent over-extension of the vessel's diameter.
- *Tunica media*: This is the middle layer, composed of smooth muscular tissue and elastic fibers, which therefore provides the key regulation of blood via muscular contraction. Due to their specific functions, the thickness of this layer controls the pressure experienced; therefore arteries will have a thicker layer than veins.
- *Tunica intima*: This inner layer is composed of elastic fibers and smooth endothelium. This layer permits the exchange of nutrients and gases, while also playing a critical role in hormonal regulation. When the intima layer is injured as a result of vascular surgery or stent placement, scar tissue might be generated in this region by abnormal migration and proliferation of smooth muscle cells from the media. Afterward, the tunica intima generates a thick scar layer described as *intimal hyperplasia* or *neointimal hyperplasia.*

What is so unique about the endothelium? The vascular endothelium is a monolayer of endothelial cells (ECs), which constitutes the inner cellular lining of arteries, veins, and capillaries. This interface is in permanent contact with the blood stream and serves as a barrier to endocrine activity (production of hormones). It regulates the degree of vasodilation or constriction in coordination with vascular smooth muscle cells.[10] Structural and compositional changes in this single layer have been correlated with pathogenesis of vascular disease, also known as endothelial dysfunction.

Vascular pathogenesis

The contraction of the heart generates pulsatile changes in blood pressure on the arterial fraction of the circulation. Large arteries show a passive response to blood pressure due to their elastic components (greater composition of elastin, as discussed in the Mechanical properties of soft tissues section); this response decreases the pressure downstream. When the vessel wall conformation or composition changes, immediate effects are seen in their physiological response. Cardiovascular disease (CVD) is the leading cause of death worldwide, with 17.8 million deaths attributed to CVD in 2017 alone.[11] Vascular disease is the main component of CVD and is linked to maladaptive remodeling of the vascular wall, controlled by the fluid shear stress, which generates occlusion or blockage that limits the ability to regulate the pressure downstream. The primary underlying cause of CVD is the lipid-driven inflammatory process, known as atherosclerosis. This condition is driven by lipidic plaque formation at the inner layer of the blood vessel (endothelium), resulting in downstream stenosis; this last term refers to the narrowing or blockage of the vessel.[12] When the blockage occurs in the arteries serving the heart, this pathology is known as coronary artery stenosis and it is caused by lipidic accumulation and calcium deposition, which is therefore conducive to calcific aortic vascular stenosis (CAVS).

Endothelial dysfunction is recognized by alterations in the endothelium response leading to reduced vasodilation, proinflammatory response, and prothrombotic properties; these alterations are associated with the development of atherosclerosis, angiogenesis in cancer, vascular leakage, and stroke. Endothelial dysfunction is associated with stenosis (illustrated in Fig. 10.9). The histological markers of CAVS are inflammation, ECM remodeling with increasing fibrosis, valve thickening, angiogenesis, and calcification (Fig. 10.10).

In a possible mechanism, the high mechanical stress on aortic valves together with atherosclerotic risk factors lead to valvular endothelial dysfunction/leakage, followed by deposition of low-density lipoprotein (LDL) particles as well as other compounds that trigger inflammation, which in turn activates valvular interstitial cells (VICs) resulting in their osteoblastic transformation, that is, calcification. On the other hand, heart disease accounts for the problems with arterial vessels, muscle tissue, or valves, all of which results in a malfunction of the heart. That said, the pathologies described previously for vessels, if not treated, result in potential heart disease.

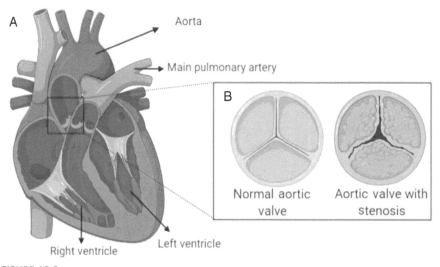

FIGURE 10.9

Graphic illustration of a heart cross-section (A) with a zoom-in of aortic valve, comparing a healthy valve vs. a valve with stenosis

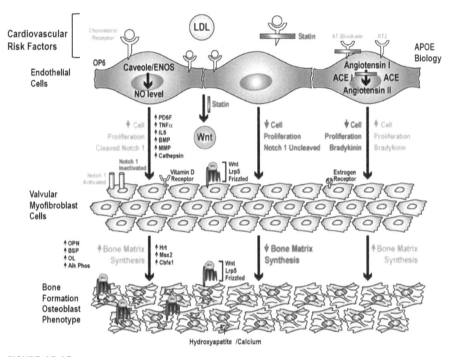

FIGURE 10.10

Basic principles of the pathogenesis of calcific aortic valve stenosis and potential therapies for this disease. *ACE*, Angiotensin-converting enzyme; *APOE*, apolipoprotein E; *BMP*, bone morphogenetic proteins; *BSP*, bone sialoprotein; *ENOS*, endothelial nitric oxide synthase; *IL6*, interleukin 6; *LDL*, low-density lipoprotein; *MMP*, matrix metalloproteinase; *OPN*, osteopontin; *TNFα*, tumor necrosis factor-*α*.

Reproduced with permission from Rajamannan.[34]

Vascular tissue repair

Apart from the general considerations for implantable biomaterials, any device that is intended to be implanted in blood vessels (blood-contacting surfaces), either for repair or replacement, must comply with the criteria summarized in Fig. 10.11.

1. **Hemocompatibility:** The materials used need to demonstrate a neutral interaction with blood, therefore preventing undesired activation and/or destruction of blood components (e.g., activation of foreign invader, coagulation factors, and immunoglobulins).
2. **Nonthrombogenic surface:** When a foreign material is exposed to blood, a significant number of thrombotic and inflammatory responses are triggered by the endothelium of the disrupted vessel. Here it is important to note again the role of EC mediating thrombosis (blood clot within a blood vessel) by activating the coagulation cascade, resulting in thrombin and finally fibrin formation. This

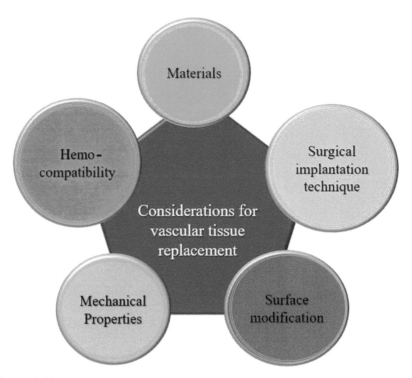

FIGURE 10.11

Design considerations for implantable devices intended for vascular tissue replacement.

last is ultimately conducesive to vessel narrowing and further blockage of the already treated vessel, which is also known as restenosis.

3. **Optimal surface modification:** The materials implanted must favor EC adhesion, migration, and proliferation, allowing the development of an endothelial layer on the material surface; this process is also known as re-endothelialization. There is plenty of evidence supporting the fact that endothelium impairment represents a major factor for restenosis.[13]

4. **Mechanical properties:** The requirements to design materials and devices for vascular repair are mainly related to wear and fatigue resistance. The device must be able to regulate the inner vascular pressure to harmonize the natural tissue remodeling.

Table 10.2 provides a summary of different approaches for surface modification, intended to optimize the interactions (reduced thrombogenicity and increased re-endothelialization) between the implanted material and blood interfaces.

Table 10.2 Selection of modification techniques employed for optimizing blood-material interactions.

Modification	Description
Physical immobilization	• Polymer gelling (growth factor mixed with the material in the liquid state and changing temperature, pH, or ion concentration to obtain a gel with nanopores) • Emulsion techniques (factors that are insoluble in aqueous solutions) • High-pressure gas foaming (incorporate growth factor into porous scaffolds without the use of solvents)
Covalent modification	• Surface distribution of ligands • Distribution of ligands through the bulk of the material
Surface adsorption	• Passive adsorption driven by secondary interactions between the molecule and the protein • Self-assembled monolayer (SAM) adsorption of the peptide (which is designed with hydrophobic tail and a spacer) from solution • Microcontact printing of alkanethiol SAMs, photolithography (on hard materials), soft lithography (on elastomeric materials) • Direct protein patterning: drop dispensing, microfluidic patterning
Cross-linking	• Photo/chemical cross-linking
Altering surface wettability	• Ion bombardment • UV irradiation • Exposure to plasma discharge
Altering surface roughness	• Deposition of polymer films/islands • Nanoparticles, metallographic paper, or diamond paste polishing • Sand blasting, photolithography, and e-beam etching

UV, *Ultraviolet.*
Reproduced with permission from de Mel et al.[38]

Vascular mechanics

The pressure on the blood vessel can be calculated by the simplified Laplace equation:

$$\sigma = P\frac{r}{t} \tag{10.9}$$

where r is the vessel radius and t is the wall thickness.

Vascular stents apply radial force (R_F) on the artery as a response to expansion (illustrated in Fig. 10.12), which originates a hoop stress associated with an expansive hoop force (H_F) on the wall of the artery.[14] The ideal performance of the stent over the artery pressure can be calculated by using the thin-walled cylinder theory (Eq. 10.10). R_F in the case of an open cylinder having thickness (t), length (l), and diameter (D) under internal pressure (P) developing hoop stress (σ_θ) can be calculated as:

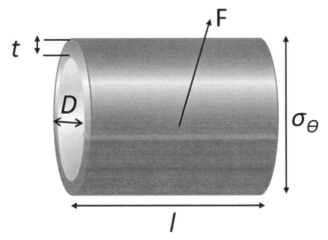

FIGURE 10.12

Vascular mechanics. Graphical illustration of the forces acting within an artery.

$$\sigma_\theta = \frac{PD}{2t} = \frac{H_F}{A_{hoop}} \tag{10.10}$$

where

$$P = \frac{R_F}{A_{radial}}, \; A_{hoop} = l.t, \; \text{and} \; A_{radial} = \pi Dt$$

Therefore,

$$R_F = 2\pi H_F \tag{10.11}$$

Example 10.2

A mouse artery has an internal diameter of 5 mm and a wall thickness of 1 mm; the artery was kept under an internal pressure of 100 mm Hg (1 mm Hg = 133 Pa), while the arterial stretch was stimulated by a longitudinal stress of 50 kPa.
a) Identify the principal stresses and direction.
b) Calculate the maximum shear stress.
c) Calculate the pressure required to produce a maximum tensile stress of 80 kPa, keeping the longitudinal stress at 60 kPa.

Solution
a) Longitudinal (σ_l) and hoop stresses (σ_θ) (assuming a cylindrical shape).

Circumference-hoop stress:

$$\sigma_\theta = \frac{PD}{2t} = \frac{13.3 \text{ kPa} \ (0.005 \text{ m})}{2 \ (0.001 \text{ m})} = 33.25 \text{ kPa}$$

The longitudinal (axial) stress (due to internal pressure) might be approximated to:

$$\sigma_{l_2} = \frac{PD}{4t} = \frac{13.3 \text{ kPa} \ (0.005 \text{ m})}{4 \ (0.001 \text{ m})} = 16.63 \text{ kPa}$$

Total longitudinal stress:

$$\sigma_l = \sigma_{l_1} + \sigma_{l_2} = 50 \text{ kPa} + 16.63 \text{ kPa}$$

$$\sigma_l = 66.63 \text{ kPa}$$

$$\sigma_l > \sigma_\theta \ \therefore \ Principal \ stress - longitudinal \ direction$$

b) Maximum shear stress:

$$\tau_{max} = \frac{\sigma_l - \sigma_\theta}{2} = 16.7 \text{ kPa}$$

c) Given that:

$$\sigma_l = \sigma_{l_1} + \sigma_{l_2}$$

$$80 \text{ kPa} = 60 \text{ kPa} + \frac{PD}{4t}$$

$$P = 16 \text{ kPa (blood pressure)}$$

Vascular stents

Vascular stents are devices used in coronary angioplasty or percutaneous coronary intervention (PCI). This nonsurgical intervention is the first response to coronary artery disease or vascular stenosis; it is performed to widen the vessel affected by blockage or narrowing, and during the procedure a stent is permanently implanted to keep the vessel open (Fig. 10.13).

Besides the criteria discussed in the previous section, the specific parameters for consideration when designing stents can be summarized as:

1. Dimensions of the stent struts
2. Full expansion of the stent (radial elasticity)
3. Extent of the balloon injury during the stent implantation
4. Ability to undergo compression from the vessel wall
5. Resistance to recoil: high R_F/hoop strength

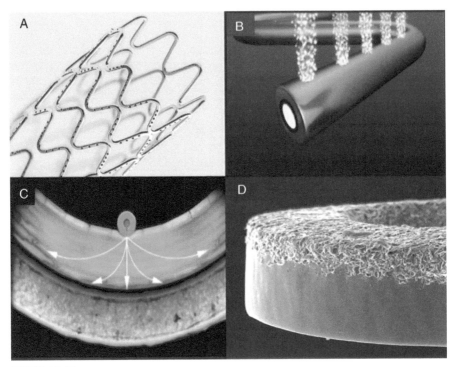

FIGURE 10.13

Next-generation polymer-free stents. The drug-filled stent (DFS, Medtronic, Santa Rosa, CA, USA) (A) exhibits a hollow design with sustained drug elution through diffusion (B) via direct interaction with the vessel wall (C). The BioFreedom drug-eluting stent platform (Biosensors Europe, Morges, Switzerland) (D) as seen with electron microscopy, illustrating the highly texturized abluminal surface.

Reproduced with permission from Harada and Kastrati.[35]

6. Compatibility with magnetic resonance imaging (MRI)
7. Fast endothelization: essential to decrease the risk of thrombosis
8. A surface that inhibits inflammatory reactions and excessive smooth cell (SMC) proliferation, while preventing intimal hyperplasia

As presented in Table 10.3, the biomaterial choices for stents are traditional metals and alloys (chemically nonreactive metals), including stainless steel (SS), tantalum, nitinol (alloy of Ni 55 wt% and Ti 45 wt%), or platinum. Due to their mechanical properties, metals can expand and be compressed with ease and are able to maintain their diameter after deployment into the vessel.

Table 10.3 History of development of coronary stents.

Device (Manufacturer)	Device description	Material(s)	Year	Status
Wallstent (Schneider-AG, Bulach, Switzerland)	Self-expanding wire-mesh structure	Stainless steel	1986	Technical limitation in using the delivery system. Withdrawn from market in 1991
Palma-Schatz (Johnson & Johnson, New Brunswick, NJ)	First FDA-approved stent in the United States. First balloon-expandable device	Stainless steel	1987[a]	Most studied and widely used stent in the 1990s. Risk of restenosis
Gianturco-Roubin Flex Stent (Cook, Bloomington, IN)	Balloon-expandable flexible coil stent	Stainless steel mesh and balloon catheter	1993[a]	Induced subacute stent thrombosis, causing the sudden occlusion of the vessel
Wiktor (Medtronic, Santa Rosa, CA)	High-pressure balloon-expandable stent	Tantalum coil	1996[a]	Induced thrombosis
Cypher (Cordis Corp, Baar, Switzerland)	First drug-eluting stent. Metallic stent coated with sirolimus	Stainless steel	2006[a]	Presented challenges as the metallic surface stimulated thrombosis
Zilver PTX (Cook, Bloomington, IN)	Self-expanding peripheral drug-eluting stent	Nitinol coated with paclitaxel using polymers as carriers	2012[a]	Postdeployment, it imparts an outward radial force upon the inner lumen of the vessel with local drug delivery
Zilver Flex (Cook, Bloomington, IN)	Self-expanding flexible stent	Bare metal nitinol	2015[a]	Designed to provide support while maintaining flexibility in the vessel
Absorb GT1 BVS (Abbott, Abbott Park, IL)	First FDA-approved bioresorbable scaffold drug-eluting (everolimus)	Poly(L-lactide) scaffold with embedded platinum markers	2016[a]	Potential allergic reactions to everolimus or the polymeric material

[a] United States Food and Drug Administration (FDA) clearance.

The first generation of commercial stents was fabricated with bare metals. Even though these devices almost eliminated the risk of the vessel collapsing, they were not effective in avoiding the renarrowing (restenosis), having the vessel closed up again in about 6 months in 25% of the cases. Drug-eluting stents (second generation) were created to mitigate vessel restenosis by introducing coating of the struts with polymer as a drug-carrier vehicle for antiproliferative agents (such as rapamycin and paclitaxel), intended to inhibit the cell cycle and avoid neointimal hyperplasia. These metallic stents coated with drugs effectively controlled the renarrowing, reducing the risk of restenosis down to less than 10%.[13]

However, there were concerns that drug-eluting stents were related to a complication called acute thrombosis. The mechanism for the late stent thrombosis is not yet clear, but it is hypothesized that the antirestenotic drug can delay the endothelialization, while the polymers remaining in the coating after drug delivery can trigger a hypersensitive reaction. For this reason, different strategies for surface modification have been introduced to improve blood compatibility. Natural polymers (presented in Table 10.4) including fibrillin and tropoelastin have been researched as surface modifiers with positive results on blood compatibility. The recognition of the natural polymers/proteins as "self" by the immune system (via conditioning

Table 10.4 Summary of vascular biological effects of biomolecule candidates.

Candidate	Effects on ECs	Effects on SMCs	Blood compatibility	Translation to date
Fibrillin-1	Proliferation enhanced	Proliferation inhibited	Not yet tested	Enhanced fibroblast attachment to PU scaffold
Fibrillin-5	Attachment enhanced Apoptosis reduced	Proliferation inhibited	Not yet tested	No translation
Tropoelastin	Attachment and proliferation enhanced	Proliferation inhibited	Hemocompatible; minimal activation of platelets	Improves EC binding, growth on steel; reduces thrombogenicity of catheters and ePTFE grafts
Perlecan	Proliferation enhanced	Proliferation and hyperplasia inhibited	Direct inhibition of thrombosis	Less thrombus present on ePTFE grafts, more ECs; stents show less neointima

EC, *Endothelial cells;* ePTFE, *expanded polytetrafluoroethylene;* PU, *polyurethane;* SMCs, *smooth cells.*
Data reproduced with permission from De Mel et al.[39]

film on the vessel surface) limits the inflammatory response, therefore reducing the risk of thrombogenesis.

The third generation of stents is characterized by polymer-free eluting systems, where the drug is introduced into the microporous surface on metallic stents. In other examples, BioFreedom (Biosensors Europe, Morges, Switzerland) stents propose the use of small reservoirs (microstructured surface) filled with Biolimus A9; this solution provides a polymer- and carrier-free alternative, especially successful for patients with high bleeding risk. Finally, bioresorbable stents offer a vascular scaffold made of resorbable polymers such as poly(L-lactic acid) (PLLA) and poly(D,L-lactic acid), or magnesium-based metallic stents. These scaffolds remain stable for a certain period of time, for further resorption. This strategy favors compatibility by avoiding an inflammatory response and improving the restoration of endothelial coverage. The long-term performance of these devices is still a matter of study.

Vascular grafts

In some cases PCI is not sufficient in the long term and a coronary artery bypass graft (CABG) intervention is necessary. This procedure is considered a good alternative for patients requiring long-term revascularization solutions.[15] A vascular graft (also known as vascular bypass) is a surgical technique that redirects the blood flow by reconnecting blood vessels and recovering the blood supply to areas affected by a diseased (blocked) artery. These bypass grafts can be either from biological or synthetic origin. Fig. 10.14 presents the conventional approaches for vascular grafting, analyzing advantages and disadvantages of each strategy.

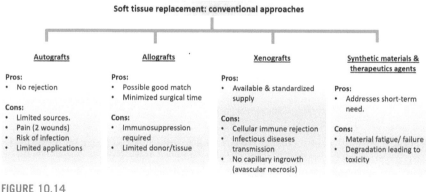

FIGURE 10.14

Conventional approaches for soft tissue repair and/or replacement.

Natural-biological grafts

- *Autografts:* Harvested from the patient. This approach avoids rejection risk although is limited by a restricted material supply and the risks associated with additional surgery procedures.
- *Allografts:* Also known as allogeneic grafts or homografts. In this case, the graft is sourced from a human donor (who is not genetically identical). This strategy relies on the donor availability and requires extended immuno-suppression to avoid rejection.
- *Xenografts:* Grafts can be obtained from donors belonging to different species. Most common examples are porcine xenografts.

Synthetic grafts

- *Synthetic materials:* Artificial vascular grafts are devices that are designed to replace a part of a damaged blood vessel. These are usually polymeric fibers that are fabricated into tubular shapes with fabric-like properties. The geometry of the fibers into the vascular graft has a significant influence on the final properties of the artificial blood vessels.[16] Because the polymers are fibers, the tensile strength of the artificial vascular grafts is high. As discussed in the previous chapters, all implants need some degree of porosity, and vascular grafts are no exception. Porosity is designed to be part of these devices to allow for good integration with the neighboring tissues.

Some specific requirements for material selection and design of vascular synthetic grafts can be summarized as follows:

1. Ensure tissue integration
2. Avoid blood diffusion (leakage) from the artificial blood vessel
3. Must be hemocompatible and nonthrombogenic

To ensure tissue integration, one common technique is the so-called "preclotting technique," which consists of deposition of proteins that initiate blood clotting, therefore inducing the integration of the prosthesis with the surrounding connective tissue. Fibrin is one of the proteins used for this purpose. Upon deposition of fibrin, a soft clot is formed that is called *pseudointima* and which seals the vessel, not allowing further blood leakage from the artificial blood vessel. Thus the biomaterials used for the vascular grafts need to be hemocompatible and nonthrombogenic.

A strategy to prevent blood clotting is using antithrombogenic molecules to coat the vascular graft materials. These include plasminogen, heparin, albumin-heparin mixtures, or phosphorylcholine. Heparin is an anticoagulant molecule that has been successfully used for coating expanded polytetrafluoroethylene (ePTFE) grafts and preventing vessel occlusion. The attachment method is via covalent bonding, which ensures long-term maintenance of the anticoagulant activity of the heparinized surface.

Another avenue to prevent vessel occlusion is via passivation of Dacron grafts with human serum albumin or heat-denatured albumin, which leads to further prevention of blood component adhesion. Combining albumin with heparin and grafting them on the surface of artificial vessels is also a successful strategy to prevent vessel occlusion. The rationale of using this molecule is that it is naturally present in the membrane of the ECs and should not precipitate the formation of thrombi. Polymers containing these functional groups in their structure seem to markedly reduce the adsorption of fibrinogen and the adhesion of platelets to the surfaces.

An ideal polymer needs to be biocompatible and inert, reducing the inflammatory response and the risk of failure by thrombosis. Back in 1992, the use of Dacron meshes (polyester threads) as a potential backbone for vessel tissue replacement was demonstrated for the first time. However, as the inflammatory response of this material is expected to resemble that encountered in its use as vascular grafts, its clinical application is still facing limitations. Currently the polymers that are most often used for these applications are Teflon (polytetrafluoroethylene) and Dacron (polyethylene terephthalate [PET]). Literature provides different examples on passivation strategies or coating with bioactive molecules, aiming to address the early rejection of the implanted prosthetic graft. For instance, Fig. 10.15 introduces a dual approach under research, by combining a PET fabric with a cryogel coating (containing biological agents) to fabricate a bioactive prosthetic graft for the effective delivery of therapeutic modulators of the anastomotic neointimal hyperplasia (ANIH) response, along with antithrombotic agents.[17]

Synthetic biodegradable polymers such as polyglycolic acid (PGA), PLLA, and polyglycolic-co-lactic-acid (PLGA) are interesting polymers reported as biocompatible materials, biodegradable (bioabsorbable), and as having favorable ingrowth rates; however, the risk of failure caused by platelet adhesion and later thrombosis development continue, imposing restrictions to their long-term implantation.

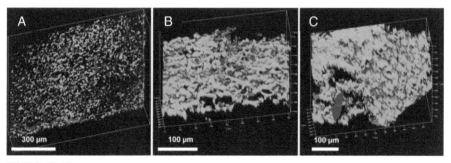

FIGURE 10.15

Endothelial cell integration into hybrid grafts (polyethylene terephthalate [PET] and alginate). Three-dimensional reconstructed confocal images display cell adhesion and integration after 24 hours in culture with hybrid electrospun PET graft (A), knitted PET (B), and woven PET (C).

Reproduced with permission from Huynh et al.[17]

Different strategies might be implemented to control and/or reduce the undesired reactions toward stent implants. Most common approaches explored to date are focused on the design of the device as a carrier for thrombolytic agents. In principle, this is advantageous since there is a significant enhancement, favoring proper performance and reducing the failure rate postimplantation; integrating a new complexity layer, such as drug delivery systems, requires to guarantee a highly controlled performance without altering the nature, composition, or mechanical response of the stent.

Despite the promising approaches described in Table 10.4, legal regulation poses a significant restriction into taking these devices to commercial stages. As drug delivery systems are introduced to enhance the polymer performance, the device will be categorized within a stricter regulatory route, leading to a discouragement of this venue and favoring the surface modification as a primary strategy to be applied when considering practical and feasible implementation. Some examples of bioresorbable polymeric scaffolds are summarized in Table 10.5.

Substitute heart valves

In general, mechanical valves are designed to have a metallic support that can be sewn to a fabric cuff or ring and fixed to the surrounding natural tissue as either a single or a bileaflet valve. Therefore tissue integration is a main requirement at the sewing cuff; PET or polytetrafluoroethylene (PTFE) is commonly used. Tissue integration rate is inversely proportional to thrombosis or sepsis risk (due to bacterial attachment and propagation), thus the materials selected must promote rapid integration. The hemodynamics of the mechanical valves is highly critical and special attention is necessary to reducing the risk of platelet adhesion, which might result in thrombosis, calcification, and final restriction of the leaflet's movement.

Table 10.5 Commercial bioresorbable polymeric scaffolds.

Bioresorbable scaffold	Strut material	Polymer type	Eluting polymer	Drug
Absorb	PLLA, Pt markers	Bioresorbable	PDLLA	Everolimus
Absorb-GT1	PLLA, Pt markers	Bioresorbable	PDLLA	Everolimus
DEsolve 100	PLLA, Pt-Ir markers	Bioresorbable	PLLA	Novolimus
MeRes	PLLA, Pt-Ir markers	Bioresorbable	PDLLA	Sirolimus
Magmaris	Mg, backbone, Ta markers	Bioresorbable	PLLA	Sirolimus

Ir, *Iridium;* Mg, *magnesium;* PDLLA, *poly-D,L-lactic acid;* PLLA, *poly-L-lactic acid;* Pt, *platinum;* Ta, *tantalum.*
Data from Thakkar and Dave.[40]

Therefore materials selection and device design are intimately linked to optimal performance.

Substitute heart valves can be classified as bioprosthetic valves (from human donors, or tissue-biological materials) or mechanical valves (Fig. 10.16). Mechanical valves are developed completely from man-made materials, therefore a cautious selection of materials will improve valve reliability and functional life.

The requirements to design mechanical valves are mainly related to:

1. Wear and fatigue resistance
2. Alteration of the natural hemodynamics and possible impact on long-term valve performance
3. Device deterioration under the continuous mechanical stresses, which can lead to the failure of the implant and release of degradation products into the circulation, with a potential risk of toxicity

Metallic alloys (commonly titanium alloys) are forged in the form of rings and cages, in order to support and guide disks and balls that are commonly made of polymers and pyrolytic carbon. The latter has been shown to result in less wear and reduced fibrous tissue formation when compared with that usually found on the surface of polymer-based systems, increasing its incorporation in vascular prosthetics.

Metallic materials used in mechanical heart valves include austenitic SSs, Co-Cr-Mo (cobalt-based alloys), tantalum, and titanium. SSs are no longer extensively used due to their passivation layer that is not as stable as other metallic alloys. However, 316 LVM (low carbon vacuum melt) is successfully used for endovascular

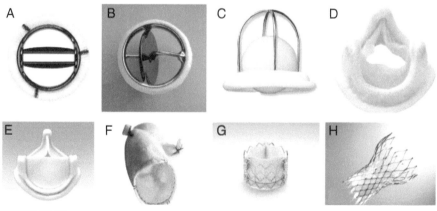

FIGURE 10.16

Different types of prosthetic valves. Bileaflet mechanical valve (St. Jude) (A); monoleaflet mechanical valve (Medtronic Hall) (B); caged ball valve (Starr-Edwards) (C); stented porcine bioprosthesis (Medtronic Mosaic) (D); stented pericardial bioprosthesis (Carpentier-Edwards Magna) (E); stentless porcine bioprosthesis (Medtronic Freestyle) (F); percutaneous bioprosthesis expanded over a balloon (Edwards Sapien) (G); self-expandable percutaneous bioprosthesis (CoreValve) (H).

Modified with permission from Dangas et al.[36]

stents and portion of valves.[18] Titanium alloys are common structural components of mechanical valves because of their low density, corrosion rate, high strength, and biocompatibility. Pyrolytic carbon is a glassy carbon used for heart valve fabrication, especially useful in the leaflets and housing of mechanical valves, due to its good biocompatibility and thromboresistance.

On the other hand, polymeric valves represent an interesting venue for tissue replacement, providing both durability and enhanced hemodynamic performance.

Silicone materials have a long history in model vascular systems; the most common silicones used for vascular replacement are casted room-temperature-vulcanizing (RTV) silicone, porous PTFE, injection molding of polyurethanes, and solvent casting and molding of polysiloxanes.[19] However, polymers have not yet reached an optimal performance compared with native tissue, due to the persistence of thrombosis and calcification, leading to early failure caused by material fatigue. A selection of examples for biomaterials employed for heart valve replacement is presented in Table 10.6.

Bioprosthetic heart valve replacement

Despite the limitations, a form of xenotransplantation has been taking place since the 1960s with relative success: bioprosthetic heart valve (BHV) replacement. The main sources of material for these valves are from bovine pericardium and porcine cusp valves. After being obtained from the donor, the tissue is generally fixed and stored in glutaraldehyde to preserve it and reduce its immunogenicity. Subsequently the bioprosthetic valves are sewn onto a polymeric ring before implantation. During implantation, the construct is sutured into the surrounding tissues. One main advantage of these valves is that they do not require anticoagulants, reducing the limitations and complexities related to surface modifications. Although these valves offer

Table 10.6 Clinically used heart prosthetics and common materials.

Valve type		Pretreatment/materials
Mechanical	Ball and cage	Titanium (Ti alloys), cobalt-based alloys, pyrolytic carbon
		Silicone rubber, elastomer
	Tilted disk	Titanium (Ti alloys), cobalt-based alloys, pyrolytic carbon, ePTFE
	Bileaflet valve	Titanium, pyrolytic carbon
Bioprosthetic	Porcine valve	Glutaraldehyde, various enzymes
	SynerGraft	DNAase, RNAase, water
	Matrix P	Deoxycholic acid
	Bovine pericardium	Glutaraldehyde

ePTFE, *Expanded polytetrafluoroethylene.*
Modified with permission from Khademhosseini and Merryman.[41]

hemodynamics similar to that of the natural valves, they often undergo calcification, and in some cases there is a structural deterioration correlated with inflammation/rejection of glutaraldehyde-fixed xenografts in human patients.[20]

More recently mechanical prosthetic valves tend to be replaced by bioprosthetic valves, although mechanical valves are still produced by some companies. Allografts can be used for the production of prosthetic valves, with bovine pericardium and porcine cusp valves being the main sources of material. After being obtained from the donor, the tissue is generally fixed and stored in glutaraldehyde and subsequently sewn onto a polymeric ring before implantation. On implantation, the construct is sutured into the surrounding tissues. The porcine valves are typically stripped of cells and sterilized with glutaraldehyde before being mounted on sewing rings. Although these valves offer hemodynamics similar to that of the natural valves, they often undergo calcification, which is termed calcific aortic valve stenosis (CAVS).

Regardless of the site of implantation, the formation of an organic layer conditioning the surface of any cardiovascular implant from the earliest phase of body fluid contact is a common feature. The organic layer is rich in blood components, which are key factors in the host response toward foreign bodies. In particular, proteins play a role in regulating the two principal systems of this host reaction, which are coagulation and inflammation.

Tissue-engineered heart valves

With the rise of tissue engineering as a research field, new polymeric materials have emerged as potential scaffold materials capable of promoting cell proliferation and regeneration of natural tissues/organs. Tissue-engineered heart valves (TEHVs) seek to develop fabrication technologies by mimicking the natural environment of native embryonic tissue as it develops in utero. Considering that one important event in the development of heart valves is epithelial-to-mesenchymal transition (EMT), during this process there is cell differentiation, and endocardium cells turn into mesenchymal cells and migrate to form developing prevalve cardiac cushions, producing the ECM proteins needed to form the leaflets. A selection of biomaterials used for tissue engineering of heart valves is presented in Table 10.7.

Several factors need to be carefully controlled in order to design and produce viable TEHV.

1. An ideal material for scaffolding needs to ensure high porosity with interconnected pore network, allowing for proper diffusion of nutrients and waste removal.
2. Scaffold materials must be biocompatible and biodegradable, providing a non-thrombogenic but compatible surface for cell seeding and propagation. The degradation rate of the scaffold material should be similar to the native tissue proliferation, to guarantee proper resorption by maintaining the support for the growing tissue.
3. The mechanical properties of the scaffold must match the mechanical properties of the native tissue and retain their natural regeneration ability.

Table 10.7 Biomaterials used for TEHV with cell lines seeded in vitro and implanted in animals.

	Biomaterials	Cell line used	Implanted into
Synthetic materials	PEG	hMSCs, porcine VICs	NA
	PGA:PLA	Ovine fibroblasts, ECs, and valve cells Human fibroblasts, bovine aortic ECs	Lamb (2 weeks)
	PGA:P4HB	Ovine myofibroblasts, ECs ovine vascular and SCs, ovine EPCs, VECs Human AFSCs	Lamb (20 weeks) Sheep (8 weeks)
	PCL	Human myofibroblasts	NA
	PGS:PCL	HUVECs	NA
	PEUU	Rat SMCs	NA
	PDO	Sheep MSCs	Sheep (1, 4, 8 months)
	PCU:POSS	NA	NA
Natural materials	Collagen	NA	Rats, beagles (84 days)
	Hyaluronic acid (HA)	Porcine VICs Neonatal rat SMCs	NA
	Collagen/HA composites	Neonatal hDF 3T3s	NA
	Gelatin	Porcine VICs	NA
	Collagen/ chondroitin sulfate composite	Porcine VICs, VECs	NA
	Fibrin/fibronectin	hDF, human aortic myofibroblasts, ovine arterial SMCs and myofibroblasts Porcine VICs	Sheep (3 months)
	Hydroxyapatite (HAp)	HUVECs	NA

Cell line abbreviations: AFSCs, amniotic fluid-derived stem cells; ECs, endothelial cells; EPCs, endothelial progenitor cells; hDF, human dermal fibroblasts; hMSCs, human mesenchymal stem cells; HUVECs, human umbilical vein endothelial cells; SCs, stem cells; SMCs, smooth muscle cells; VECs, valvular endothelial cells; VICs, valvular interstitial cells.
NA, *Not applicable; P4HB, poly-4-hydroxybutyrate; PCL, polycaprolactone; PCU, poly(carbonate-urea) urethane; PDO, polydioxanone; PEG, poly(ethylene glycol); PEUU, poly(etherurethane urea); PGA, polyglycolic acid; PGS, poly(glycerol sebacate); PLA, polylactic acid; POSS, polyhedral oligomeric silsesquioxane.*
Reproduced with permission from Khademhosseini and Merryman.[41]

The field of tissue engineering is in continuous development and new advances are being sought to overcome its limitations and improve the functions of the tissue engineering constructs. In addition to implantable applications, work is being performed toward using these approaches to create sensing strategies of various toxicities or medications. Despite the promising impact of this technology, the clinical application of TEHVs still faces major challenges, mainly represented by the accurate prediction of the tissue performance, the ability of the matrix to follow remodeling after implantation, and high costs related to laborious processes for production.

Intraocular lens implants

The use of biomaterials for ophthalmological applications dates back to the 19th century, when Adolf Fick invented the glass contact lens. Since then the development of ocular materials has quickly expanded, achieving to date the successful implantation of intraocular devices. Among all the intraocular devices, intraocular lens implants (IOLs) are commonly used to permanently replace the natural lenses and improve visual acuity in patients undergoing cataract surgery. The early use of polymethyl methacrylate (PMMA) as a biocompatible implantable material has a unique connection to World War II, when the favorable reaction of canopy material accidentally implanted in pilots' eyes was repeatedly observed.[21] From these observations, the first IOL was developed.

PMMA has been for a long time the most popular material used for development of IOLs due to its outstanding optical properties and relatively good biocompatibility. However, this material might trigger in some cases the damage of tissue surrounding the implant, with possible effects as serious as iris adhesion to the implant and loss of vision. Due to its low tolerance to temperature and rigidity, new materials with foldable performance are now being explored.

Nowadays, different polymeric materials are used to fabricate IOLs—silicone, acrylic compounds, and hydrogels being the most common, as shown in Table 10.8. The surface properties of the implantable lenses have been extensively studied, due

Table 10.8 Examples of biomaterials for IOLs.

Manufacturer	Lens type	Material	Refractive index
Advanced Medical Optics	Rigid	PMMA	1.49
ALCON	ACRYSOFT foldable	PEA/PEMA	1.55
Bausch & Lomb	Hydro-view foldable	HEMA/HEXMA	1.47
Calhoun Vision	Multifocal foldable	PDMS	1.41
STAAR Surgical	Collamer foldable	Collagen/HEMA	1.45

HEMA, *Hydroxyethylethacrylate;* HEXMA, *6-hydroxyhexyl methacrylate;* PEA, *poly(ethyl acrylate);* PEMA, *poly(ethyl methacrylate);* PMMA, *polymethyl methacrylate.*
Modified from Lloyd et al.[42]

to the need to guarantee biocompatibility, and avoid adhesion, proliferation, and migration of lens epithelial cells (LECs). Different strategies to prevent later deterioration of the refractive ability of the implanted lens are reported, going from the controlled growth of a LEC monolayer, thus avoiding further cell proliferation and lens opacification,[22] to immobilization of bioactive compounds such as fibronectin, vitronectin, laminin, and type IV collagen.[23] From these latest experiments, fibronectin and laminin were found to perform better when bonded to hydrophobic acrylate IOLs in terms of capsule attachment and reduced opacity.

The vision process heavily relies on the ability of the eye to refract (bend) light. Since the cornea possesses a spherical surface, it reflects the light while acting as a converging lens (Fig. 10.17). Most of the eye's refractive power comes from this process, where light passes through two different media (air and aqueous humor) interacting with different refractive indexes (n) (1.00 and 1.34, respectively). The same phenomenon occurs when light passes through the pupil and hits the lens. However, the vitreous humor has a refractive index (n) of 1.34 and the lens has a refractive index (n) of 1.44, hence providing a fine-tuning of imaging.

In addition to the general requirements for implantable biomaterials already covered, IOLs involve specific optical properties that allow a clear path for light to provide appropriate optical imaging. Hence an ideal IOL must fulfill the following conditions:

1. Transparent material with a smooth surface
2. Appropriate reflective index
3. Sharp-edged optic design
4. Hydrophobic surface (capsular biocompatibility)
5. Hydrophilic anterior surface (uveal biocompatibility)
6. Enhanced mechanical properties (controlled unfolding)

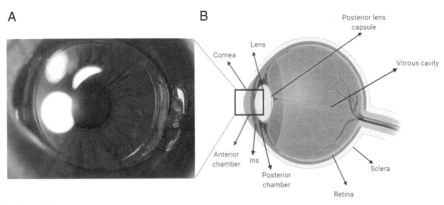

FIGURE 10.17

Intraocular lens implants. Close up of the anterior segment of a human eye, containing a three-piece silicone lens (A); illustration of the eye's anatomy (B). (Fig. 10.17A reproduced with permission from Isaac K CC BY-SA 4.0 (https://commons.wikimedia.org/w/index.php?curid=55046374).)

IOLs are composed of optics (lens) with side structures (haptics) to keep the implant fixed within the capsular bag (inside the eye) (Fig. 10.18A). According to the material used and hence its properties, IOLs are classified as rigid (PMMA), flexible (silicone), foldable (hydrophobic acrylic, hydrophilic acrylic), and collamer. Based on the refractive correction they perform when implanted (optics design), they can be classified as monofocal, Crystalens, toric, and multifocal (Fig. 10.18B). Expanded classification of IOLs is presented in Table 10.9. Monofocal IOLs traditionally used for cataract surgery use a standard single-focus lens, providing excellent vision but only at a fixed distance. Crystalens are flexible lenses designed to reestablish the eye's accommodation functionality, providing a shift of focus between nearby and distant objects after cataract surgery; their operation relies on the eye's natural muscle control of focus. Toric lenses, on the other hand, provide a solution for patients who need cataract and astigmatism intervention; their unique design provides both spherical and astigmatic correction. Finally, multifocal implants deliver improved vision from near and distance; however, these lenses do not follow focus accommodation.

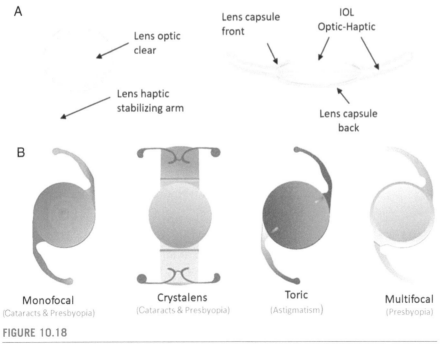

FIGURE 10.18

Illustration of the structure of intraocular lens implants (IOLs) and view within the lens capsule (A); main types of IOLs commercially available (B).

Reproduced with permission from Jung et al.[37]

Table 10.9 Classification of IOLs.

Destination	Capsular bag, ciliary sulcus, scleral fixation, iris fixation, angle supported
Overall design	3 piece/1 piece
Overall length	10−13 mm
Optics material	Rigid (PMMA), flexible (silicone), foldable (hydrophobic acrylic, hydrophilic acrylic), collamer
Refraction index	1.42−1.55
Optics shape	Biconvex, planoconvex, meniscus
Optics diameter	5−7 mm
Optics design	Spherical, aspheric, toric multifocal, multifocal toric
Optics color	Transparent, tinted
Haptic properties	3 piece/1 piece (PMMA, PVDF, polyamide, 2, 3, 4, 6 haptics)
Type of implantation	Injectable, not injectable
Type of packaging	Preloaded, not preloaded

PMMA, *Polymethyl methacrylate*; PVDF, *polyvinylidene fluoride.*
Reproduced with permission from Bellucci.[43]

As an effort to improve the postoperative performance of IOLs, different materials, surface modifications (hydrophobic, hydrophilic, passivated, bioactive), geometries, and haptic design have been tested. Recent studies suggest that the most critical factor associated with optimal performance of IOLs resides in the haptic design rather than the material itself, based on the optical performance of the tested IOLs under dynamic compression conditions.[24] However, considering biocompatibility and the potential risk of posterior capsule opacification (PCO) development, the material is the most important in terms of critical variables that affect the long-term performance of the IOL.[23]

Optical performance of IOLs is analyzed in terms of the refractive power of a lens; this power is measured in diopter units, and is defined as:

$$D = 1/f \tag{10.12}$$

where D is the power in diopters and f accounts for the local length in meters.

Now, the surface power can be calculated as a function of the refraction index and radius of the curvature as follows:

$$D_s = (n - 1)/r \tag{10.13}$$

where n is the refractive index and r describes the curvature radius in meters.

Therefore the lens power can be calculated by Lens-Maker's equation:

$$D_n = (n - 1)/r_1 + (n - 1)/r_2 \tag{10.14}$$

In the range of 4 diopters, thickness must be accounted for by compensation, applying the effective power formula:

$$D_e = D_1 + D_2 + \frac{t}{u}(D_1)_2 \tag{10.15}$$

Example 10.3

A prescription for a corrective lens is 3.0 diopters; you selected a material with a refractive index of 1.56. If the radius of the curvature on the front surface of the lens is 27 cm, what must the radius of the curvature be for the other lens surface?

Solution Applying the lens maker's equation (Eq. 10.14):

$$D = (n-1)\left(\frac{1}{r_1} + \frac{1}{r_2}\right)$$

$$\frac{1}{r_2} = \left(\frac{D}{(n-1)} - \frac{1}{r_1}\right)$$

$$\frac{1}{r_2} = \left(\frac{3.0}{(1.56-1)} - \frac{1}{0.27}\right) = \frac{1}{1.653}$$

$$r_2 = 0.60 \text{ cm}$$

Biomaterials with a high refractive index are preferred for fabrication of IOLs, since this allows for a thinner IOL configuration, enhancing the mechanical properties of the device, which favors a slow and controlled unfolding thereby contributing to a lower risk of PCO, which is the most common complication after cataract surgery, also known as "secondary cataract."

PCO is caused by propagation and migration of residual (postsurgery) LECs. LECs adhere to the lens, inducing PCO, which implies a progressive reduction of visual acuity; since LECs arrange into multilayer islets, light is scattered and therefore vision is reduced.[25] The development of PCO might be influenced either by natural wound-healing mechanisms and/or IOL surface properties. As previously discussed in this book, there is strong evidence that demonstrates how hydrophilic surfaces favor epithelial cell adhesion and proliferation, hence increasing the risk of PCO. IOL surface modification is an approach largely explored by immobilizing a variety of molecules or treating the material (such as carbon, titanium, heparin, PTFE, gas plasma, polyethylene glycol, and 2-methacryloyloxyethyl phosphorylcholine among others),[26] preventing the adsorption of cells and proteins.

Breast implants

Breast implants are medical devices designed to replace or augment mammary tissue for cosmetic or reconstructive purpose. As for any other implantable devices, biocompatibility is a major challenge and silicone-based materials have led to the development of new generations of breast implants during the last decades. The most common complications postimplantation are infections, rejections, capsular ruptures, contractures, and more recently, anaplastic large cell lymphoma.[27] Great attention has been directed to understand the mechanisms leading to capsular contracture as a major response to the collective reactions occurring at the implant surface after implantation, such cascade reaction is also known as foreign body response (illustrated in Fig. 10.19). As for any foreign object, the host creates a capsule surrounding the implant as a strategy to isolate the nearby tissue. Right after implantation, the implant is coated with proteins from the blood and neutrophils that rapidly reach the wound site. Monocytes differentiate into macrophages, creating foreign body giant cells that trigger the proliferation of fibroblasts, and it is here that the synthesis of collagen surrounding the implant is initiated, isolating it from the tissue, thereby creating a fibrous capsule. Finally, the capsule contracts and becomes stiff, causing capsular contracture.

Although immune response is necessary for a proper healing process, an excessive inflammatory response limits the viable life of the implant and increases the risk for capsular contraction, posing an important challenge for succesful long-term implantation.

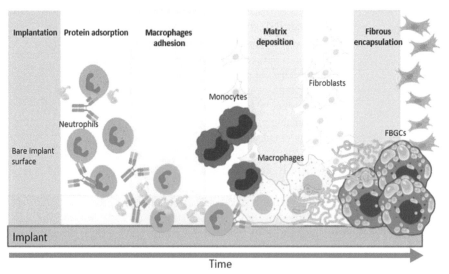

FIGURE 10.19

Foreign body response to biomaterial implant. Illustration of the immune response following implantation and ending with fibrous capsule formation. *FBGCs*, Foreign body giant cells.

Different strategies have been implemented in order to improve the implant performance, focusing on three major fronts: (1) reducing the risk of bacterial contamination, (2) optimizing the implant surface texture (roughness, porosity), and (3) designing implants that avoid the leakage of the gel filling content.

Biomaterials for breast implants

The global market for breast implants has been dominated by silicone materials since its early origins in 1961 (Table 10.10), when thick shell implants were designed, containing viscous silicone gel and including patches of PET (Dacron). In the 1970s the shell material became thin, containing a less viscous gel, and Dacron patches were removed. Different materials were rapidly introduced, such as perforated silicone (to favor tissue ingrowth) and polyurethane coating (implant fixation), providing textured surface to the implants. More recently (1993), breast implants evolved into nonbleed cohesive gels.[28]

One of the main challenges with breast implants is the risk of leaking of silicone fluids into the surrounding tissue. In order to calculate the sustained leakage in a period of time, and assuming the leakage is governed only by diffusion, the flux (units of mass per unit area per time) can be expressed using Fick's first law:

$$F = -D\frac{dc}{dx}$$

where D is the diffusion coefficient and c accounts for concentration.

Example 10.4

You are requested to calculate the leaking of a silicone membrane for 2 years. The silicone gel has a molecular weight (MW) of 1200 amu; the thickness of the membrane is 2 mm with a surface area of 350 cm^2; the diffusion coefficient is constant ($D = 4.6 \times 10^{-18}$ cm^2 s^{-1}). The overall volume of the implant is 750 cm^3 with a density of 1.05 g cm^{-3}.

Solution The flux is expressed in units of mass per unit area per time. Therefore,

$$\frac{\text{Mass}}{\text{Time}} = \text{Flux} * \text{area} = FA = D\frac{dc}{dx}\,350\ \text{m}^2$$

$$= 4.6 \times 10^{-18}\text{cm}^2\text{s}^{-1}\frac{1.05\ \text{g cm}^3}{0.2\ \text{cm}}350\ \text{cm}^2$$

$$\frac{\text{Mass}}{\text{Time}} = \text{Flux} * \text{area} = FA = D\frac{dc}{dx}\,\text{cm}^2$$

$$= 8.45 \times 10^{-15}\text{g s}^{-1}\ \text{or}\ 2.67 \times 10^{-7}\text{g year}^{-1}$$

$$= 5.3 \times 10^{-7}\text{g in 2 years}$$

Table 10.10 Evolution of materials used for breast implants.

Approach	Materials	Year	Advantages	Disadvantages
Solid allopathic implants	Polyurethane Polytetrafluoroethylene (Teflon) Expanded polyvinyl alcohol formaldehyde (Ivalon sponge)	1950s and 1960s	Lack of antigenicity Non-risk of infection	Local tissue reactions Firmness, distortion of the breast Significant discomfort
Semisolid materials injected into breast parenchyma	Liquid silicone (polydimethyl siloxane)	1961	Softness Inert nature	Infections Chronic inflammation Drainage, granulomas Necrosis
Two-component implant: silicone elastomeric shell filled with stable liquid material	Shell: thick, smooth silicone elastomer with Dacron fixation patches Filling: viscous silicone gel	1962	Fixed position of the implant	High capsular contracture rate Patches promoted a stress point where the elastomer could break
Two-component implant	Shell: silicone elastomer Filling: saline solution	1965	Softer implant	2%–3% deflation rate per year High failure rate
Polyurethane foam–covered implants	Coating: polyurethane Core: silicone elastomer	1967	Reduced contracture rates	Delamination of foam coating Breast pain/infections Difficulty of removal
Silicone shell without Dacron patches	Shell: thin, smooth silicone elastomer Filling: less viscous silicone gel	1970s	Natural feel	Silicone bleeding into periprosthetic intracapsular space
Multilayer implant shell (barrier-coat material sandwiched between two layers of silicone elastomer)	Shell: multilayer silicone elastomer (monomers, and copolymers: Me_2SiO, Ph_2SiO, $CF_3CH_2CH_2SiMeO$) Filling: silicone gel/saline solution	1980s	Reduced silicone bleeding	Rupture Migration of silicone gel and platinum to surrounding tissue
Anatomical form stable implants		1994	Anatomically shaped implants	

Regulatory constraints

Although the regulatory requirements are very specific to each of the technologies discussed, the authors consider that early efforts should be made to understand the needs of final users and market drivers. This approach might improve the chances to take the lab-bench research into sustainable, cost-effective, and scalable commercial solutions. Also, early testing in human cell culture, along with digital accurate modeling, would improve the design iterations, thus reducing the R&D timeline and optimizing resources.

Summary

Regardless of the site of implantation, the formation of an organic layer conditioning the surface of any implant from the earliest phase of body fluid contact is a common feature. The organic layer is rich in blood components, which are key factors in the host response toward foreign bodies. In particular, proteins play a key role in regulating the two principal systems of this host reaction, which are coagulation and inflammation response.

Inspired by native ECMs, new advances for tissue engineering and targeted regeneration are expected to flourish. The introduction of surface features in a nanoscale and controlled fashion will play a key role in mimicking the natural uniformity of porosity in scaffold materials, leading to micro- and macroscale gradient structures, interconnectivity, pore distribution, and topological features.[29] Finally, the growth of powerful digital tools opens the possibility to create predictive models that permit simultaneous testing of a number of different variables, allowing for data-driven design and decision-making. New approaches need to be developed in order to offer not only avenues that are not only scientifically appropriate, but also cost-efficient, sustainable, and safe, while in compliance with current regulation.

Problems

1. You need to design a vascular graft. List and explain the advantages and disadvantages of using ceramics, metals, or polymers for such application.
2. What is the main component of arterial tissue?
   a) From your response, what synthetic material should be used to design an arterial graft?
   b) Elaborate on the mechanical properties of the natural tissue and describe the main property that will determine the success of such an arterial graft.
   c) Would you choose a natural or synthetic material for this device?
   d) What special characteristics must the inner/outer surfaces of the arterial graft have? Explain.

3. What are the common complications/malfunctions of natural and synthetic replacement valves postsurgery?

4. Soft tissue is characterized by its mechanical response. Its mechanical performance is described by its elastic behavior.

    a) Draw a typical stress-strain curve for soft tissue, showing the differences between high-collagen and high-elastin content.

    b) Describe the mechanical testing performed (lab tests) for hard tissue and contrast with the testing performed in soft tissue. Why do these protocols differ?

    c) Discuss the viscoelastic behavior of soft tissues. How does it relate to their biomechanical function?

5. Describe the main structures and functions of the ECMs.

    a) How does the cell signaling operate through the ECMs?

    b) What is the role of ECMs on cellular differentiation? List the most important steps.

    c) How is the matrix contact related to cell division cycles?

6. What are the main challenges encountered with autografts, allografts, and xenografts as potential strategies for soft tissue replacement?

7. You are characterizing a group of specimens with potential to be used as vascular graft materials. According to the mechanical test results presented below, which candidate material would you select? Explain.

Sample ID	Ultimate tensile strength (MPa)	Yield strength (MPa)	Ultimate tensile strain (%)
[1]	100	30	10
[2]	80	27	15
[3]	1	5	98

8. From the specimens listed in the last problem, make an educated guess about the sample's composition. Which one shows a typical behavior of highly concentrated collagen fibers?

    a) From your response, discuss what kind of soft tissue would be ideally replaced by this material.

    b) What synthetic material might mimic the mechanical properties of natural vascular tissue?

9. What is the difference between cohesive forces and surface energy?

    a) How does surface tension impact the interaction of blood with the implanted material's surface? Elaborate on the potential triggering of iatrogenic reactions.

    b) List some of the common measures taken to control the wettability of metallic implanted materials.

    c) Explain the effect of topographical changes on the interaction between blood and metallic surfaces when the materials are implanted.

10. Mention at least two different methods used to measure the surface energy of an implantable material. Describe the advantages/disadvantages of using either high surface energy or low surface energy materials.
11. What is the relationship between surface roughness, total surface energy, and wettability of metallic implants?
12. According to the metallic surface designed and fabricated (flat and homogeneous or rough), describe the model applied to calculate their surface energy.
13. You are requested to design a metallic implant for a cardiovascular stent. When considering wettability, would you prefer a hydrophilic surface or a hydrophobic one? Explain the long-term implications of each scenario.
14. Describe the biochemical pathways activated postimplantation.
    a) What is the role played by M2 macrophages in collagen synthesis?
    b) Why does implant encapsulation increase the risk of rejection? Elaborate on the immunological cascade response involved.
15. Describe at least two of the fundamental differences between silicone gel and saline solution breast implants.
16. How much tension might the blood vessel walls be able to endure to maintain the positive pressure difference in equilibrium?
17. A titanium-alloyed dental implant with outstanding mechanical properties is designed. You are requested to reduce its surface energy in order to improve its biocompatibility. Do you agree with this? If so, what would be your strategy? Explain.
18. A new polymer is developed providing an optimal refractive index, transparency, and mechanical properties (folding). You are requested to test the wettability and determine its viability as an implantable intraocular lens prototype. After characterization you learn that both faces of the lenses' prototype are hydrophobic. Would you approve its implantation? Explain your response.

References

1. Holzapfel GA. *Handbook of Material Behavior: Nonlinear Models and Properties*. Graz, Austria: Biomech Preprin; 2000.
2. Shoulders MD, Raines RT. Collagen structure and stability. *Annu Rev Biochem*. 2009;78: 929—958.
3. Henriksen K, Karsdal MA. Type I collagen. In *Biochemistry of Collagens, Laminins and Elastin: Structure, Function and Biomarkers*, Karsdal MA, Ed. London: Academic Press; 2016:1—11.
4. Buckley MR, Evans E, Satchel LN, et al. Distributions of types I, II and III collagen by region in the human supraspinatus tendon. *Connect Tissue Res*. 2013;54(6):374—379.

5. Hoeve CAJ, Flory PJ. The elastic properties of elastin. *J Am Chem Soc*. 1958;80(24): 6523—6526.

6. Stecco C, Hammer W, Vleeming A, De Caro R, Connective tissues. In *Functional Atlas of the Human Facial System*, Stecco C, Hammer W, Vleeming A, De Caro R, Eds. London: Churchill Livingstone; 2015:1—20.

7. Ebrahimi M, Ojanen S, Mohammadi A, et al. Elastic, viscoelastic and fibril-reinforced poroelastic material properties of healthy and osteoarthritic human tibial cartilage. *Ann Biomed Eng*. 2019;47(4):953—966.

8. Chanda A, Callaway C. Tissue anisotropy modeling using soft composite materials. *Appl Bionics Biomechanics*. 2018;2018:7—12.

9. Özkaya N, Nordin M, Goldsheyder D, Leger D. *Fundamentals of Biomechanics: Equilibrium, Motion, and Deformation*. 3rd ed. New York: Springer-Verlag; 2012:1.

10. Krüger-Genge A, Blocki A, Franke RP, Jung F. Vascular endothelial cell biology: an update. *Int J Mol Sci*. 2019;20(18):1—22.

11. Lyle AN, Taylor WR. The pathophysiological basis of vascular disease. *Lab Invest*. 2019; 99:284—289.

12. Hahn C, Schwartz MA. Mechanotransduction in vascular physiology and atherogenesis. *Nat Rev Mol Cell Biol*. 2009;10(1):53—62.

13. Bedair TM, ElNaggar MA, Joung YK, Han DK. Recent advances to accelerate re-endothelialization for vascular stents. *J Tissue Eng*. 2017;8:1—14.

14. Cabrera MS, Oomens CWJ, Baaijens FPT. Understanding the requirements of self-expandable stents for heart valve replacement: radial force, hoop force and equilibrium. *J Mec Behav Biomed Mat*. 2017;68:252—264.

15. Al-Lamee R, Thompson D, Dehbi HM, et al. Percutaneous coronary intervention in stable angina (ORBITA): a double-blind, randomised controlled trial. *Lancet*. 2018; 391(10115):31—40.

16. Kucinska-Lipka J, Gubanska I, Janik H, Sienkiewicz M. Fabrication of polyurethane and polyurethane based composite fibres by the electrospinning technique for soft tissue engineering of cardiovascular system. *Mater Sci Eng C*. 2015;46:166—176.

17. Huynh C, Shih T-Y, Mammoo A, et al. Delivery of targeted gene therapies using a hybrid cryogel-coated prosthetic vascular graft. *PeerJ*. 2019;7:1—19.

18. Bailey SR. DES design: theoretical advantages and disadvantages of stent strut materials, design, thickness, and surface characteristics. *J Intervent Cardiol*. 2009;22:S3—S17.

19. Coulter FB, Faber JA, Zilla P, et al. Bioinspired heart valve prosthesis made by silicone additive manufacturing. *Matter*. 2019;1(1):266—279.

20. Manji RA, Menkis AH, Ekser B. Porcine bioprosthetic heart valves: the next generation. *Am Heart J*. 2012;164(2):177—185.

21. He W, Benson R. Polymeric biomaterials. In: Kutz M, ed. *Applied Plastics Engineering Handbook*. Oxford, UK: Elsevier; 2017:145—164.

22. Linnola RJ. Sandwich theory: bioactivity-based explanation for posterior capsule opacification. *J Cataract Refract Surg*. 1997;23(10):1539—1542.

23. Pérez-Vives C. Biomaterial influence on intraocular lens performance: an overview. *J Ophthalmol*. 2018;2018:1—17.

24. Remón L, Cabeza-Gil I. Influence of material and haptic design on the mechanical stability of intraocular lenses by means of finite-element modeling. *J Biomed Opt*. 2018; 23(3):1—10.

25. Yuen CHW, Williams R, Batterbury M, Grierson I. Modification of the surface properties of a lens material to influence posterior capsular opacification. *Clin Exp Ophthalmol.* 2006;34(6):568−574.

26. Tan X, Zhan J, Zhu Y, et al. Improvement of uveal and capsular biocompatibility of hydrophobic acrylic intraocular lens by surface grafting with 2-methacryloyloxyethyl phosphorylcholine-methacrylic acid copolymer. *Sci Rep.* 2017;7:1−13.

27. De Faria Castro Fleury E, D'Alessandro GS, Wludarski SCL. Silicone-induced granuloma of breast implant capsule (SIGBIC): histopathology and radiological correlation. *J Immunol Res.* 2018;2018:1−9.

28. Heitmann C, Schreckenberger C, Olbrisch RR. A silicone implant filled with cohesive gel: advantages and disadvantages. *Eur J Plast Surg.* 1998;21:329−332.

29. Lam MT, Wu J. Biomaterial applications in cardiovascular tissue repair and regeneration. *Expert Rev Cardiovasc Ther.* 2012;10(8):1039−1049.

30. Domene C, Jorgensen C, Abbasi SW. A perspective on structural and computational work on collagen. *Phys Chem Chem Phys.* 2016;18:24802−24811.

31. Weihermann AC, Lorencini M, Brohem CA, de Carvalho CM. Elastin structure and its involvement in skin photoageing. *Int J Cosmet Sci.* 2017;39(3):241−247.

32. Green EM, Mansfield JC, Bell JS, Winlove CP. The structure and micromechanics of elastic tissue. *Interface Focus.* 2014;4(2):1−9.

33. Muiznieks LD, Keeley FW. Molecular assembly and mechanical properties of the extracellular matrix: a fibrous protein perspective. *Biochim Biophys Acta.* 2013;1832(7): 866−875.

34. Rajamannan NM. Calcific aortic stenosis: lessons learned from experimental and clinical studies. *Arterioscler Thromb Vasc Biol.* 2009;29(2):162−168.

35. Harada Y, Kastrati A. Polymer-free drug-eluting stents—a safe and effective option for ACS. *Nat Rev Cardiol.* 2016;13:447−448.

36. Dangas GD, Weitz JI, Giustino G, Makkar R, Mehran R. Prosthetic heart valve thrombosis. *JACC (J Am Coll Cardiol).* 2016;68(24):2670−2689.

37. Jung GB, Jin KH, Park HK. Physicochemical and surface properties of acrylic intraocular lenses and their clinical significance. *J Pharmaceut Invest.* 2017;47:453−460.

38. de Mel A, Jell G, Stevens MM, Seifalian AM. Biofunctionalization of biomaterials for accelerated in situ endothelialization: a review. *Biomacromolecules.* 2008;9(11): 2969−2979.

39. De Mel A, Cousins BG, Seifalian AM. Surface modification of biomaterials: a quest for blood compatibility. *Int J Bio.* 2012;2012:707863.

40. Thakkar AS, Dave BA. Revolution of drug-eluting coronary stents: an analysis of market leaders. *Eur Med J.* 2016;1(4):114−125.

41. Khademhosseini A, Merryman WD. EMT-inducing biomaterials for heart valve engineering: taking cues from developmental biology. *J Cardiovas Trans Res.* 2011;4: 658−671.

42. Lloyd AW, Faragher RGA, Denyer SP. Ocular biomaterials and implants. *Biomaterials.* 2001;22(8):769−785.

43. Bellucci R. An introduction to intraocular lenses: material, optics, haptics, design and aberration. *Cataract.* 2013;3:38−55.

Materials and devices for sensors and detectors: biocatalysts, bioimaging, and devices with integrated biological functionality

Learning objectives

The contents of this chapter will provide students with the fundamental concepts in the study of biosensors. At the end of the chapter, the reader will be able to:

- Understand the main principles, common materials used for fabrication, detection strategies, and representative applications of the major biosensing platforms.
- Know the types and principles of electrochemical biosensing and biocatalysis.
- Recognize the types and principles of optical sensing.
- Gain a working knowledge of the design, fabrication, and performance evolution of biosensors while discussing the main challenges and drawbacks that this growing technology presents to date.
- Understand the materials property requirements for effective signal transduction in biosensors and as contrast agents in bioimaging.

Introduction

By definition, a detector is a device with the ability to register a specific signal or physical phenomena whereas a sensor is capable of analyzing the input and retrieving a confirmative response and/or quantifying the analyte tested.[1] When we define biosensors, there is a reference to biological strategies for recognition. The most common types of biorecognition interactions are antibody/antigen, nucleic acid hybridization, enzyme/ligand, synthetic nucleic acids (aptamers)/whole-cell analyte, and molecular imprinted polymers.[2] A typical sensing system consists of three main components, as illustrated in Fig. 11.1: the recognition element (A), which specifically interacts with the target in order to generate an output known

Introductory Biomaterials. https://doi.org/10.1016/B978-0-12-809263-7.00011-1

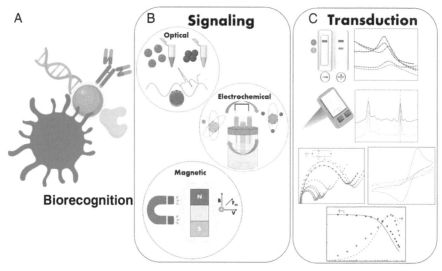

FIGURE 11.1

Typical components of a sensing system. Biorecognition elements: nucleic acids, antibodies, and enzymes (A). Examples of signaling: optical (colorimetric, plasmonic), electrochemical, and magnetic (B). Transduction systems: naked-eye visible response, ultraviolet-visible (UV-Vis) spectroscopy, Raman spectroscopy, impedance spectroscopy, cyclic voltammetry, and alternating current (AC) magnetic susceptibility (C).

as the signal (B), which might have a physical or chemical origin. Last, the transduction system (C) provides a readout which can be analyzed and, in some cases, amplified for enhanced sensitivity.

Nanotechnology is a research field dedicated to study and apply matter on a near-atomic scale. It allows for the fabrication of novel, highly controlled structures, materials, and devices, while unveiling unique advantageous properties only accessible at such a tiny scale; when you find the prefix "nano" is used, keep in mind that we are dealing with manipulation of matter at a scale equivalent to one-billionth of a meter. The unique performance displayed by nanosized materials is inherent to scale and can be harnessed to impart significant enhancement to the overall performance of biosensors.

Among the advantages of nanomaterials applied to biosensors, we can find enhanced surface area-to-volume ratio, superior electron transfer (high electrical conductivity), label-free transduction, multiplexing (simultaneous detection of more than one analyte), and integration with lab-on-a-chip systems, therefore enabling the development of sensitive, reliable, portable platforms.[3] The development of novel and enhanced biosensors as a research field is rising rapidly, due to the significant advancement in nanotechnology during the last decades, along with the increasing commercial need for rapid, low-cost, on-site, and portable sensing strategies. All these conditions have nurtured the exploration of novel nanomaterials and sophisticated fabrication techniques, delivering sensing platforms that might be used for a variety of diverse applications such as drug discovery, food and agriculture (e.g., soil health and produce freshness monitoring, foodborne pathogen detection), early disease detection (human, animal, and plant pathogens), and environmental monitoring (e.g., detection of toxic ions and pesticides in soil and bodies of water). Humanity has been relying on bioanalysis instinctively since the dawn of time; just think for a second about the ability of your sensory nervous system to recognize scents, or the detection of a foreign agent leading to the inflammatory response.[2] Biosensing is a research field inspired by nature and compiles a vast diversity of technical approaches for material design and fabrication, detection strategies, and a continuously expanding number of applications. Based on recent market research, the global market for biosensors is expected to continue growing annually by 8%, reaching a value of USD 31.5 billion by 2025; such growth is driven mainly by the increasing inclination toward rapid and portable diagnostic testing. According to the biosensors forecast, when the market is segmented by detection technique, electrochemical biosensors dominate with a market share of 55%, while optical biosensors account for a share of nearly 38% of the global market.[4] The scope of this chapter considers the extensive body of research on electrochemical and optical biosensing, their promise to be scaled up at low cost, and their commercial relevance. This chapter will cover the main strategies within those specific detection technologies, along with current challenges. We will explore the most common nanomaterials incorporated in successful portable approaches for optical and electrochemical biosensing, as we provide examples of promising applications in medicine, environmental and health monitoring, and food safety.

Ideal criteria and common challenges for biosensors

When working on designing a biosensor, there are some general principles that can be used as a guide for material selection, detection approach, fabrication, and optimization.

1. **Compatibility with biomolecules:** The selection of materials, design, and fabrication methods employed must prove to be nonharmful for the biocapture molecules and/or biological analytes contained in the samples. Any risk of denaturation, structural change, or degradation needs to be assessed and avoided as much as possible.
2. **Multiplexing:** This refers to the capability of a sensing platform to incorporate multiple detection events (signals) in a single device. This feature allows for simultaneous testing of multiple analytes or multiple independent measurements of the same analyte (replication with statistical value).
3. **Sensitivity:** This term refers to the smallest absolute amount of change versus a reference value (signal response) that can be detected by a testing event or measurement. The optimal concentration threshold for a biosensor strictly

depends upon the specific target or analyte. This range is ruled out according to the relevant range of concentration (which is stipulated by either regulation or medical guidelines).

4. **Specificity:** This is the ability of the sensing arrangement (usually based on antibodies, nucleic acids, or enzymes) to recognize only a specific single analyte. In an ideal specific design, the recognition elements will not bind to any other target, even when tested in highly complex samples.

5. **Selectivity:** When specific strategies are unavailable or a set of analytes is required for certain testing, the term selectivity comes into place. By designing arrays of biorecognition elements to target specific sets of analytes, selectivity is introduced within the targets, obtaining characteristic profiles (also known as fingerprints). It enables discrimination of each analyte due to the cross-reactivity of the array components.[5] This approach allows to control analyte-sensor interactions without the need for characterizing individual components.

6. **Precision:** This is a statistical measurement of repeatability. It refers to the sensor's ability to provide the same measurement consistently.

7. **Accuracy:** This is a measure of the deviation from the sensor output compared with the ideal value. In other words, this term refers to the uncertainty or intrinsic error of the measurement.

8. **Cost-efficiency and manufacturability:** From the device design perspective, materials and processes need to be optimized, in order to maximize the chances of adoption and further revenue, considering commercialization as the ultimate goal. Likewise the fabrication methods proposed since the early stage should be validated for proper scalability in an efficient, repeatable, and sustainable fashion.

Putting aside the great potential that biosensors represent, let us now consider the main drawbacks or challenges of this technology when assessed as high-throughput alternatives for reliable screening.

1. **Nonspecific adsorption of biological materials (NSA):** When the active surface of the biosensors comes in contact with the sample, there is a risk of having atoms, ions, or molecules adsorbed on the surface. In NSA, molecules are adsorbed as a result of intermolecular forces (hydrophobic forces, ionic interactions, Van der Waals forces, and hydrogen bonding). NSA increases background signals, affecting the biosensor performance by decreasing sensitivity, specificity, and reproducibility. It can result in different types of NSA:
   a. Molecules adsorbed on vacant spaces (restricting proper unfolding of proteins, affecting stability)
   b. Molecules adsorbed on non immunological sites (intensifying background signals)
   c. Molecules adsorbed on immunological sites (preventing access to antigens)

2. **Biofouling:** In the context of electrochemical sensors, biofouling indicates any deposition or accumulation of either proteins, minerals, fats, cells, target

analyte, or electrochemical reaction products on the sensing surface. It generally involves NSA, polymerization, or precipitation of fouling agents.[6] This process is associated with progressive passivation of the transducer, inhibiting the contact of the target with the electrode and is the major cause of sensor malfunctions in vivo.

3. **Biocapture molecules' stability:** Despite the great advances in the biosensors field, the great majority of technologies developed at the laboratory scale get lost during the translation to prototyping and industrial scaling-up. This restriction is imposed mainly by the sensitive nature of the capture biomolecules. The extensive use of enzyme-linked and antibody-based assays implies high cost of sourcing, extraction, isolation, and purification, along with a number of challenges such as limitations to survive normal conditions of traditional processing and manufacturing.[7]

In this chapter we will explore different strategies to improve the specificity of biosensors, reduce the risk of NSA, and increase their sensitivity, by developing novel materials, unique arrangements, and exciting approaches for controlled manufacturing on a larger scale (Fig. 11.2).

FIGURE 11.2

Common criteria for material selection and design of biosensors.

Modified with permission from Bhalla et al.[60] [CC-BY-4.0 License].

Metallic nanoparticles for applications in imaging and sensors

The development of multifunctional nanomaterials is a focus of attention, given their versatility, tunability, and tailored surface functionalization. Among them, metallic nanoparticles (MNPs) represent the most studied nanomaterials with great potential for cancer detection (bioimaging), sensing (optical and electrochemical), and thera-peutics (targeted drug release). The particles' size, shape, configuration, and surface properties determine the viability of their application as biomedical probes or drug delivery systems.

In the context of topical drug delivery, *particle size* has a direct effect on skin penetration through the outer layer of the epidermis, also known as *stratum corneum* (SC) barrier. As a rule of thumb, particles sized below 10 nm seem to have a positive performance in drug absorption, while smaller particles (below 1 nm) would act as single molecules.[8] Moreover, when thinking in terms of oral delivery, particle size is a property of nanoparticles (NPs) that relates to the surface area; by reducing the par-ticle size, the surface area-to-volume ratio increases, which therefore makes the nanomaterial surface more reactive to interact with the surrounding environment/ media.

On the other hand, *particle shape* has repeatedly been reported as a relevant parameter for tissue penetration and cellular uptake.[9] According to the literature, rods tend to show the best uptake, followed by spheres, cylinders, and cubes.[10] This response might be influenced by the different local curvatures and hydrody-namic behavior, which is unique to each shape/configuration. Also, there is an important effect of geometric shape on critical physiological interactions such as opsonization, internalization, and circulation half-life.[11]

In addition to careful tuning of physical features such as size, geometry, and configuration, additional criteria must be met in order to achieve successful perfor-mance of MNPs in vivo:

1. **Water dispersibility:** The stability of MNP dispersion is a key criterion to determine the NP performance in biological systems. Most of the physiological and biological environments are water based. Therefore MNPs must be designed to be stable and well dispersed in aqueous systems; this behavior is mainly driven by the particle's size and concentration, surface properties, and media composition (ionic strength, pH, and concentration of circulating pro-teins and fats).[12]
2. **Tolerance to relatively high ionic strength:** Ideally the surface of MNPs must be designed with a robust surface charge able to induce electrostatic repulsion, thus keeping a steady interparticle distance among the NPs, even when dispersed in samples with high ionic pressure.[7]
3. **Resistance to oxidation in biological environments:** As discussed in Chapter 3, physiological environments are highly corrosive for metallic materials and MNPs are not an exception. Although, MNPs have unique physical and

chemical properties, the use of noble metals with outstanding resistance to corrosion and oxidation is required to avoid material degradation and negative effects on both performance and toxicity.

4. **Low toxicity:** The materials selected must guarantee low toxicity toward the environment and biological tissues, as well as high stability. The biological impact of metals can be completely changed either by their oxidation states, association with specific ligands, or concentration in solution. An ideal material therefore needs to be stable under different pH, ionic strength, and changing redox conditions.[13]

Surface optical properties of MNPs: localized surface plasma resonance

Nanotechnology has provided the grounds for successful development of imaging probes with an outstanding performance. Among nanomaterials, inorganic NPs are widely studied due to their unique properties derived from their dimensions. Plasmonic nanomaterials such as noble metals show a unique surface plasmon resonance (SPR) as a result of the proton confinement at nanometric size, thereby showing enhanced radiative and non radiative properties.[14]

In the context of NPs, when the incident light passes through the NP, there occurs a displacement of electrons that results in a dipolar oscillation of electrons[15] described as localized surface plasmon resonance (LSPR), illustrated in Fig. 11.3. When assuming the electric field of the light to be constant, the polarizability of the sphere is given by the Clausius-Mossotti relation as follows:

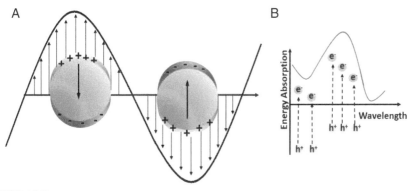

FIGURE 11.3

Localized surface plasma resonance excitation and extinction on spherical nanoparticles. (A) Illustration of the localized electron oscillation, and (B) representation of the absorption spectrum of plasmonic metal.

Modified with permission from Chou and Chen.[61]

$$\alpha = 3\varepsilon_0 V \left(\frac{\varepsilon - \varepsilon_m}{\varepsilon + 2\varepsilon_m} \right) \tag{11.1}$$

where ε_0 is the permittivity of vacuum, ε_m is the dielectric constant of the surrounding media, ε is the dielectric function of the MNP, and V is the spherical volume of the particles, which is calculated as follows for spherical geometries:

$$V_{sphere} = \frac{4}{3}\pi r^3 \tag{11.2}$$

The propagation constant of the surface plasma towards the interface between dielectric and metal is given by:

$$\beta = k \left(\frac{\varepsilon_m \eta s^2}{\varepsilon_m + \eta s^2} \right)^{1/2} \tag{11.3}$$

where k describes the wavenumber, ε_m ($\varepsilon_m = \varepsilon_{mr} + I_{emi}$) is the dielectric constant of metal, and η_s denotes the refractive index of the dielectric. The SPR propagation is possible if $\varepsilon_{mr} < -\eta_s^2$. The most common metals that comply with this condition at optical wavelengths (wavelength range between 100 nm and 1 mm) are silver and gold.[16]

MNPs are the basis of plasmonic sensing nanoplatforms. Specifically, the surface properties of Au (or Ag) NPs become dominant and the electrons from the NP together with the photons of an incident beam of light come together to form quasi-particles called surface polaritons. These surface polaritons have oscillating frequencies known as LSPR. Because the LSPR displays wavelength dependence, the plasmonic nanoplatforms exhibit a very specific color in the visible spectrum. This property can be used in a diversity of platforms tuned to detect different targets, such as disease marker molecules, small ions, or whole cells, by registering a change of wavelength when the target binds to the surface of the NP-based sensor.

Example 11.1

A radiation of a 1-μm wavelength is applied to a silver/glass interface and surface plasmon polaritons are generated. Given the permittivity values are 1.5 for glass and -10 for silver, calculate the propagation constant for the surface plasmon polaritons.

Answer The propagation constant is given by:

$$\beta = \frac{\omega}{c} \left(\frac{\varepsilon_{Ag}\varepsilon_{glass}}{\varepsilon_{Ag} + \varepsilon_{glass}} \right)^{1/2}$$

$$\omega = 2\pi\vartheta = 2\pi\frac{c}{\lambda} = 6\pi 10^{14}$$

After introducing the values in the first equation, the value for the propagation constant is found to be 8.342×10^6 m^{-1}.

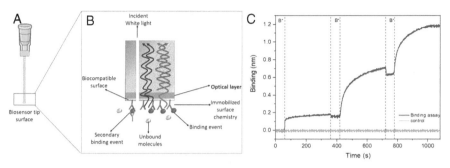

FIGURE 11.4

Bilayer interferometry (BLI): illustration of single biosensor tip (A); representation of the BLI mechanism of detection (B), and typical sensorgram (response vs. time plot) (C).

Label-free optical biosensors: surface plasma resonance

Based on SPR principles, in 1984 Pharmacia (a Swedish pharma-biotech company) founded Pharmacia Biosensors AB. This company developed the first commercial SPR-based instrument with engineered biosensors and microfluidics cartridge, becoming the first Biacore SPR commercially released in 1990. In 1996, the company transitioned to be Biacore AB Corp, which was acquired by General Electric (GE Healthcare) in 2006. In April 2020, GE Healthcare became Cytiva, which is now a leading global provider of technologies for development of therapeutics. Such a technological success with expanding commercial potential!

In principle, SPR-based biosensors operate by applying white light to the functionalized metallic surface, and track in real time the refractive index/mass changes in the presence of the ligand of interest. When plotting the response versus time, a binding curve is obtained (sensorgram), as illustrated in Fig. 11.4. This technology is all about recognizing subtle changes in electron oscillations, allowing for a label-free detection of molecular binding; this approach avoids the need for fluorescent agents typically used for labeling purposes, which significantly reduces the testing complexity, along with the toxicity risks associated with fluorescent dyes. Very quickly this technique became essential for drug discovery and validation, and a popular method to characterize molecular interaction in real time. To date, SPR represents a critical resource to understand and validate drug function, immune-based therapeutics, and binding events in general; it remains the gold standard in life sciences and the pharmaceutical industry for real-time detection of antibody-antigen binding and kinetics, as summarized in Table 11.1.

Label-free optical biosensors: bilayer interferometry

Bilayer interferometry (BLI) is another popular approach to run analysis of molecular binding and kinetics. It is an optical analytical technique that measures

Table 11.1 Examples of label-free (SPR and BLI) commercial biosensors.

Manufacturer	Technique	Description	Validated applications
Pharmacia Biosensor (Biacore SPR, Uppsala, Sweden)	SPR	First functional SPR biosensor launched to the market Manual sampling	Lead characterization and preclinical drug discovery
GE Healthcare (Biacore, Barrington, IL)	SPR	First label-free system to receive FDA approval. As most of the state-of-the-art SPR sensors it applies a hydrophobic, self-assembled monolayer (SAM) as adhesion promoter	Immunogenicity testing and drug discovery
Pioneer (FE System, Santa Fe Springs, CA)	SPR	Multi-assay (three channels): competition/inhibition mechanisms	Screening of fragment-based drug discovery
Xantec Bioanalytics (SPR-PLUS, Düsseldorf, Germany)	SPR	Two-channel SPR platform. The sensor's surface chemistry promotes lower level of nonspecific interactions by using hydrophilic adhesion promoters	Miniaturization of chromatographic approaches
Bruker (Sierra SPR-32 Pro, Billerica, MA)	SPR	Eight flow cells with 32 or 24 independent sensors; continuous flow microfluidics	Small-molecule screening and early detection of nonspecific binding proteins
Fortebio (Octec HTX, Menlo Park, CA)	BLI	Monitors up to 96 biosensors simultaneously at unmatched speed	ELISA conversion, epitope binning, and protein purity testing
Fortebio (Next Gen OneStep, Menlo Park, CA)	SPR and BLI	Next-generation SPR technology. More information and eliminates sample dilutions	Fragment screening and characterization of high-affinity biologic interactions

BLI, *Bilayer interferometry;* ELISA, *enzyme-linked immunosorbent assay;* FDA, *United States Food and Drug Administration;* SPR, *surface plasmon resonance.*

interference patterns from white light reflected by the biosensor tip.[17] To guarantee a successful detection event, close proximity between the two molecules analyzed is required. Most of the commercial instruments available for BLI rely on microfluidics to run the sample over the functionalized metallic film's surface; although this approach has proved to be effective in overcoming diffusion limitations in many cases it represents challenges for certain samples, leading to clogging and fouling. Since 2010, new designs have come to light (e.g., dip-in BLI biosensors), preventing these issues from happening, and facilitating the exploration of multiples matrices and complex samples.

The example illustrated in Fig. 11.4 shows a typical dip-in BLI biosensor that incorporates a fiber optic tip that is dipped in the sample of interest. In this case, the biofunctionalized surface moves to reach the sample, which reduces the risk of contamination and avoids the potential clogging, which is persistent in systems relying on microfluidics.

The shifts in the interference pattern are analyzed and correlated to the increase of thickness of molecules on the sensor's surface (as illustrated in detail in Fig. 11.4B,C, which show a typical sensorgram of a double-binding event). The example presents a preliminary immobilization of the biocapture molecules (i), followed by a first binding event (association) (ii), which is correlated to a mass transfer event, and a secondary binding (iii).

Overall, in the last decade the use of BLI has increased and continues gaining acceptance due to its capability for label-free and real-time analysis with high throughput. Table 11.1 summarizes examples of commercially available BLI instruments.

Spectroscopic applications: SERS-based biosensors

The second effect of the polariton formation is observed when there is an enhancement of the electromagnetic field at the Au/Ag NP surface. This high electromagnetic field helps couple the molecular vibrations of a target analyte (molecule to be detected) that is up to 3 nm away from the metallic surface to the LSPR, leading to an extraordinary increase (10^6-10^8) of the molecular signal intensity. This is called surface-enhanced Raman scattering and is used as signal amplifier in a technique we call surface-enhanced Raman spectroscopy (SERS), as illustrated in Fig. 11.5. SERS is a powerful ultrasensitive spectroscopic technique that can be used for molecular sensing, including for in vivo diagnostics. The sensitivity of this technique can reach single molecules. Many times, SERS sensors involve the deposition of a metallic film on a glass substrate. In this case, the propagation constant of the surface plasmon towards the metal/glass interface is given by:

$$\beta = \frac{\omega}{c} \left(\frac{\varepsilon_{metal}\varepsilon_{glass}}{\varepsilon_{metal} + \varepsilon_{glass}} \right)^{1/2} \tag{11.4}$$

In summary, SPR is the resonant oscillation of electrons at the interface between negative and positive permittivity materials, stimulated by an incident light. The surface plasmon polariton is an electromagnetic surface wave that propagates in the direction parallel to the negative permittivity material (a dielectric, such as glass). These waves are very sensitive to any surface changes, such as molecular adsorption of proteins on the surface of the NP-based sensor.

Raman scattering spectroscopy provides structural and chemical information on molecules based on the specific vibrational shifts that occur when changes in their

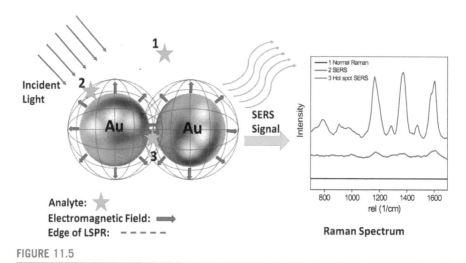

FIGURE 11.5

Illustration of surface-enhanced Raman spectroscopy *(SERS)* phenomenon. *LSPR,* Localized surface plasma resonance.

Reproduced with permission from Wei et al.[62]

polarizability are registered upon exposure to infrared (IR) light, which generates a chemical fingerprint of the sample analyzed. Spectroscopy tools, such as mid-IR, have shown important potential toward the development of simple optical methods that require minimal sample preparation and provide cost-efficient sensitivity, fast response, and quantitative analysis.[18] A chemical enhancement is introduced by the modification of the polarizability of a molecule as a result of its interaction with the substrate; this enhancement requires molecule-metal interactions to induce charge transfer between energy levels of the adsorbate molecules (also referred to as labels) and the MNPs.[19] According to research findings, SERS enhancement is possible in metals with high optical reflectivity (such as silver, gold, copper, and aluminum). Among candidate materials, gold and silver are ideal for their yielded SERS enhancement and their chemical stability. In particular for biosensors, gold is the most popular substrate, due to its non reactive nature.[20]

The LSPR-enabled SERS phenomenon on MNPs coupled with IR spectroscopic tools results in the ultrasensitive detection of molecular analytes. The analyte binding on the surface of an SERS-based sensor leads to detection events that can be registered without the need for additional sample preparation. For enhanced performance, a specificity layer can be conjugated to the metallic surface (e.g., biomolecular recognition ligands), resulting in ultrasensitive and specific detection events. Some examples of chemical enhancers and applications are presented in Table 11.2.

MNPs: effect of size on optical (colorimetric) properties

When an MNP is in contact with white light, its plasmonic surface properties will determine the color we can observe by the naked eye. The SPR can be tuned by

Table 11.2 Examples of research applications of SERS-based sensors.

Target	Biorecognition element	Metallic material	Chemical label	Reference
Whole-cell bacteria (*Escherichia coli* O157: H7)	DNA aptamers	Gold nanoparticles	4-ATP (aminothiophenol)	20
Cancer marker: prostate-specific antigen (PSA)	Antibody	Gold-coated zinc oxide nanoparticles	Toluidine blue (TB)	67
Trace microRNA	Stimuli-responsive DNA microcapsule	Silver	TB	68
Respiratory virus (H1N1 and HAdV)	Antibody	Gold-coated iron oxide particles	2-DTNB (nitrobenzoic acid)	69
Biological pH detection	4-Mercaptopyridine (4-MPy)	Gold nanorod array	Label-free	70

HAdV, *Human adenovirus;* SERS, *surface-enhanced Raman spectroscopy.*

changing the size and shape of the NPs, as well as the dielectric medium surrounding them, becoming a simple, stable, and low-cost sensing platform for colorimetric-based detection.[21] As we covered in the Metallic nanoparticles for applications in imaging and sensors section (and illustrated in Fig. 11.3), when incident light passes through the MNP, a dipolar oscillation occurs described as LSPR. In particular, the LSPR of NPs strongly depends on the interparticle distance upon assembly, progressively shifting from the monodisperse state.[22]

The most common materials used as nanoprobes for colorimetric-based assays are noble metals: silver (Ag) and gold (Au). In order to identify the detection event, different strategies might be followed to modulate a shift on the particle's SPR that might be reflected as a progressive color change (Fig. 11.6). The optical properties of NPs are quantified in terms of their absorption, scattering efficiency (Q_{abs} and Q_{sca}), and optical resonance wavelength (i_{max}). A full analysis of the mathematical model for calculating the absorption and scattering properties of gold nanoparticles (AuNPs) is covered in Jain et al.[23]

Optical density

This optical property commonly expressed as OD is a measure of the optical attenuation per centimeter of a material measured and is typically specified within a wavelength range (cm^{-1}). This metric is directly proportional to particle

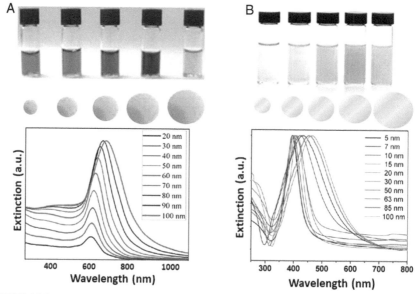

FIGURE 11.6

The tunable colorimetric response of gold nanoparticles (A) and silver nanoparticles (B) as an effect of particle size modification.

Modified with permission from Starowicz et al.[63] and Agnihotri et al.[64]

concentration and is often used instead of molar concentration due to the convenience of reporting the optical absorption in the wavelength of interest. The optical absorbance is a measure of loss or attenuation of light through a material. Generally, absorbance is the ratio of the amount of light absorbed by the NP solution to the amount of incident light directed, as expressed in Eq. 11.8 (Beer-Lambert law). Therefore absorbance correlates to optical density by the following equation:

$$OD = A/L \tag{11.5}$$

where A is the absorbance and L accounts for the thickness of the sample.

MNPs: effect of interparticle distance on optical (colorimetric) properties

Most of the predictive models of the colloidal stabilization of NPs are based on the DLVO (Derjaguin-Landau-Verwey-Overbeck) theory[24]; this proposes that the net interaction potential (G) between two adjacent particles is determined by both Van der Waals attraction (G_{vdW}) and electrostatic repulsion (G_{elec}).

$$\Delta G = \Delta G_{vdW} \text{ attraction} + \Delta G_{elec} \text{ repulsion} + \Delta G_{non-DLVO} \tag{11.6}$$

However, an extended model has been recently proposed, accounting for the non-DLVO forces (hydrophobic, steric, and solvation), as presented in the following equation:

$$\Delta G = \Delta G_{vdW} \text{ attraction} + \Delta G_{elec} \text{ repulsion} + \Delta G_{non-DLVO} \qquad (11.7)$$

Achieving colloidal stabilization (electrostatically or sterically) is a major concern when developing colorimetric assays, due to the need to guarantee a color change that is modulable and predictable; colloidal aggregation in the absence of the analyte must be avoided in order to guarantee the test reliability. Aggregation in this context is defined as the shrinking of the interparticle distance leading to formation of particle clusters.[25] It occurs when equilibrium between Van der Waals attractive and electrostatic repulsive forces is disrupted[26] and can be fine-tuned by modifying either the ionic strength or pH, or by adding organic modifiers (e.g., thiourea, cysteamine hydrochloride, and ethanol amine).[26] Different strategies have been successfully applied to achieve a sensible colorimetric response for diverse targets. Among the detection approaches, particle aggregation is extensively used due to its low cost, chemical versatility, and enhanced stability.

Example 11.2

Why are AuNPs the best candidates for biomedical applications compared with silver NPs (AgNPs)?

Answer AuNPs are not only inert under the harsh biological environment compared with AgNPs, which tend to oxidize even when exposed to air, but Au offers much needed chemical properties for functionalization. Au has six free valence electrons that are available to form covalent bonds with amines and thiols. Thus Au can form covalent bonds with biomolecules that have amino acids in their composition, such as proteins, nucleic acids, antibodies, or aptamers. By contrast, it is more difficult to form these bonds if Ag is used. Thus AuNPs can be functionalized in many ways with a variety of biomolecular ligands, making them ideal to be used in cellular imaging, biosensing, or targeted drug delivery.

MNPs: surface modifications
Surface functionalization

The primary goal of ligand functionalization is to control agglomeration and guarantee solubility of AuNPs in biological environments. Binding of ligands provides steric stabilization as an effect of repulsive forces (electrostatic interactions, steric exclusion, or hydration layer).[27] AuNPs are typically synthesized in organic solvents; however, their incorporation into most of the biomedical applications requires

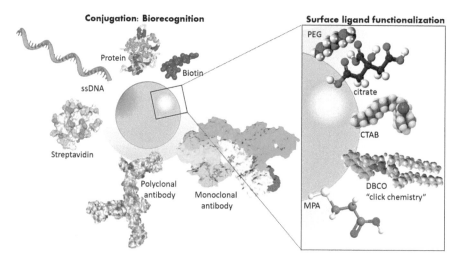

FIGURE 11.7

Chemical functionality of gold nanoparticles: surface functionalization and targeting. Illustration of a ligand shell with a gold nanoparticle core. Conjugated biomolecules for biorecognition and immobilization of capture molecules (A). Zoom-in of the grafting shell showing different ligand molecules (B): polyethylene glycol *(PEG)*, citrate, cetrimonium bromide *(CTAB)*, dibenzo cyclooctyne-amine *(DBCO)*, and 3-mercaptopropionic acid *(MPA)*. *ssDNA*, Single-stranded DNA.

3D structures obtained from https://3dprint.nih.gov [CC-BY] and http://www.3dchem.com [CC-BY].

stable suspension in aqueous solvents. Thus hydrophilic ligands are commonly used to maintain the interparticle distance, protecting from aggregation and favoring the labels' mobility through tissue.[28] Moreover, by increasing the size of the contrast agent by using either polymeric backbones, dendrimers, or core of inorganic materials, the blood half-life of the label might be incremented, with the drawback of increasing risk of bioaccumulation and toxicity[29] (see Fig. 11.7).

Biorecognition: NPs-bioconjugates

Conjugation of inorganic particles with biomolecules creates hybrid functional materials, which provide highly specific interaction with the biological system targeted. This hybridization process brings together the unique properties and functionality of the materials conjugated (Fig. 11.7). The strategies to bind biological molecules to metallic NPs can be summarized into four classes:

- Ligand-like binding (commonly chemisorption, e.g., thiol groups)
- Electrostatic adsorption of layers with an opposite net charge
- Covalent binding (taking advantage of the functional groups in biological molecules)
- Noncovalent, affinity-based receptor (e.g., avidin-biotin system)

Depending upon the synthesis route of the NPs and the nature of the colorimetric detection mechanism, the NPs might need to go through preliminary surface modification in order to incorporate the desired chemical functionalities that will act as anchor points for the recognition biomolecules.

Among some examples of polymeric materials grafted on AuNPs we can find branched polyethylenimine (B-PEI), used for covalent immobilization of carboxyl/DNA aptamers on the surface of Au-decorated polystyrene particles[7,30]; polyethylene glycol (PEG), used to fabricate highly stable AuNP-antibody conjugates[31]; and copolymers such as copoly(DMA-PMA-MAPS), for covalent binding of immunoglobulin G (IgG) antibody on AuNPs.[32]

The orientation of biomolecules plays a critical role in the overall sensor performance due to its direct impact on the probability of achieving a selective and sensible interaction with the desired analyte (Fig. 11.8). It is desired to chemically orient

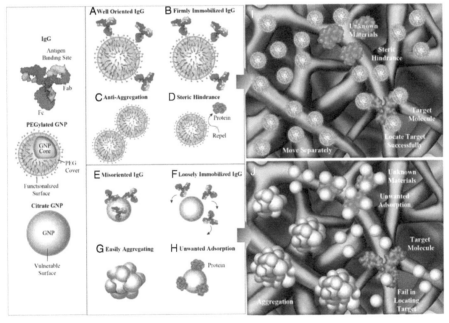

FIGURE 11.8

Schematic illustration of the comparison of PEGylated gold nanoparticles (GNPs) and citrate GNPs. (A) Functionalized surfaces orient the antibodies on the surfaces. (B) Antibodies are firmly fixed on the surfaces by EDC/NHS chemistry. (C) PEG layer prevents the particles from aggregating. (D) Steric hindrance decreases nonselective adsorption. (E, F) Antibodies are loosely and randomly attached to the surfaces by electrostatic forces. (G) GNPs easily aggregate under different conditions. (H) Undesirable adsorptions occur frequently. (I) The customized particles move separately in a nitrocellulose membrane and successfully conduct target locating and decrease the possibility of false negatives/positives. (J) Citrate GNPs aggregate easily in the membrane and increase the possibility of false negatives/positives.

Reproduced with permission from Lin et al.[31].

the biomolecule (e.g., via covalent bonding, or affinity-based receptor interactions) to expose the specific antigen-binding site and facilitate hybridization and adaptive folding as required for aptamers.[33] PEG remains the most widely used compound for MNPs' surface ligand functionalization. It provides four major advantages versus nonfunctionalized particles:

1. **Surface charge:** Ligand modification offers a high positively charged surface, which increases the electrostatic stability of colloidal materials over time, avoiding undesired aggregation in complex samples, and prevents nonspecific adsorption (Fig. 11.8D,J).
2. **Tunable chemical functionality:** The exposed functional groups serve as anchor points for oriented conjugation of biorecognition molecules via covalent bonding (illustrated in Fig. 11.8A,B).
3. **Increased hydrophilicity:** This property improves the flow in chromatographic assays (Fig. 11.6) and enhances bioavailability (better solubility) and retention time when PEGylation is used in drug delivery systems (Fig. 11.8I).
4. **Reduced complement activation:** The complement system is a pool of about 40 proteins that activates the cascade responses in both innate and adaptive immunity. Complement elicitation is associated with hypersensitivity, inflammation, and cancer, therefore biomaterials in general must avoid its activation as much as possible.[34] There is evidence of a link between hydrophilicity of PEGylated AuNPs and reduction of phagocytosis (compared with naked AuNPs). This effect is mediated by a reduced adsorption of complement proteins and their downstream complement activation, offering an indicator of enhanced biocompatibility.[35]

Colorimetric applications

This section is dedicated to analyzing the interaction of white light with metallic NPs in the visible spectrum, changes upon interparticle distance, and effect on colorimetric response.

AgNPs and AuNPs have been extensively used as colorimetric probes for detection of diverse analytes due to the tunability of their electrical, optical, and chemical properties. AuNPs are of special interest for their biocompatibility, their ease for functionalization, and the dimensional-geometric compatibility with biomolecules.[36] Typically NPs are functionalized with ligands or conjugated with biomolecules to achieve recognition of a specific analyte; following the detection event, there will be changes in surface composition and/or conformation, leading to environmental modifications that will trigger the reduction of interparticle distance. Therefore NPs need to be designed to analytically reduce their interparticle distance as a response to selective interactions; this distance requires to be less than 2.5 times their own diameter in order to provide an effective and traceable color change.[36]

Paper-based chromatographic assays

Taking advantage of the versatility of paper-based platforms, lateral flow assays (LFAs) appear as one of the most popular and widely commercialized colorimetric assays. This colorimetric setup uses nitrocellulose capillary action as a chromatographic matrix. The flow is controlled to create two interrogation zones along the porous path (test line and control line) in order to provide a detection confirmation along with internal control. MNPs play a critical role as label materials. In LFA, AuNPs provide a color signal as visualization of the analyte/cell detection when interacting with the bioconjugates. This technology has found its foundation in the fundamental understanding of the human immune response, which led to the early development of pregnancy tests and the subsequent expansion of a variety of applications in the last decade, as presented in Table 11.3.

This assay may either follow a sandwich-like or a competitive mechanism (Fig. 11.9A,D). The sandwich format is applied for large targets (e.g., whole cells), immobilizing the target within the test line by a double trapping event, therefore a positive test will be identified with two colored lines (Fig. 11.9B). On the other hand, the competitive approach will be used for small targets (e.g., small molecules, proteins, ions). In this case, antigen-carrier conjugates are previously immobilized on the test line, showing a red line anytime the target is not detected. A detection event will be identified by the absence of color due to the interaction of label particles with the target, avoiding the particles to react with the immobilized carrier, identifying a positive result with a single line (control line) (Fig. 11.9E).

As for color labels, the particles are expected to maintain their colloidal stability in order to provide the maximal optical density as a metric for color intensity, hence improving the sensitivity threshold. Polymeric materials have been extensively used to provide steric hindrance and particle stability against agglomeration. Some examples presented in Fig. 11.9 show transmission electron microscopy (TEM) microstructures of B-PEI-Au—decorated polystyrene particles (Fig. 11.9C) and PEGylated AuNPs (Fig. 11.9F) applied for enhanced detection of *Escherichia coli* O157:H7[30] and bisphenol A (BPA),[31] respectively.

Particle aggregation as a detection strategy

Non—cross-linking aggregation

This approach involves the fabrication of electrostatically stabilized AuNPs by ligand modification. The aggregation is controlled by the strong affinity between the ligand and the analyte, removing the ligands from the particle surface as a function of analyte concentration. Different examples of targets and biorecognition molecules employed are summarized in Table 11.4.

Table 11.3 History of lateral flow assays (LFAs).

Milestone	Impact	Type of analysis	Company/author	Year
First therapeutic use of antibodies	This knowledge served as the basis for immunization and other therapeutics	NA	Emil Adolf vin Behrig	1890
The pregnancy hormone hCG in urine was used for diagnosis for the first time	First immunoassay for pregnancy testing (laboratory test)	Qualitative/quantitative	Aschheim and Zondek	1927
The first pregnancy test at home was approved by the FDA in 1970	First commercial point of care (POC) developed by using antibodies	Qualitative	Drs. Vaitukaitis, Braunstein, and Ross	1970
Enzyme-linked immunosorbent assays (ELISA) were first described	This discovery led to a series of developments, as it provided a combination of high sensitivity in extremely complex sample matrices	Quantitative	Engvall and Perlmann	1971
The first automated ELISA device was launched	Fully automated device that was quickly integrated into the clinical and laboratory environment	Quantitative	Boehringer Mannheim now Roche Diagnostics	1980
First LFA was launched (Clearblue)	The platform has been re-engineered over the years to adopt new technologies. To this date, Clearblue is an over-the-counter hCG test with an integrated digital reader	Qualitative (sandwich assay)	Unilever Company	1988
Immunofluorescence-LFA to detect influenza A and B (Sofia Influenza A + B)	Commercial test that allows differential detection, fully integrated with an optical reader system	Qualitative (sandwich assay)	Quidel Corp	2009
FDA cleared multipanel drug screen test (Chemtrue)	Introduces chemically modified drugs (drug-protein conjugates) to compete for limited antibody binding sites and drugs present in the sample	Qualitative (competitive assay)	Chemtron Biotech Inc.	2011

FDA-approved diagnostic test to diagnose inflammation around a prosthetic joint (Synovasure LF test Kit)	FDA-clinical trials demonstrated that 89.5% of subjects with an infection diagnosis based on standard-of-care criteria were also identified as positive for alpha defensin by the Synovasure Lateral Flow Test Kit	Qualitative (sandwich assay)	Zimmer Biomet	2019
FDA-approved semiquantitative detection of cryptococcal meningitis (CrAg LFA)	Interpretations based upon the semiquantitative methodology can be indicative of prognosis and response to treatment	Qualitative and semiquantitative	Immuno-Mycologics, Inc.	2020
First LFA approved by FDA antibody test for COVID-19 (ASSURE SARS-CoV-2 IgG/IgM Rapid Test)	In vitro immunochromatographic test to detect and differentiate IgC/IgM antibodies against SARS-CoV-2	Qualitative (multipanel sandwich assay)	MP Biomedicals	2020

COVID-19, *Coronavirus disease 2019*; FDA, *United States Food and Drug Administration*; hCG, *human chorionic gonadotropin*; Ig, *immunoglobulin*; SARS-CoV-2, *severe acute respiratory syndrome coronavirus 2.*

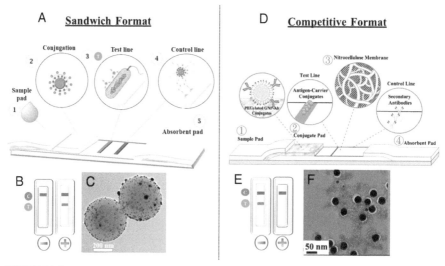

FIGURE 11.9

Colorimetric lateral flow assay (LFA). Sandwich format (A), sandwich test interpretation (B), fabrication strategy for particle stability (C), competitive format (D), competitive test interpretation (E), fabrication strategy for particle stability (F).

Reproduced with permission from Díaz-Amaya et al.[30] and Lin et al.[31]

Table 11.4 Summary of AuNP-based LSPR assays (NCL aggregation).

Target	Biorecognition element	Reference
Hg^{2+}	T-rich ssDNA	37
As^{2+}	S-layer	71
Pb^{2+}	DNAzymes	72
Thiocyanate	Cetyltrimethylammonium bromide (CTAB)	73
As^{2+}	Polymers and aptamers	74
Progesterone	CTAB	75

AuNP, *Gold nanoparticle;* LSPR, *localized surface plasma resonance;* NCL, *non–cross-linked;* ssDNA, *single-stranded DNA.*

AuNPs have the ability to interact with short single-stranded DNA (ssDNA) sequences and antibodies, showing spontaneous adsorption on their surface with high affinity, creating a stable complex that overcomes the ionic strength introduced by high concentrations of salts. As illustrated in Fig. 11.10A, a thymine-rich ssDNA-AuNP complex remains undisturbed when Hg^{2+} is not present in the samples tested (red color), whereas the selective affinity of thymine bases for Hg^{2+} (via thymine-Hg(II)-thymine base-pair stabilization) triggers the removal of ligands, leaving the particles vulnerable to agglomeration by ionic-induced compression of the

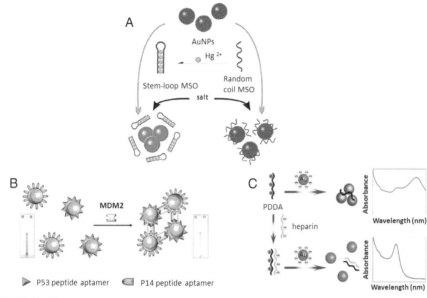

FIGURE 11.10

Aggregation-mediated colorimetric detection. DNA aptamer-based colorimetric detection of Hg^{2+} ions (A); peptide aptameric detection of MDM2 (B); label-free heparin colorimetric detection (C). *AuNPs*, Gold nanoparticles; *MSO*, mercury-specific oligonucleotide; *PDDA*, polydiallyldimethylammonium chloride.

Reproduced with permission from Ma et al.,[38] Retout et al.,[40] and Wang et al.[65]

interparticle double layer (purple color).[37] Label-free strategies have been proposed, as in the case of heparin detection via induced aggregation by electrostatic capping of AuNPs with polydiallyldimethylammonium chloride (PDDA) (purple color) (Fig. 11.10C). Due to the high affinity of PDDA to heparin, the presence of the analyte (heparin) forms a stable complex, reverting the aggregation and progressively (as a function of concentration) recovering a pink color.[38]

Interparticle cross-linking aggregation

Interparticle aggregation (cross-linking) is one common mechanism designed to bring the particles together via linkage formation among individual particles. This effect can be reached either by cross-linker molecules or direct interaction (e.g., antibody-antigen or DNA hybridization).[36] In either case, specific binding forces overcome the interparticle electrostatic repulsive forces leading to aggregation.[39] Cross-linker aggregation can be mediated by DNA aptamers and peptides,[40] as presented in Table 11.5 and illustrated in Fig. 11.10.

The extension of the fields where colorimetric-based detection technologies might be applied is under continuous expansion; while it is exciting, there is still

Table 11.5 Summary of AuNP-based LSPR assays (CL aggregation).

Target	Biorecognition element	Reference
Cu^{2+}	Protein	76
Restriction endonuclease DNA methyltransferase	DNA	77
T4 polynucleotide kinase	5′OH-DNA	78
Acetyl choline esterase	Rhodamine-B	79
ALP	Peptide	80
Aminopeptidase N	Thermo-responsive co-polymer	81
MDM2	Peptide aptamers	40

ALP, *Alkaline phosphatase; AuNP, gold nanoparticle; CL, cross-linked; LSPR, localized surface plasma resonance; MMDM2, mouse double minute 2 homolog (Independendet negative cancer marker; NCL, non–cross-linked; ssDNA, single-stranded DNA.*

a need to overcome the stability issues when incorporating complex samples (e.g., food, blood), and develop simple and scalable fabrication methods that are cost-effective at high scale, reliable, reproducible, and stable under manufacturing conditions.

MNPs as contrast agents

AuNPs are emerging nanomaterials with great potential theranostics; this last term refers to the dual action of using certain radioactive materials to identify (diagnose) and deliver a therapy. AuNPs are very effective in scattering the visible light and have a high X-ray attenuation coefficient that is in line with the ranges applied at clinical procedures such as X-ray and computerized tomography (CT), which turns the attention to their potential application as X-ray contrast agents and radio-enhancers (Fig. 11.11).

CT is based on a measure of loss of detection or beam intensity (high-energy protons) when passing through biological tissue. This phenomenon is also known as *attenuation* and is caused by the beam absorption by the body segment or due to scattering out of the detector field of view.[41] During acquisition, X-rays are projected at different angles across the tissue, and a detector collects the transmitted protons[29]; later, the numerical analysis allows to reconstruct three-dimensional (3D) images representing the attenuation of X-rays in tissues, which is highly dependent on their respective attenuation coefficient (μ). When traveling through the tissue, X-ray undergoes attenuation (loss of beam intensity) due to photoelectric absorption or scattering. During acquisition, the limited X-ray absorption by tissues plays an important factor in diminishing the contrast, resulting in low-quality images. The degree of mass attenuation obeys the Beer-Lambert law:

$$I = l_o e^{-\mu x} \tag{11.8}$$

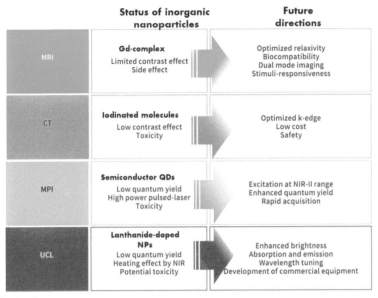

FIGURE 11.11

Current use of inorganic nanoparticles *(NPs)* as contrast agents and future outlook. Imaging modalities: magnetic resonance imaging *(MRI)*, computed tomography *(CT)*, myocardial perfusion imaging *(MPI)*, and upconversion luminescence *(UCL)*. *NIR*, Near infrared; *QDs*, quantum dots.

where l_o is the initial beam intensity, μ is the material's adsorption coefficient, and x accounts for the thickness of the tissue. Therefore the attenuation coefficient can be calculated by:

$$\mu = \frac{\rho Z^4}{A E^{3'}} \tag{11.9}$$

where ρ is the density, Z is the atomic number, A is the atomic mass, and E accounts for the X-ray energy.

Besides the modulation of X-ray properties and optimization of the optical setup, contrast agents are used to enhance the image quality. In biomedical applications, AuNPs have been studied to offer an alternative to overcome the well-known limitations of traditional iodine-based chemical platforms, such as reduced contrast in large subjects, toxicity, and fast degradation.[42] To date the use of AuNPs has demonstrated enormous opportunities in extensive biomedical imaging techniques due to their unique and tunable plasmonic properties. AuNPs can be designed to be used with multiple modalities (multimodal imaging probes). The capability to introduce these multimodal NPs in bioimaging allows gathering a variety of information with a single contrast agent, increasing their clinical value and the possibilities to be later transferred to the market. Multimodality can be incorporated by modification of

Table 11.6 Imaging approaches using AuNPs by functionalized moieties or formulation methods.

Imaging modality	Formulation method	Reference
X-ray/CT	Gold nanoparticles	82
MRI	Inclusion of heavy metals (gadolinium, iron oxide)	49
Ultrasound	Core-shell structure with high acoustic impedance	83
Fluorescent	Inclusion of fluorescent dyes	84
Nuclear imaging	Radioisotopes (^{64}Cu, ^{111}In)	84

AuNPs, *Gold nanoparticles;* CT, *computerized tomography;* MRI, *magnetic resonance imaging.*

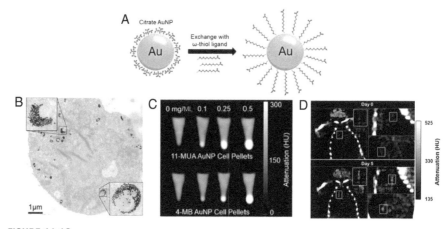

FIGURE 11.12

Gold nanoparticles *(AuNPs)* as bioimaging labels. Schematic representation of ligand exchange with citrate-capped AuNPs (A). A transmission electron microscopy image of monocyte cells after 24 hours of incubation with AuNP labels (B). Computed tomography (CT) images of cell pellets labeled with 11-mercaptoundecanoic acid *(11-MUA)* and 4-mercaptobutanol *(4-MB)* (C), and CT scans of an atherosclerotic mouse before and after injection of AuNP-labeled monocytes (D).

Reproduced with permission from Chhour et al.[44]

particles' design by introducing different moieties that provide added functionality. Different modalities have been explored such as magnetic resonance imaging (MRI), imaging fluorescence, ultrasound, photoacoustic imaging, and nuclear imaging[28] as presented in Table 11.6.

Due to their strong optoacoustic signal, AuNPs have found an opportunity as contrast agents as well, with successful performance as effective and specific uptake of targeted cells without disruption of cell viability or inflammatory function.[43,44] Fig. 11.12 presents AuNPs used to label monocytes and monitor their migration into atherosclerotic plaques noninvasively using CT. In addition, the encapsulation

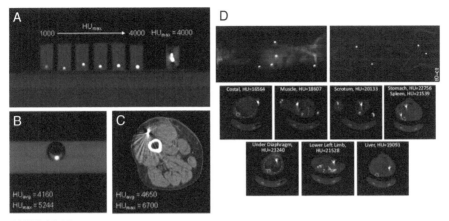

FIGURE 11.13

Computed tomography (CT) images of the gold nanoparticle (AuNP) fiducial markers. AuNP fiducial markers with increasing nanoparticle (NP) concentration (from left to right), compared with the commercial bulk gold fiducial cylinder (A). CT image of a single fiducial marker in a tissue-equivalent phantom (B). Implanted fiducial marker with the report of maximum attenuation efficiency (C). In vivo evaluation of the AuNP fiducial markers implanted in different body regions in different mice (D).

Reproduced with permission from Maiorano et al.[66]

of AuNPs in polymeric matrices is explored as well as an opportunity for the development of fiducial (reference) markers for X-ray imaging and radiosurgery (Fig. 11.13), providing a simple, low-cost fabrication that might provide increased clinical value.

Functionalization and targeting of AuNPs for cancer detection

The selective detection of cancer cells is achieved by the incorporation of an additional layer of specificity provided by biocapture molecules conjugated to the AuNPs' surface. Application of bioconjugates as tagged contrast agents allows for the effective recognition and cell internalization for further detection via bioimaging. Different strategies can be used for selective tagging, such as immune-targeted gold nanoprobes for recognition of tumor-specific antigens,[45] monitoring of tumor microenvironment by conjugation of peptide substrates that are specifically cleaved by overexpressed proteases (matrix metalloproteinase-2 [MMP-2]),[46] tracking the acidic angiogenic tumor microenvironment using ultra pH-sensitive fluorescent nanoprobes,[47] or DNA aptameric sequences highly selective for epidermal growth factor receptors (EGFRs)[48] among others. The surface chemistry of AuNPs offers a versatile platform that can be used for a variety of biomedical

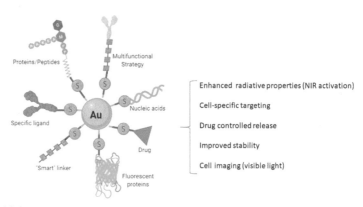

FIGURE 11.14

Gold nanoparticles as multifunctional nanomaterials. Graphical illustration of common surface modifications applied for cancer detection and imaging. *NIR,* Near infrared.

applications (Fig. 11.14). To date, the introduction of MNPs as tagged conjugates for bioimaging detection of cancer offers the ability for combined approach (detection and therapeutic), as well as the possibility to incorporate personalized therapy protocols that might result in an improved prognosis.[49]

Electrochemical biosensors

Electrochemical biosensors, a subclass of analytical chemical sensors, combine the high sensitivity of electrical transduction systems with the outstanding specificity of organic biomolecules[50]; they are widely used for diverse applications because of their rapid response, ease of use, enhanced performance, versatility, and portability.[51] Electrochemical biosensors operate by a reaction of the recognition element with the analyte while retrieving an analytical electric signal. The biochemical reactions are induced by applying an external voltage or current, using a setup known as *electrolytic cell;* the most commonly used electrocyclic cell is the three-electrode system, which consists of a working electrode (WE), where the reaction of interest will be taking place, a reference electrode (RE), and a counter electrode (CE), as represented in the schematic illustration in Fig. 11.15. After the WE comes in contact with the electrolyte, the transfer of electrons between the WE and analyte takes place. The current will pass through the CE for balance, and since the RE has a known potential, it is only adopted as a reference value to control the WE potential. In most cases a reversible redox couple (e.g., ferricyanide/ferrocyanide) is introduced to enhance the signal, along with the supporting electrolyte.

The electrochemical cell potential is calculated based on the electrode potentials. As a general convention, it is considered that the half-reaction on the left is oxidation and that on the right is reduction. Therefore we have:

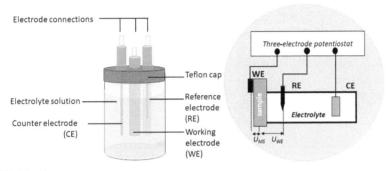

FIGURE 11.15

Schematic illustration of a three-electrode electrolytic cell.

$$E_{cell} = E_{reduction} - E_{oxidation}$$

The potential on an electrode will depend on the nature of the ions present in solution and their concentration, which is described by the Nernst equation:

$$E = E_0 - \frac{RT}{nF} \ln Q \tag{11.10}$$

where E is electrode potential, E_0 is standard potential of the electrode, R is universal gas constant (8.314 J/K·mol), F is Faradaic constant (96,485 C/mol), T is absolute temperature (K), n is charge of the ions or number of electrons, and Q is dynamic version of equilibrium constant, which is defined by:

$$Q = \frac{[C]^c [D]^d}{[A]^a [D]^d} \tag{11.11}$$

where $[C]$ and $[D]$ are molar concentrations of product C or D, or partial pressure in the atmosphere (for gas).

Example 11.3

Use the Nernst equation to calculate the electrode potential E for a copper concentration cell under standard conditions. The copper concentration at the anode is 0.1 M and the concentration of copper at the cathode is 2.0 M.

Answer The Nernst equation:

$$E_{cell} = E^0_{cell} - \frac{0.0592}{n} \log Q = E^0_{cell} - \frac{0.0592}{2} \log \frac{C_{anode}}{C_{cathode}}$$

The standard cell potential for a concentration cell—a cell where both the anode and the cathode are made of the same material—is zero. Thus E^0_{cell} is zero. Two electrodes are exchanged by the copper ion, therefore $n = 2$.

Thus substituting the copper concentrations at the anode and cathode, we calculate:

$$E_{cell} = 0\ V - \frac{0.0592}{2} \log \frac{0.1\ M}{2.0\ M}$$

Thus the theoretical cell potential for the given cell is $E_{cell} = 0.0385\ V$.

Basic concepts

AC circuit theory:

a) The electrode potential determines the kinetics of the analyte, and the current is usually reported as the rate of response. The cell potential (E) difference governs the free energy change for the chemical reaction.

b) The electrochemical cell. This concept is covered in Fig. 11.15.

c) The concentration of the analyte at the electrode surface might vary when compared with a bulk solution. It is important to account for the analyte gradient, which creates the concept of interfacial concentration.

d) The analyte might participate in other reactions. It is key to calculate the diffusion coefficient to identify mass transport due to unspecific adsorption on the electrode surface.

e) Current and potential cannot be controlled simultaneously.

Classification by biochemical recognition approach

From their biochemical recognition mechanism, electrochemical biosensors can be classified into two groups: affinity and biocatalytic sensors (Fig. 11.16). The affinity biosensors rely on the strong and selective binding of capture biomolecules, such as antibodies, aptamers, nucleic acids, or membrane receptors, with the analyte. As an effect of the detection event, the electric properties of the WEs are modified, thus such changes are tracked down and correlated with the target analyte concentration. On the other hand, biocatalytic biosensors take advantage of the high specificity of compounds that recognize the target analyte, producing electroactive species or detectable outputs that can be monitored as a function of analyte concentration. These sensors primarily use enzymes, due to the specificity provided by their complex structure and their ability to selectively detect in complex biological matrices such as blood, urine, and food.[50]

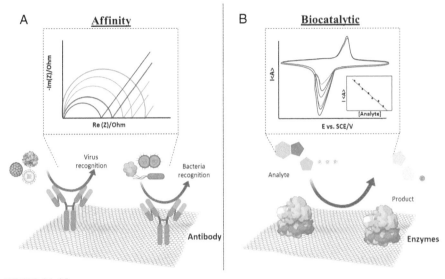

FIGURE 11.16

Illustration of an affinity electrochemical biosensor for whole-cell pathogen detection (A), and biocatalytic electrochemical biosensor for selective detection of hydrogen peroxide (B). Moreover, this illustration highlights the versatility of graphene oxide as an electrochemical platform.

The incorporation of biomolecules introduces an additional layer of complexity when considering the proper interaction between the bio-entity and the surface. Next, a summary of the most relevant affinity systems explored to date is presented.

Classification by electrochemical method

Interfacial electrochemical methods can be classified into other basic groups: voltammetric and potentiometric.[52] In this section, we will briefly discuss the most common techniques applied for characterization and evaluation of electrochemical biosensors: potential electrochemical impedance spectroscopy (PEIS), cyclic voltammetry (CV), square wave voltammetry (SWV), and differential pulse amperometry (DPA).

Potentiometric methods

Potentiometric methods measure the potential difference between the RE and the WE with a negligible current applied. The most common technique applied to biosensors is PEIS. It performs impedance measurements in potentiostat mode while applying a sinus direct current (DC) potential. This technique analyzes the

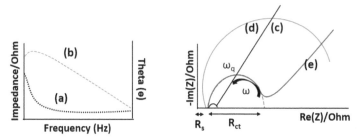

FIGURE 11.17

A Bode plot *(a)* and a phase angle plot *(b)* for a typical alternating current impedance experiment. Nyquist plots for *(c)* a system that is diffusion controlled, *(d)* a system that is controlled by electron transfer, and *(e)* an intermediate system that is controlled by electron transfer (high frequencies) and diffusion (low frequencies).

electron-transport properties of the modified electrode and provides kinetic information. It uses alternating current (AC) to induce an electrochemical reaction on the electrode surface, thus when the frequency of AC varies, the overall impedance value is registered as a function of frequency. Impedance results can be plotted either as a Bode plot or a Nyquist plot, as represented in Fig. 11.17.

Nyquist plots are of special interest because they provide kinetic details of the different process occurring at the electrode surface. When the electron transfer between the electrolyte and the electrode is very fast, it will be reflected as a quasi-linear response, as presented in Fig. 11.17 (c). Fig. 11.17 (d) shows a curve with a slower electron-transfer rate, while Fig. 11.17 (e) shows a system governed by both factors evidencing electron transfer at high frequencies, and diffusion of redox species controlling the low-frequency range.

The PEIS technique has been used for highly sensitive detection of different target analytes moving from breast cancer—specific gene sequences,[53] dengue viral RNA,[54] to whole-cell bacterial detection.[55] Fig. 11.18A shows a Nyquist plot with the impedimetric analytical response of an ink-jet printed gold platform under increasing concentrations of Hg^{2+} ions. In this case, the gold surface was functionalized with thiol-DNA aptamers for efficient detection of Hg^{2+} ions in phosphate-buffered saline (PBS) and organic solvents (dimethyl sulfoxide [DMSO]), achieving a limit of detection of 10 ppb and 5 ppb, respectively.[56]

Voltammetric biosensors

The sensors grouped within this category apply a transducing technique where the current is measured as a function of the applied potential, and the current peaks are proportional to the chemical species concentration.[52]

Fig. 11.19A shows the characteristic result for a CV experiment. The potential is scanned negatively (cathodically) following a reduction (from point A to D); at point

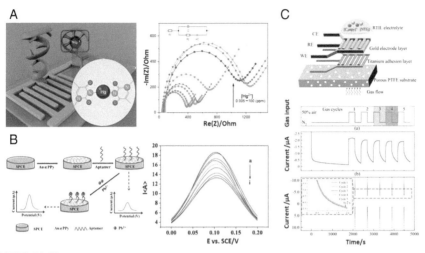

FIGURE 11.18

Examples of electrochemical techniques applied for the detection of different targets. Aptamer-based detection of Pb^{2+} using differential pulse amperometry for signal quantification (A). Ink-jet printed approach for aptamer-based detection of Hg^{2+} using potential electrochemical impedance spectroscopy (PEIS) (B). Amperometric detection for gas sensing (C). *Au@PPy*, Gold@polypyrrole; *CE*, counter electrode; *PTFE*, polytetrafluoroethylene; *RE*, reference electrode; *RTIL*, room temperature ionic liquid; *SPCE*, screen-printed carbon electrode; *WE*, working electrode.

From Diaz-Amaya et al.,[56] Wan et al.,[59] and Ding et al.[58]

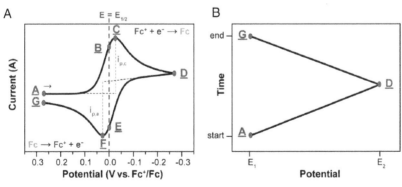

FIGURE 11.19

Voltammogram (cyclic voltammetry) of a reversible reduction of a 1-mM Fc^+ solution to Fc (A); applied potential as a function of time, showing the initial, switching, and end potentials (B).

From Elgrishi et al.[57]

C where a peak is recognized, the current is controlled by additional diffusion of the redox label from the bulk solution, producing the growth of the diffusion layer that is evidenced by a decrease in the current as the scan reaches point D. By switching potentials, the scan direction reverses to positive (anodic), going through oxidation. The separation of the peaks is a metric of the diffusion of the analyte to and from the electrode.[57]

Example 11.4

Do we need two or three electrodes to run a successful experiment via voltammetry?

Answer For voltammetric detection we need a system of three electrodes. Voltammetric techniques measure the current as a function of voltage, which is in fact the electrode potential. In the simplest electrochemical cell, minimum two electrodes are needed. However, it is not possible to extract analytical, quantitative information about the concentration of target analytes in solution if both the anode and the cathode measure the flowing current. Generally one of the two electrodes has a smaller dimension than the other, and this is the electrode that limits the flowing current. This is what we call the working electrode (WE). It is useful to use a third electrode in such a cell, called a reference electrode (RE). The RE has a known and constant electrode potential and no current passes through it. This electrode is used to control the potential of the WE by measuring the voltage between the WE and RE. The counter electrode (CE) completes the circuit and carries the current so that the potential of the RE does not change.

Among the voltammetry techniques, DPA is a popular and highly sensitive method for quantification of biologically active species at a delimitated potential range. Fig. 11.18B presents the use of gold@polypyrrole (Au@PPy) composites for detection of lead (Pb^{2+}) ions. A carbon screen-printed electrode was modified with Au@PPy composite as a strategy for signal amplification by increasing the surface area and promoting the electron transfer, followed by a DNA aptamer, electrostatically immobilized on the modified surface for selective interaction. The as-fabricated platform reached a limit of detection of 0.36 nM.[58]

Fig. 11.18C shows amperometric sensing of gas. A gold interdigitated electrode was fabricated by photolithography on a porous polytetrafluoroethylene substrate aiming to enhance the gas diffusion, and an interlayer of titanium was introduced to enhance adhesion of the integrated electrodes to the porous substrate. The detection is mediated by the reverse reaction of oxygen and superoxide, allowing to detect oxygen in 4 seconds, with a sensitivity of 0.2863 $\mu A/[\%O_2]$.[59]

Summary

Overall, promising advancements have been made on the high precision detection of a wide range of analytes. The exploration of novel materials, arrangements, and configurations, along with the exploration of nonconventional biomolecules, is expected to push the field forward. The growth of these technologies is consistently leading to the development of a next generation of healthcare and environmental monitoring devices. The current limitations for the advancement of the biosensors field are mainly encountered by the instability of biorecognition elements repeatability issues, manufacturing challenges, and regulatory restrictions. An appropriate rationale for materials selection, design, and fabrication will provide new avenues to improve the chances to translate emerging technologies into real commercial applications.

Problems

1. What is surface plasmon resonance (SPR)? What are the principles of operation of SPR biosensors?
2. What are the factors that control the plasmon resonance frequency? Why is the plasmon resonance for Ag and Au different, despite these elements having very similar lattice parameters?
3. Explain the reason for the difference in the absorption spectrum of the aggregated AuNPs compared with well-dispersed ones.
4. Spherical AuNPs with a 12.5-nm diameter were synthesized via the citrate reduction method. Calculate the mass of one perfectly spherical NP. The density of gold is 19.3 g/cm^3. How many Au atoms are in one NP?
5. A colloidal solution of NPs shows an absorbance band in the ultraviolet-visible (UV-Vis) spectrum of 0.750 at 680 nm in a 1.00-cm cuvette. For a solution of 1.00 mL and a molar extinction coefficient of 3.58×10^4 M^{-1} cm^{-1}, calculate how many NPs are in your dispersion.
6. How can AuNPs be used to detect bacteria and how can this test be made in such a way that it has specificity to different types of bacteria?
7. Name one drawback to the use of AuNPs in colorimetric biosensors.
8. What are the general principles of operation of electrochemical biosensors? List at least three types of electrochemical biosensors and compare their advantages and disadvantages. Are there any commercially available electrochemical biosensors? If yes, which one and what are its operating principles?
9. Why would someone use gold electrodes for the design of an electrochemical biosensor?
10. Which biomolecular recognition elements are based on an organism's immune response?
11. Describe a scenario when an electrochemical biosensor would be more appropriate to use for the detection of a target analyte than a colorimetric biosensor.
12. Foodborne bacteria have a zero tolerance standard imposed by regulatory agencies. However, on-site single-cell detection in real samples and without time-consuming sample preparation and laboratory expertise have yet to be realized. What would be some limiting factors to reaching single-cell bacteria detection with an electrochemical biosensor?
13. Would a competitive or a noncompetitive LFA be more appropriate to use for the detection of bacteria? For what kind of targets would you recommend the use of competitive assays? Explain.

14. You are asked to design an LFA for the detection of *E. coli* O157:H7 in milk based on antibody recognition. Describe how you envision the design of this assay, including the number of different types of recognition elements involved in successful detection.

References

1. Bhalla N, Jolly P, Formisano N, Estrela P. Introduction to biosensors. *Essays Biochem.* 2016;60(1):1—8.
2. Vo-Dinh T, Cullum B. Biosensors and biochips: advances in biological and medical diagnostics. *Fresenius J Anal Chem.* 2000;366:540—551.
3. Sposito AJ, Kurdekar A, Zhao J, Hewlett I. Application of nanotechnology in biosensors for enhancing pathogen detection. *Wiley Interdisciplinary Reviews: Nanomedicine and Nanobiotechnology.* 2018;10(5):1—22.
4. Markets and Markets. *Biosensors Market by Type (Sensor Patch and Embedded Device), Product (Wearable and Nonwearable), Technology (Electrochemical and Optical), Application (POC, Home Diagnostics, Research Lab, Food & Beverages), and Geography—Global Forecast to 2024.* Jacksonville: Close-Up Medi; 2020.
5. Peveler WJ, Yazdani M, Rotello VM. Selectivity and specificity: pros and cons in sensing. *ACS Sens.* 2016;1(11):1282—1285.
6. Campuzano S, Pedrero M, Yáñez-Sedeño P, Pingarrón JM. Antifouling (bio)materials for electrochemical (bio)sensin. *Int J Mol Sci.* 2019;20(2):1—19.
7. Diaz-Amaya S, Zhao M, Allebach JP, Chiu GTC, Stanciu LA. Ionic strength influences on biofunctional Au-decorated microparticles for enhanced performance in multiplexed colorimetric sensors. *ACS Appl Mater Interfaces.* 2020;12(29):32397—32409.
8. Lemos CN, Pereira F, Dalmolin LF, Cubayachi C, Ramos DN, V Lopez RF. Nanostructures for the engineering of cells, tissues and organs: from design to applications. In: Grumezescu AM, ed. *Nanostructures for the Engineering of Cells, Tissues and Organs.* Oxford: Elsevier; 2018:187—248.
9. Sobiech M, Lulinski P. Imprinted polymeric nanoparticles as nanodevices, biosensors and biolabels. In: Grumezescu AM, ed. *Nanostructures for the Engineering of Cells, Tissues and Organs.* Oxford: Elsevier; 2018:331—374.
10. Albanese A, Tang PS, Chan WCW. The effect of nanoparticle size, shape, and surface chemistry on biological systems. *Annu Rev Biomed Eng.* 2012;14:1—16.
11. Nejati S, Vadeghani EM, Khorshidi S, Karkhaneh A. Role of particle shape on efficient and organ-based drug delivery. *Eur Polym J.* 2020;122(109353):1—13.
12. Labille J, Brant J. Stability of nanoparticles in water. *Nanomedicine.* 2010;5(6):985—998.
13. Auffan M, Rose J, Wiesner MR, Bottero J. Chemical stability of metallic nanoparticles: a parameter controlling their potential cellular toxicity in vitro. *Evironment Pollut.* 2009; 157(4):1127—1133.
14. Chang WS, Willingham B, Slaughter LS, Dominguez-Medina S, Swanglap P, Link S. Radiative and nonradiative properties of single plasmonic nanoparticles and their assemblies. *Acco Chem Res.* 2012;45(11):1936—1945.
15. Shafiqa AR, Abdul Aziz A, Mehrdel B. Nanoparticle optical properties: size dependence of a single gold spherical nanoparticle. *J Phys Conf.* 2018;1083(012040):1—6.

16. Panchapakesan B, Book-Newell B, Sethu P, Rao M, Irudayaraj JJ. Gold nanoprobes for theranostics. *Nanomedicine*. 2008;23(1):1−7.
17. Petersen RL. Strategies using bio-layer interferometry biosensor technology for vaccine research and development. *Biosensors*. 2017;7(4):1−15.
18. Díaz-amaya S, Lin L, Deering AJ, Stanciu LA. Aptamer-based SERS biosensor for whole cell analytical detection of *E. coli* O157: H7. *Anal Chim Acta*. 2019;1081:146−156.
19. Kneipp K. Chemical contribution to SERS enhancement: an experimental study on a series of polymethine dyes on silver nanoaggregates. *J Phys Chem C*. 2016;120(37): 21076−21081.
20. Chong NS, Donthula K, Davies RA, Ilsley WH, Ooi BG. Significance of chemical enhancement effects in surface-enhanced Raman scattering (SERS) signals of aniline and aminobiphenyl isomers. *Vib Spectrosc*. 2015;81:22−31.
21. Simovski C, Tretyakov S. Plasmonics. In: *An Introduction to Metamaterials and Nanophotonics*. Cambridge: Cambridge University Press; 2020:154−173.
22. Yao Y, Tian D, Li H. Cooperative binding of bifunctionalized and click-synthesized silver nanoparticles for colorimetric Co^{2+} sensing. *ACS Appl Mater Interface*. 2010;2(3): 684−690.
23. Jain PK, Lee KS, El-Sayed IH, El-Sayed MA. Calculated absorption and scattering properties of gold nanoparticles of different size, shape, and composition: applications in biological imaging and biomedicine. *J Phys Chem B*. 2006;110(14):7238−7248.
24. Kim T, Lee CH, Joo SW, Lee K. Kinetics of gold nanoparticle aggregation: experiments and modeling. *J Colloid Interface Sci*. 2008;318(2):238−243.
25. Ganesh AN, Donders EN, Shoichet BK, Shoichet MS. Colloidal aggregation: from screening nuisance to formulation nuance. *Nano Today*. 2018;19:188−200.
26. Chegel V, Rachkov O, Lopatynskyi A, et al. Gold nanoparticles aggregation: drastic effect of cooperative functionalities in a single molecular conjugate. *J Phys Chem C*. 2012; 116(4):2683−2690.
27. Sperling RA, Parak WJ. Surface modification, functionalization and bioconjugation of colloidal inorganic nanoparticles. *Phil Trans Math Phys Eng Sci*. 2010;368:1333−1383.
28. Mahan MM, Doiron AL. Gold nanoparticles as X-ray, CT, and multimodal imaging contrast agents: formulation, targeting, and methodology. *J Nanomater*. 2018; 2018(5837276):1−15.
29. Fortin M-A, Simao T, Laprise-Pelletier M. Gold nanoparticles for imaging and cancer therapy. In: Gonçales G, Tobias G, eds. *Nanooncology*. Cham, Switzerland: Springer International Publishing; 2018:1e50.
30. Díaz-Amaya S, Zhao M, Lin L-K, et al. Inkjet printed nano patterned aptamer-based sensors for improved optical detection of foodborne pathogens. *Small*. 2019;15(1805342): 1−14.
31. Lin LK, Uzunoglu A, Stanciu LA. Aminolated and thiolated PEG-covered gold nanoparticles with high stability and antiaggregation for lateral flow detection of bisphenol A. *Small*. 2018;14(1702828):1−10.
32. Finetti C, Sola L, Pezzullo M, et al. Click chemistry immobilization of antibodies on polymer coated gold nanoparticles. *Langmuir*. 2016;32(29):7435−7441.
33. Ricci F, Lai RY, Heeger AJ, Plaxco KW, Sumner JJ. Effect of molecular crowding on the response of an electrochemical DNA sensor. *Langmuir*. 2007;23(12):6827−6834.
34. Szeto GL, Lavik EB. Materials design at the interface of nanoparticles and innate immunity. *J Mater Chem B*. 2017;4(9):1610−1618.
35. Quach QH, Kong RLX, Kah JCY. Complement activation by PEGylated gold nanoparticles. *Bioconjugate Chem*. 2018;29(4):976−981.

36. Aldewachi H, Chalati T, Woodroofe MN, Bricklebank N, Sharrack B, Gardiner P. Gold nanoparticle-based colorimetric biosensors. *Nanoscale*. 2018;10:18−33.
37. Chansuvarn W, Tuntulani T, Imyim A. Colorimetric detection of mercury (II) based on gold nanoparticles, fluorescent gold nanoclusters and other gold-based nanomaterials. *Trac Trends Anal Chem*. 2015;65:83−96.
38. Ma X, Kou X, Xu Y, Yang D, Miao P. Colorimetric sensing strategy for heparin assay based on PDDA-induced aggregation of gold nanoparticles. *Nanoscale Adv*. 2019; 1(2):486−489.
39. Weian Z, Brook MA, Li Y. Design of gold nanoparticle-based colorimetric biosensing assays. *Chembiochem*. 2008;9(15):2363−2371.
40. Retout M, Valkenier H, Triffaux E, Doneux T, Bartik K, Bruylants G. Rapid and selective detection of proteins by dual trapping using gold nanoparticles functionalized with peptide aptamers. *ACS Sens*. 2016;1(7):929−933.
41. Blanc-Durand P, Khalife M, Sgard B, et al. Attenuation correction using 3D deep convolutional neural network for brain ^{18}F-FDG PET/MR: comparison with Atlas, ZTE and CT based attenuation correction. *PloS One*. 2019;14(10):e0223141.
42. Ahn S, Jung SY, Lee SJ. Gold nanoparticle contrast agents in advanced X-ray imaging technologies. *Molecules*. 2013;18(5):5858−5890.
43. Kim J, Chhour P, Hsu J, et al. *Bioconjugate Chem*. 2017;28:1581−1597.
44. Chhour P, Naha PC, Neill SMO, et al. Labeling monocytes with gold nanoparticles to track their recruitment in atherosclerosis with computed tomography. *Biomaterials*. 2016;87:93−103.
45. Reuveni T, Motiei M, Romman Z, Popovtzer A, Popovtzer R. Targeted gold nanoparticles enable molecular CT imaging of cancer: an in vivo study. *Int J Nanomed*. 2011; 6:2859−2864.
46. Chen WH, Xu XD, Jia HZ, et al. Therapeutic nanomedicine based on dual-intelligent functionalized gold nanoparticles for cancer imaging and therapy in vivo. *Biomaterials*. 2013;34(34):8798−8807.
47. Wang Y, Zhou K, Huang G, et al. A nanoparticle-based strategy for the imaging of a broad range of tumours by nonlinear amplification of microenvironment signals. *Nat Mater*. 2014;13(2):204−212.
48. Melancon MP, Zhou M, Zhang R, et al. Selective uptake and imaging of aptamer- and antibody-conjugated hollow nanospheres targeted to epidermal growth factor receptors overexpressed in head and neck cancer. *ACS Nano*. 2014;8(5):4530−4538.
49. Guo J, Rahme K, He Y, Li L-L, Holmes JD, O'Driscoll CM. Gold nanoparticles enlighten the future of cancer theranostics. *Int J Nanomed*. 2017;12:6131−6152.
50. Ronkainen NJ, Halsallb HB, Heinemanb WR. Electrochemical biosensors. *Chemical Society Reviews*. 2010;39:1747−1760.
51. Zhang S, Wright G, Yang Y. Materials and techniques for electrochemical biosensor design and construction. *Biosens Bioelectron*. 2000;15(5−6):273−282.
52. Silva NFD, Magalhães JMCS, Freire C, Delerue-Matos C. Electrochemical biosensors for *Salmonella:* state of the art and challenges in food safety assessment. *Biosens Bioelectron*. 2018;99:667−682.
53. Xu H, Wang L, Ye H, et al. An ultrasensitive electrochemical impedance sensor for a special BRCA1 breast cancer gene sequence based on lambda exonuclease assisted target recycling amplification. *Chem Commun*. 2012;48:6390−6392.
54. Jin SA, Poudyal S, Marinero EE, Kuhn RJ, Stanciu LA. Impedimetric dengue biosensor based on functionalized graphene oxide wrapped silica particles. *Electrochim Acta*. 2016;194:422−430.

55. Braiek M, Rokbani KB, Chrouda A, et al. An electrochemical immunosensor for detection of *Staphylococcus aureus* bacteria based on immobilization of antibodies on self-assembled monolayers-functionalized gold electrode. *Biosensors*. 2012;2(4):417−426.

56. Diaz-Amaya S, Lin LK, DiNino RE, Ostos C, Stanciu LA. Inkjet printed electrochemical aptasensor for detection of Hg^{2+} in organic solvents. *Electrochim Acta*. 2019;316: 33−42.

57. Elgrishi N, Rountree KJ, McCarthy BD, Rountree ES, Eisenhart TT, Dempsey JL. A practical beginner's guide to cyclic voltammetry. *J Chem Educ*. 2018;95(2): 197−206.

58. Ding J, Liu Y, Zhang D, et al. An electrochemical aptasensor based on gold@polypyrrole composites for detection of lead ions. *Microchi Acta*. 2018;185(12):1−7.

59. Wan H, Yin H, Mason AJ. Rapid measurement of room temperature ionic liquid electrochemical gas sensor using transient double potential amperometry. *Sen Actuators B Chem*. 2017;242:658−666.

60. Bhalla N, Pan Y, Yang Z, Payam AF. Opportunities and challenges for biosensors and nanoscale analytical tools for pandemics: COVID-19. *ACS Nano*. 2020;14:7783−7807.

61. Chou C, Chen F. Plasmonic nanostructures for light trapping in organic photovoltaic devices. *Nanoscale*. 2014:8444−8458.

62. Wei H, Abtahi SMH, Vikesland PJ. Plasmonic colorimetric and SERS sensors for environmental analysis. *Environ Sci: Nano*. 2015;2:120−135.

63. Starowicz Z, Ozga P, Sheregii EM. The tuning of the plasmon resonance of the metal nanoparticles in terms of the SERS effect. *Colloid Polym Sci*. 2018;296(6):1029−1037.

64. Agnihotri S, Mukherji S, Mukherji S. Size-controlled silver nanoparticles synthesized over the range 5−100 nm using the same protocol and their antibacterial efficacy. *RSC Adv*. 2014;4:3974−3983.

65. Wang L, Zhang J, Wang X, et al. Gold nanoparticle-based optical probes for target-responsive DNA structures. *Gold Bulletin*. 2008;41(1):37−41.

66. Maiorano G, Mele E, Frassanito MC, Restini E, Athanassiou A, Pompa PP. Ultra-efficient, widely tunable gold nanoparticle-based fiducial markers for X-ray imaging. *Nanoscale*. 2016;8:18921−18927.

67. Xie L, Yang X, He Y, Yuan R, Chai Y. *Polyacrylamide gel-contained zinc finger peptide as the "lock" and zinc ions as the "key" for construction of ultrasensitive prostate-specific antigen SERS immunosensor. ACS Appl Mater Interfaces*. 2018;10(17):15200−15206.

68. Yang X, Wang S, Wang Y, He Y, Chai Y, Yuan R. Stimuli-responsive DNA microcapsules for SERS sensing of trace microRNA. *ACS Appl Mater Interfaces*. 2018;10(15): 12491−12496.

69. Wang C, Wang C, Wang X, et al. Magnetic SERS strip for sensitive and simultaneous detection of respiratory viruses. *ACS Appl Mater Interfaces*. 2019;11(21):19495−19505.

70. Bi L, Wang Y, Yang Y, et al. Highly sensitive and reproducible SERS sensor for biological pH detection based on a uniform gold nanorod array platform. *ACS Appl Mater Interfaces*. 2018;10(18):15381−15387.

71. Lakatos M, Matys S, Raff J, Pompe W. Colorimetric as (V) detection based on S-layer functionalized gold nanoparticles. *Talanta*. 2015;144:241−246.

72. Liu J, Lu Y. Accelerated color change of gold nanoparticles assembled by DNAzymes for simple and fast colorimetric Pb^{2+} detection. *J Am Chem Soc*. 2004;126(39): 12298−12305.

73. Song J, Huang PC, Wan YQ, Wu FY. Colorimetric detection of thiocyanate based on anti-aggregation of gold nanoparticles in the presence of cetyltrimethyl ammonium bromide. *Sensor Actuator B Chem*. 2016;222:790−796.

74. Wu Y, Zhan S, Wang F, He L, Zhi W, Zhou P. Cationic polymers and aptamers mediated aggregation of gold nanoparticles for the colorimetric detection of arsenic(III) in aqueous solution. *Chem Commun.* 2012;48(37):4459−4461.

75. Du G, Wang L, Zhang D, et al. Colorimetric aptasensor for progesterone detection based on surfactant-induced aggregation of gold nanoparticles. *Analytical Biochemistry.* 2016; 514:2−7.

76. Chen Y, Xianyu Y, Wu J, Yin B, Jiang X. Click chemistry-mediated nanosensors for biochemical assays. *Theranostics.* 2016;6(7):969−985.

77. Kanaras AG, Wang Z, Hussain I, Brust M. Site-specific ligation of DNA-modified gold nanoparticles activated by the restriction enzyme StyI†. *Small.* 2006;3(1):67−70.

78. Wang G, He X, Xu G, et al. Detection of T4 polynucleotide kinase activity with immobilization of TiO_2 nanotubes and amplification of Au nanoparticles. *Biosens Bioelect.* 2013;43:125−130.

79. Liu D, Chen W, Tian Y, et al. A highly sensitive gold-nanoparticle-based assay for acetylcholinesterase in cerebrospinal fluid of transgenic mice with Alzheimer's disease. *Adv Healthcare Material.* 2012;1(1):90−95.

80. Choi Y, Ho N-H, Tung C. Sensing phosphatase activity by using gold nanoparticles. *Angew Chem Int Ed.* 2007;46(5):707−709.

81. Uehara N, Fujita M, Shimizu T. Colorimetric assay of aminopeptidase N activity based on inhibition of the disassembly of gold nano-composites conjugated with a thermoresponsive copolymer. *Anal Sci.* 2009;25(2):267−273.

82. Sokolov K, Follen M, Aaron J, et al. Real-time vital optical imaging of precancer using anti-epidermal growth factor receptor antibodies conjugated to gold nanoparticles. *Canc Res.* 2003;63(9):1999−2004.

83. Li W, Chen X. Gold nanoparticles for photoacoustic imaging. *Nanomedicine.* 2015; 10(2):299−320.

84. Li H, Wang P, Deng Y, et al. Combination of active targeting, enzyme-triggered release and fluorescent dye into gold nanoclusters for endomicroscopy-guided photothermal/photodynamic therapy to pancreatic ductal adenocarcinoma. *Biomaterials.* 2017;139: 30−38.

Biodegradable materials for medical applications

Learning objectives

Most classes of biomaterials that are subsequently used for implantation are those that are permanently present in the body. Hip implants, elbow implants, or vascular stents are some examples of such permanent implants, and they have been discussed in previous chapters. However, in some cases it is desirable that an implant be removed from the body without a second surgery. After reading this chapter on biodegradable materials, the readers will:

- Gain an understanding of concepts related to the main types of biomaterials that are used in medical applications where biodegradability is designed as a desired characteristic of a particular biomedical implant, drug delivery, or tissue engineering system.
- Learn the types of polymeric, metallic, and ceramic materials that are typically designed for biodegradable applications, as well as what are their pertinent properties concerning performance requirements.
- Understand concepts related to degradation rates, degradation mechanisms, degradation stages, as well as characterization methods to aid their understanding.

Introduction

Biodegradable materials can find applications as temporary implants (e.g., stents or orthopedic screws or wires), surgical sutures, and also in tissue engineering or drug delivery. We will start by discussing the biodegradable biomaterials that are used for implantation and will continue to the other types of applications.

The performance of permanent biomedical implants (e.g., total joint replacements, fracture fixation devices, artificial heart valves) can be affected by issues related to inflammation, thrombus formation, stress shielding, and subsequent device removal surgeries. Other disadvantages of permanent biomedical devices are long-term migration of permanent implants, fracture-induced pain, as well as interferences stemming from the use of magnetic fields of the material with standard imaging equipment, and the restriction posed on the development of new tissue for children[1] Permanent implants are typically operational in the body for many years or even decades since they are meant to replace or repair function of an anatomical

Introductory Biomaterials. https://doi.org/10.1016/B978-0-12-809263-7.00012-3

body part. When a biomedical implant is meant to be only temporary, the time that the biodegradable biomaterial will be designed to be present in the body will be highly dependent on the type of tissue where the repair or replacement is located. For example, for bone, a device needs to mechanically support healing for at least 6 weeks, after which slow degradation of the biomaterial is desired to occur at a velocity that is as close as possible to that of new bone synthesis.

In other applications, such as drug delivery, the degradation needs to be designed in such a way that the drug is released in a timely manner at the right anatomical location, and this time can vary from minutes or hours to days. At the same time, biodegradable materials can be used in tissue engineering, mainly as scaffolds guiding the formation of new tissue or organs. For tissue engineering, biodegradable scaffolds are porous to allow cell migration. Next, cells work to produce new tissue, while the biodegradable scaffold slowly dissolves and gives way for the new tissue to form in its place.

Before going further, it is useful to define three different terms that are often used: biodegradable, bioresorbable, and bioabsorbable. Most biomaterials discussed here are biodegradable. That is, these are materials that are able to break down, or degrade, when either partially or fully exposed to the physiological environment, where a biological agent (such as an enzyme) is responsible for the degradation. This term should not be confused with "bioresorbable." A certain material is bioresorbable in the context of biological reactions, such as bone resorption by osteoclasts, and has the ability to grow back. Since artificial biomaterials are not living tissues, they lack the potential of regrowth and are thus biodegradable but not bioresorbable. Bioabsorbable refers to how biodegradation products of a foreign material are affected by the host metabolic processes. Often sutures are called absorbable, and they tend to be absorbed after approximately 90 to 120 days.

For temporary biomedical implants, it is a desirable direction to substitute biomedical polymers or alloys that are used in functional vascular and orthopedic implants with biodegradable biomaterials. This chapter focuses mainly on orthopedic and vascular devices because these are the most commonly encountered devices where biodegradable biomaterials are used.

Performance characteristics of biodegradable implants

The main requirements to achieve a successful biodegradable implant[1] for their most well-known applications, that is, vascular and orthopedic, are shown in Table 12.1. Although it seems that many of the reported ideal mechanical properties have been based on current clinically employed stents such as stainless steels (SSs), this may prove in the future to be different for biodegradable applications. Currently there is no vascular stent or orthopedic implant with every ideal property listed. However, depending on the environment and application, the most appropriate implant can be chosen.

Table 12.1 Desired properties for biodegradable implant materials based on application.

Properties	Biodegradable materials for vascular stents		Biodegradable materials for orthopedics	
	Constraints	Ref.	Constraints	Ref.
Cell response	Encourage endothelial cell attachment but not smooth muscle cell attachment, as that may have a negative effect on vessel patency	8,147	Encourage new bone formation through both osteoblast and osteoclast attachment and proliferation, but also avoid fibrous capsule formation	148
Mechanical integrity	>8 months	11,101,104	>6 months (based on longest healing time for neck of femur)	104
Yield strength	>200 MPa	101	>230 MPa	149
Ultimate tensile strength	>300 MPa	101	>300 MPa	101
Elongation to failure (% strain)	>15%–18% Higher ductility is ideal for higher flexibility while expanded while in arteries, but still needs enough radial force to open lesions	23,101	>15%–18%	101
Elastic modulus	Low elastic modulus to be able to bend around the human circulatory system, but still stiff enough to retain necessary hoop and radial strengths for artery support	150–152	As close to cortical bone as possible to avoid stress shielding (10–20 GPa)	20
Fatigue strength at 10^7 cycles	>256 MPa Strength must be sufficient to prevent acute recoil and negative remodeling	23,101,153	>256 MPa	23,101
Elastic recoil on expansion	<4%	101	N/A	N/A
Hydrogen evolution	<10 µL/cm²/day (though blood flow may increase this maximum tolerance)	101,107	<10 µL/cm²/day	107,109

N/A, *Not applicable.*

1. The material must be biocompatible and not produce any negative local or systemic side effects in vivo, thereby requiring immediate elimination from the body.
2. The material must fully mechanically support the tissue under reconstruction for the full period of healing. For hard tissue applications in particular, the material should have an elastic modulus as close as possible to that of natural bone to minimize or eliminate severe stress concentrations while supporting the surrounding hard tissue. A material with too high an elastic modulus denies the bone its normal stress, delaying the natural production of bone.[2] Furthermore, the compressive and tensile strengths of the material should be greater than the tissue that it supports to protect from further fracture. High fatigue strength and fracture resistance are also essential in absorbing the natural stresses and loads experienced by the skeletal system.
3. To increase the likelihood of the implant integrating well with the surrounding tissue, the surface chemistry, roughness, and topography all need to be carefully tailored to promote suitable cell adhesion and proliferation in each particular environment.
4. The degradation parameters need to be carefully controlled in order for the material to be completely resorbed in a timely manner (generally accepted as within a few years).[1] After degradation of the implant, there should not be any residual material left in the system that could produce detrimental effects to the healthy tissue. Furthermore, the degradation mode of the implant must not cause tissue damage; flakes or large chunks of metal must be avoided during degradation, which could detrimentally obstruct or block normal cellular functions.

Existing hard tissue biomaterials are employed as permanent fixation devices, which are used primarily in load-bearing applications. These could be bone plates, staples, suture anchors, screws and pins, as well as dental implants.[2] However, most fracture fixation devices are removed after healing, requiring invasive procedures and additional costs.

Late stent thrombosis and restenosis with permanent metal and drug-eluting stents are persistent problems.[3,4] Permanent stenting can also result in jailing of side branches of a blood vessel, creating problems with intervention in nearby affected sites.[4] A successful vascular stent needs to support the vessel with adequate force to avoid elastic recoil following angiography and resorb at the same speed as the vessel heals.[3] Biodegradable polymers have been investigated for transient (temporary) vascular stents. Biodegradable materials did not find many applications for vascular grafts, since these typically need to serve a more permanent function, rather than just opening up the natural blood vessel like stents. Polymeric stents require a larger strut size compared with their metal counterparts to attain the required mechanical properties.[3] However, most other proposed biodegradable metallic stents have also been reported as having thicker strut sizes than the current permanent metallic vascular stents.[5,6] Depending on the design, the larger strut size can influence the vessel hemodynamics.[7] Early biodegradable polymer stents have been shown to have poor control over the degradation rate.[8] In newer designs, degradation rates via hydrolysis (refer to Chapter 5) are better controlled by altering hydrophilic or hydrophobic end groups, adding copolymers or catalysts.[9] Methods that increase surface area, such as electrospinning polylactic acid (PLA) with polyglycolic acid (PGA), enhance the degradation rate (see Chapter 5 for details on these polymers' degradation mechanisms via hydrolysis).[10] The main drawback for biodegradable polymeric stents is a rapid loss of radial strength upon the start of hydrolysis in the body.[9]

There is thus a high level of interest in biodegradable metallic vascular stents and orthopedic fracture fixation devices, which may be a solution to the drawbacks that permanent biomedical devices have shown.[11]

We will only briefly discuss ceramic-based biodegradable biomaterials due to their brittle nature and low degradability. However, we note that hydroxyapatite and calcium phosphates have been used in composites with biodegradable metals and polymers.[12,13] Most work on transitory implants has focused on biodegradable polymers.[1,14–19] Additional drawbacks of biodegradable polymers include stress relaxation and creep, which decrease the material's ability to encourage tissue regrowth through mechanical support.[1]

The low hardness of polymer orthopedic screws has been found to be inadequate to withstand screw insertion, where threads can be damaged or the screw's head can be torn off due to the high amount of torque.[20] One other potential challenge of biodegradable polymers is that the degradation products from some of them could cause negative tissue reactions in the patient, generating inflammatory responses, swelling, or cytotoxicity.[1,19] However, there are many types of polymers, some of which with degradation products that have been found to be safe as far as tissue interactions are concerned. These include PLA and PGA, as well as poly(lactic-co-glycolic acid) (PLGA), and others, which will be discussed later in this chapter.

The degradation speed of all biodegradable implants needs to be optimized to ensure that an appropriate balance between tissue support during healing and degradation is maintained. Drogset et al. conducted an experiment on 19 human patients with anterior cruciate ligament ruptures and used poly-L-lactic acid (PLLA) interference screws to fix them in place for reconstructive healing. After 2 years, one-third of the volume of the screw was remaining in the bone tunnels, which is considered excessive.[14] In order for advances to be made in the field of biodegradable polymers, the mechanical, thermal, and viscoelastic properties are currently the focus of efforts aiming to improve outcomes.

Metallic biomaterials similarly need to overcome several obstacles concerning optimizing the degradation rate and eliminating any toxic effects that could occur in the body before becoming practical supportive materials for clinical applications.[21] The advantages of using metals as biodegradable materials include their typically high impact strengths, wear resistances, ductility, and toughness. It has been stated as an advantage of biodegradable metals that some, such as magnesium, iron, and zinc, already exist in varying quantities within the human body, which highlight them as highly biocompatible.[22] However, it is important for scientists and engineers to carefully monitor the concentration of metal ions diffusing into the surrounding tissues. The fact that an element is present in the physiological range of concentration in the body does not make that element safe when its concentration exceeds this range. Another challenge is the fact that several metal implants interfere with radiologic images of the underlying tissues, making it difficult to follow the evolution of the device within the body.[23] The current state on the viability of employing specific biodegradable polymers and metals in hard and soft tissue applications is discussed below.

Performance characteristics of biodegradable materials for tissue engineering and drug delivery applications

Biodegradable materials can also be used in tissue engineering. Tissue engineering is a field that aims to regenerate biological substitutes (such as tissues and organs) by using synthetic biomaterials, cellular components, and growth factor molecules. In most cases the biomaterials are used as scaffolds that need to be biodegradable. More often, such biomaterials are either biodegradable polymers or ceramics. In this case, they can be used as scaffolds, or matrices, which are then seeded with living cells and growth factors. Ideally the tissue-specific cells and growth factors produce new tissue, while the scaffold material degrades and is replaced with this newly created tissue. For example, research in tissue engineering is striving to grow blood vessels by seeding biopolymers (e.g., gelatin, collagen) with endothelial cells. Another example would be biodegradable calcium phosphate scaffolds being seeded with osteoblasts to create tissue engineering bone.

Scaffolds are three-dimensional porous materials, which need to meet the following criteria: (i) biodegrade at a rate that can be designed to match the rate of tissue regeneration; (ii) allow cell migration, adhesion, and extracellular matrix (ECM) deposition; (iii) allow facile transport of nutrients and regulatory factors that are necessary for cell survival, proliferation, and differentiation; and (iv) do not provoke negative tissue reactions and immune response.

These requirements suggest that, in designing scaffolding materials, one critical variable to control is porosity, along with shape and size, degradation rate, and mechanical properties. Again, as discussed throughout this book, the design of biomaterials requires a constant balancing and trade-off of materials properties to achieve the best performance once implanted. Maximum porosity is not desirable for mechanical properties, but it is desirable for increased degradation speed; in addition, appropriate pore size improves cell migration and nutrient diffusion, which guarantees tissue ingrowth and favors the degradation of the temporary support material.

Not all scaffold materials used for tissue engineering are degradable. However, here we will only discuss the biodegradable scaffolds. Polymers such as PLA, PGA, and PLGA are among the most commonly used.

Among ceramics, calcium phosphates with various compositions are of most interest as biodegradable materials. While commercially available hydroxyapatite ($Ca_{10}(PO_4)_6(OH)_2$) forms close contact with the bone tissue, it resorbs very slowly upon implantation. On the other hand, biodegradable beta-tricalcium phosphate (TCP, with the chemical formula $Ca_3(PO_4)_2$) resorbs within 6 weeks from implantation. Because of some limitations imposed by the rate of biodegradability combined with mechanical strength, sometimes biodegradable polymers or ceramics cannot be used clinically on their own. In these cases, polymer composites and ceramics are considered, where ceramic particles act as reinforcers and improve mechanical strength.

Drug delivery is the field that aims to design systems that can be used to introduce therapeutic substances (drugs) into the body to target and/or treat disease. In this case, certain drugs are either encapsulated in or linked to a biodegradable matrix material, usually a polymer. The drug-material complex is then used to deliver the drug molecules at a chosen site (e.g., a tumor site) and the system is designed to degrade the matrix material and release the drug at the location where it is needed. The matrix material in this case is used as a carrier for the drug. For this application, often the carriers could be natural biodegradable polymers such as chitosan, polysaccharides, hyaluronic acid, arginine, or dextrin. At the same time, they can also be synthetic biodegradable polymers used for drug delivery, such as dendritic polymers, PLA, poly(2-hydroxyethyl methacrylate), poly(N-isopropyl acrylamides), or poly (ethylenimines). The biodegradable polymers used for drug delivery applications need to be water soluble, nontoxic and nonimmunogenic, biodegradable, and safe for all stages of drug delivery.

Biodegradable polymers
Overview

Although the low strength and high ductility of polymers could pose certain drawbacks for their use in biomedical applications, there are, however, other characteristics that offer advantages. These include their easy biodegradation via hydrolysis, availability of polymers with low toxicity and high biocompatibility, and facile fabrication methods. Polymer chemistry has the advantage of versatility, which can help tune properties to meet usage criteria in biomedical applications. Polymers could also be incorporated into composites with other biocompatible metals or ceramics, which could mitigate some of their low strength disadvantage. Biodegradable polymers could be used in vascular, nerve, or orthopedic soft tissue (e.g., cartilage) repair.[24–30] Examples of temporary biomedical devices that feature polymers are surgical sutures, vascular stents, composite fixation screws, or bone grafts.[31–38] It is well known that surgical sutures are composed of polymeric materials that are biodegradable. Degradable polymeric sutures started to be used in surgery from the 1970s. To date, the applications of such polymers are wide. PLA and PGA, as well as

PLGA copolymer are the most well-known and widely used degradable polymers for biomedical applications.[39–45] Generally speaking, the classes of polymers that display biodegradable characteristics include poly(amides), poly(phosphazenes), poly(amino acids), poly(ortho esters), aliphatic polyesters, as well as poly(anhydrides). These polymer classes found applications in tissue reconstruction or replacement, including bone replacement materials, dental materials, or wound care dressings, as well as in surgery. Surgical and disposable applications for degradable polymers include sutures, sealants, adhesives, blades, catheters, or blood bags.

Both natural and synthetic biodegradable polymers have been incorporated into biodegradable devices and scaffolds. Natural polymers with appropriate properties include collagen, cellulose and other polysaccharides, and natural, microbial polyesters.[46–48] In addition to PLA and PGA, biodegradable polymers that are widely used include polycaprolactone (PCL) and polydioxanone (PDO).[47–57]

The criteria for bioresorbability of a polymer restrict the choice to use such biomaterials in temporary implants to those that contain chemical linkages in their structure that can be cleaved under physiological conditions, and at the same time meet the biocompatibility and mechanical properties criteria. At the same time, care needs to be taken with respect to the toxicity of the degradation products, which need to be either metabolized safely or eliminated by the body.

Biodegradable polymers are used not only in temporary implants but also for drug delivery applications. The first class of polymers that became of interest for both implantation and drug delivery was that of lactic acid derivatives. In subsequent sections we will present examples of natural and synthetic polymers and their use for applications where biodegradability is the main performance criterion. Similarly, with the requirements stated in the beginning of this chapter, for polymers to be used in transient implants or in surgery they need to meet the criteria for nontoxicity and biocompatibility, which covers the mechanical property requirements for the specific application they are intended for.

Depending on the specific application, the mechanical properties of the polymer will need to be customized. Parameters to be included in this customization include molecular weight, hydrophilicity, or the level of crystallinity, as well as surface area (higher surface area correlates with faster degradation), pore size, and level of porosity. These will also determine the rate at which resorption occurs under physiological conditions. Lower-molecular-weight polymers will resorb faster than those with higher molecular weights. Similarly, an increase in the polymer crystallinity will generally lead to a lower resorption rate; however, this rule may not always be valid. For example, although PGA has a slightly higher crystallinity than PLA, it will degrade faster due to its more hydrophilic character, which makes it more easily hydrolyzable. During the polymer degradation process, generally the long polymer chains break down into smaller fragments that can be subsequently metabolized by cells. During degradation, chemical reactions such as hydrolysis and/or enzymatic attack lead to breakage of chemical bonds. Next, either phagocytosis is activated to engulf the fragments, or, if small enough (under 50 μm), they are eliminated via metabolic routes. It is clear, therefore, that the degradation products must show low toxicity to make one specific polymer amenable for use in biomedical applications. One concern that has been brought up with some biodegradable polymers, including PLGA, is that degradation leads to changes in the local pH, affecting surrounding tissues.

Example 12.1

The typical density of PLGA is 1.3 g/cm^3. A porous scaffold made of PLGA is 3.21 cm^3 and weighs 2.47 g. Calculate the porosity of the scaffold.

Solution In order to calculate the porosity of the scaffold, we use the following equations:

$$\rho = \frac{m}{v} \tag{12.1}$$

Therefore,

$$P = \left(1 - \frac{\rho_{scaffold}}{\rho_{bulk\ material}}\right) * 100 \tag{12.2}$$

$$P = \left(1 - \frac{0.77\ \text{g/cm}^3}{1.3\ \text{g/cm}^3}\right) * 100$$

$$P = 40.77\ \%$$

The PLGA scaffold has a porosity of approximately 41%.

Any biocompatibility and toxicity testing performed on biodegradable polymers has to follow standards put forward by the American Society for Testing and Materials (ASTM), US Pharmacopeia, and the International Organization for Standardization (ISO 10993). These include both in vitro cytotoxicity studies and in vivo testing.

Polymer biodegradation

The physiological environment is high in water content and also contains a large variety of enzymes and other chemicals. Polymers can degrade in vivo through two main avenues, either enzymatic degradation or hydrolytic degradation, which can act either separately or at the same time.

Hydrolytic degradation occurs via hydrolysis, which is one of the mechanisms through which polymers can break down by interaction with water. Hydrolysis is a chemical reaction with water, and a molecule is said to be hydrolyzable if it contains chemical bonds that can be broken down by reaction with water. For this to happen, a polymer needs to contain hydrolyzable covalent bonds, which are typically part of functional groups including, but not limited to, ester, ether, anhydride, amide, carbamide (urea), or ester amide (urethane). To control hydrolysis rate, and thus the rate of its degradation of a particular polymeric material of interest for a certain application, molecular polymer design has to center around the amount and type of such hydrolyzable functional groups in its composition.

PLA is a very well-known biodegradable polymer, a polyester, and is hydrolyzable. PLA as well as PGA and PLGA can be synthesized by ring opening polymerization or direct condensation polymerization (Fig. 12.1).

Because it is a polyester, PLA is hydrolyzable. The pH value also plays a role in the degradation rate. PLA tends to degrade more slowly at neutral pH and faster in either basic or acidic conditions. Moreover, the speed of degradation for PLA is the highest at basic pH compared with acidic pH. Other polymers that can degrade in a hydrolytic manner include polyanhydrides, polyesters (e.g., PLA), polyamides, polyethers, and polycarbonate (Fig. 12.2). Fig. 12.2 shows reactions of how hydrolyzable polymers react with water. The chemical structure of each polymer will have an influence on the degradation rate via hydrolysis. The hydrolysis rate is the highest for polyanhydrides, and then polyesters such as PLA, and decreases for polyamides and further for polyethers and polycarbonates. The degradation occurs first in the amorphous regions of the polymers and later in the crystalline regions. Aliphatic polyanhydrides or esters, including PLA and PCL, degrade easily via hydrolysis.

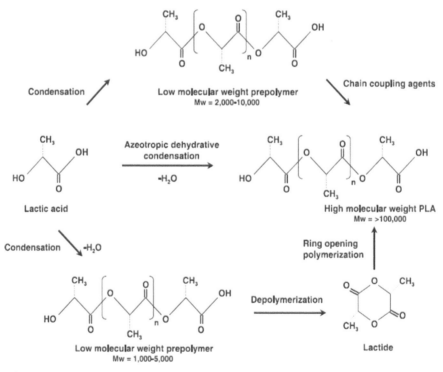

FIGURE 12.1

Mechanisms of polylactic acid *(PLA)* synthesis.

Reproduced with permission from Kenny.[143]

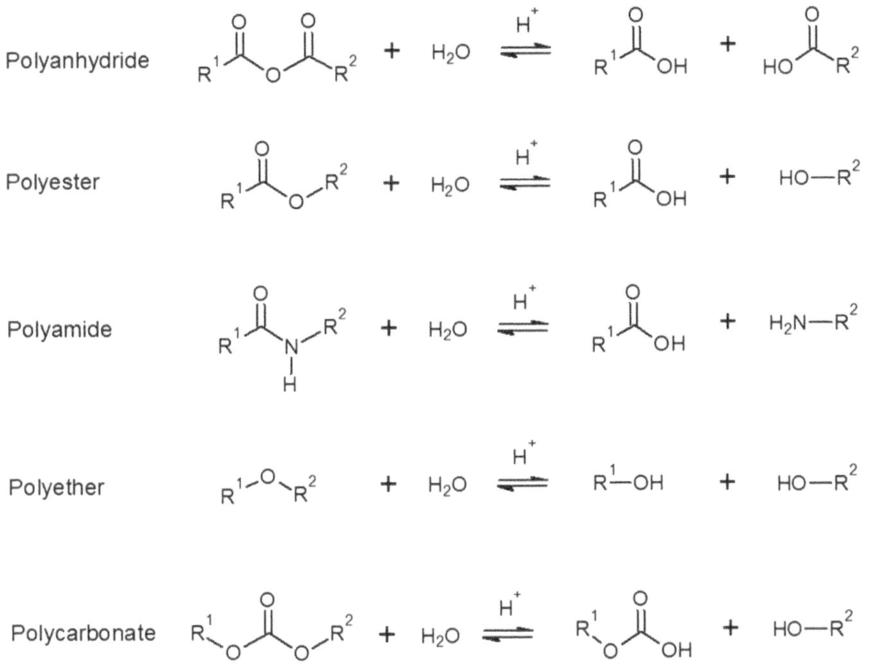

FIGURE 12.2

The hydrolytic degradation reactions of polymers, depending on their functional groups.[144]

Natural proteins as well as cellulose and starches are also in the category of readily hydrolyzable polymers.

Various factors influence the rate of hydrolytic degradation. Polymers can be designed to degrade as fast as in hours and as slow as in years, by varying the functional groups, the structure of the polymer backbone, crystallinity, morphology, pH, and molecular weight.

There are two types of degradation via hydrolysis that polymers undergo. The first one, bulk hydrolysis, is the type of erosion that polyesters and polyamides undergo. However, polymers less prone to hydrolysis, such as polyethers and polycarbonates, erode by surface erosion rather than bulk erosion. The degradation rate of bulk-eroding polymers increases as the degradation advances because of the formation of pores, which increases the hydrolyzable area, as well as leading to an increase in chain mobility and water solubility.

While we showed previously that the polymers can hydrolyze in the absence of enzymes if they have certain hydrolyzable functional groups and bonds in their chemical structure, enzymes can accelerate this process. Enzymes are natural catalysts and they can become involved in the hydrolysis of polymers in vivo by going through several steps: (i) diffusion of enzyme from the biological environment to the polymer's surface; (ii) enzyme adsorption at the polymer's surface, forming an enzyme-substrate (polymer) complex; (iii) catalysis of the hydrolysis reaction; and (iv) diffusion of the hydrolysis degradation products from the polymer substrate to the biological environment.

The factors that affect the enzyme adsorption to the polymer surface, as well as the hydrolysis rate, include the polymer's molecular weight and composition, surface area, and crystallinity. Lower crystallinity, lower molecular weight, higher surface area, and the presence of hydrolyzable groups all result in a higher hydrolysis and thus a higher degradation rate. Other factors that will play a role in the hydrolysis rate are the pH and temperature, as well as the enzymatic activity, its concentration, stability, or conformation. There is an enzyme saturation point in such a reaction, which is the enzyme concentration beyond which there will not be an increase in the hydrolysis rate. That typically occurs when the surface area of the polymer is saturated with enzyme molecules. When a polymer is introduced in the body either as an implant, a tissue engineering construct, or a drug delivery vehicle, other proteins immediately adsorb at its surface. The protein adsorption can have an effect on the rate of enzymatic hydrolysis and thus polymer degradation since adsorption sites typically available for hydrolysis-catalyzing enzymes would now become less available, limiting access to the hydrolyzable bonds. Another factor could be that the adsorbed proteins could prevent the release of the hydrolysis products back into the surrounding biological fluid, thus slowing down degradation. Examples of enzymes that can catalyze in vivo hydrolysis and degradation of polymers include, but are not limited to, lysozyme, alpha-amylase, lipase, alkaline phosphatase, proteolytic enzymes, collagenase, trypsin, cholesterol esterase, or alkaline phosphatase.

For example, cholesterol esterase has been shown to be involved in the hydrolysis of polyurethanes, while the enzymatic hydrolysis of PLLA has been shown to be catalyzed by the enzyme proteinase K. Biodegradation of polymers may also lead to activation of other enzymes that are typically involved in the immune response.[58]

Poly(lactic) and poly(glycolic) acids in medicine

PLA and PGA and their copolymers are aliphatic polyesters and are the most widely used biodegradable polymers. Their clinical applications include fracture fixation devices such as pins, screws, and nails, as well as temporary scaffolds for soft tissue repair.[59–68] The mechanism of the in vivo degradation for PLA and PGA is that of hydrolysis, with lactic and glycolic acids being the main degradation byproducts. The degradation process comes with changes in the implant structure and properties, as well as with recognizable, albeit generally safe, changes to the surrounding tissues. Implant mechanical properties that are altered include a reduction in stiffness and strength. Factors that also influence degradation rate include enzymatic activity (esterase enzymes), host-tissue interactions, functional loading, and the geometry of the implant.

Early on after its discovery, PGA was known to have the disadvantage of hydrolytic instability. This limitation was addressed by copolymerization with polyethylene terephthalate (PET). Subsequently, in 1962 PGA was the first to be used by American Cyanamid Co. for absorbable sutures. Around the same time period, Du Pont introduced PLA for similar sutures. The commercial name for surgical sutures made of PGA was introduced as Dexon in the 1970s. In 1975 Vicryl was introduced as a bioresorbable suture as well. This is a copolymer containing 90 wt% glycolic acid and 9 wt% lactic acid. Later, PLA, PGA, and their copolymers were used in biodegradable applications for drug delivery and orthopedic and dental devices.

The homopolymers of PGA and PLA have different levels of crystallinity, with PGA having an approximately 50% crystallinity and PLA 35% to 40%. This implies that PGA has better initial mechanical properties than PLA. However, because the carbonyl groups in PGA are hydrolyzable either enzymatically or hydrolytically, this polymer loses its mechanical properties within 6 weeks after implantation and is completely biodegraded at 6 months. On the other hand, PLA, despite its lower crystallinity, is a hydrophobic material, which makes it less exposed to hydrolysis and thus is slower to degrade after implantation compared with PGA. Because of its lower water absorption potential, PLA maintains its mechanical properties for months after implantation and does not fully biodegrade until 10 months. PLA has a low melting range of 180° to 220°C and a glass transition temperature of 60° to 65°C. PGA, on the other hand, has a higher melting range of 225° to 230°C and a glass transition temperature of 35° to 40°C.

To control degradation rate for biomedical applications, PGA-PLA copolymers are often designed. The degradation and mechanical properties are modulated to meet usage criteria for specific applications by variations in the ratios of each polymer that is present in the copolymer composition.

During synthesis of any biodegradable polymer, toxic solvents or reactants need to be completely eliminated for safe use in implantable devices. Biocompatibility and toxicity are some of the concerns related to the use of PLA, PGA, and their copolymers and composites with other materials. Naturally, their implantation should not lead to abnormal tissue changes or be toxic or carcinogenic in any way. The degradation products have to be nontoxic and not have a negative effect on tissue repair. These materials have been shown by multiple studies[69–74] to have reasonable biocompatibility and lack of significant toxicity. However, there are reports that indicate some problems related to reduction of cellular proliferation, inflammatory responses resulting from the accumulation of particulate debris, as well as some reported in vitro toxicity effects. Despite these, the clinical performance of these devices has been shown to be satisfactory and they continue to be used in biodegradable implants.

In orthopedic applications, PLA-PGA copolymers have been used for bone fixation, ligament and tendon repair, as well as osteochondral defect repair and as bone substitutes. For orthopedic applications, especially for fracture fixation devices, the PLA-PGA implants have often to be fiber-reinforced with carbon or fibers of the polymer itself (self-reinforcing) to meet the mechanical property demands. This is true especially when used for fracture fixation in the long bones, which withstand more functional loading than other areas. Self-reinforced PGA rods have been reported to be used for malleolar fracture fixation. These rods can be fabricated via sintering together of PGA fiber bundle in the 205° to 232°C range. Reports of inflammatory effects and aseptic sinus formation, however, have been made in some instances with use of these rods. These negative effects have been transitory in nature and did not lead to additional surgical interventions, which is why they are considered to be minor and not a deterrent to the rods' use.

One of the most challenging problems in orthopedics is that of cartilage repair. An attempt to use PGA rods for cartilage repair in a rabbit led to foreign body reaction and was therefore a failure. On the other hand, PLA was reported to be used in a rabbit knee repair, with satisfactory results. PLA-PGA implants have also been used for the delivery of growth factors to repair osteochondral defects and cartilage regeneration in rabbit knees.[75,76] Other successful applications in cartilage repair include polymer-chondrocyte grafts for cartilage regeneration in rats, as well as allogenic demineralized freeze-dried bone composite with PLA-PGA that has been used for calvarial defect repair in rabbits.[77] Overall, PLA-PGA copolymers have a reasonably good performance when used in orthopedic applications. Unwanted side effects are only occasionally reported and often are only transitory.

Scaffolds containing PGA, PLA, or PLGA can also be fabricated into porous tissue engineering scaffolds and further mixed with natural polymers such as collagen. This construct can be used for the fabrication of biodegradable scaffolds and used for cartilage, bone, skin, or ligament tissue engineering. Porous scaffolds allow for tissue-specific cells to migrate within their pores and start producing the desired tissue. After the scaffold biodegradation, only newly formed tissue, free of artificial materials, is ideally left behind and used for damage repair at the site of implantation.

The biodegradable PLGA copolymer has also found applications in drug delivery.[78] The main requirements that need to be met for this type of application include the ability to release the stored drug within the required timeline, while meeting the criteria for biodistribution and the right concentration necessary for therapeutic effectiveness. PLGA's biodistribution and pharmacokinetic profile is nonlinear and dose dependent. Drug dose and exact composition of PLGA also have an effect on the mechanism of immune clearance and phagocytosis. For example, PLGA nanoparticles have been reported to accumulate in various organs (e.g., liver, bone marrow, lymph nodes). These challenges were addressed in studies that aimed at surface modification leading to higher blood circulation half-life of PLGA. The drug release rate in PLGA drug delivery devices can be increased by an increase in hydrophilicity, a decrease in crystallinity, and a high volume-to-surface ratio. For an application requiring short-term drug release, using a PLGA with low crystallinity (amorphous) and high molecular weight would be the right choice. By contrast, for long-term drug release, higher crystallinity would be appropriate because of the lower drug release rate that this would afford. PLGA can be formulated as microspheres, nanospheres, or even macrospheres (in the millimeter range) that incorporate drugs, proteins, or peptides that can subsequently be released via different routes. Examples of applications include delivery of cisplatin to prostate cancer cells,[79] release of the anticancer drug doxorubicin,[80,81] delivery of paclitaxel for cancer targeting,[82,83] and release of the vascular endothelial growth factor (VEGF).[82,84–87]

PLA, PLLA, PGA, and PLGA polymers have also been used in various combinations as drug-loaded coatings in drug-eluting stents. The therapeutic agents that are delivered to the vascular tissue through the slow degradation of the polymers include angiopeptin, lovastatin, colchicine, as well as growth factors. PLGA has also been used for the delivery of DNA, RNA, and peptides. These are part of either antithrombogenic agents, antiangiogenesis agents, antiproliferative agents, or radiochemical groups. The polymer-drug mixtures are either sprayed on the stent surface or deposited via dipping processes. The polymer and the therapeutic agent are initially together in a mixture that includes a solvent, which is subsequently evaporated. PLGA showed good performance, allowing sustained drug release at required doses in these applications.

Biodegradable polymeric stents are also pursued, in addition to the drug-eluting stents. The stent materials include PLLA, poly(ethyl acrylate) (PEA), as well as tyrosine-derived polycarbonate.

Polycaprolactone and polydioxanone in medicine

PCL and PDO are synthetic polymers that are biodegradable and have also been used in biomedical implants either alone or as part of composite materials. Poly(ε-caprolactone) found applications in contraceptive devices (Capronor, Research Triangle Institute, Research Triangle Park, NC), wound dressings, repair of maxillofacial defects (Osteopore, Spinecraft, Elmhurst, IL), and other scaffolds for bone tissue engineering.[88–94] PCL, however, has a highly crystalline character, which results in a lower degradation rate compared with PLA, PGA, and PLGA. Cellular compatibility with respect to surface proliferation and mechanical properties has also been reported to be inferior to that of PLGA. The challenge for polymer applications in bone tissue engineering is that the scaffold needs to have similar mechanical properties to those of the natural bone. The cortical bone has a tensile strength between 122.9 and 175.2 MPa, compressive strength between 102.7 and 111.3 MPa, and a Young's modulus between 17.5 and 18.9 GPa. Meeting these requirements often mandates that biodegradable polymeric scaffolds be reinforced with an inorganic material.

PCL is an aliphatic linear polyester (Fig. 12.3) that can be synthesized through different methods, the most often cited being the catalyzed reaction (e.g., catalyst stannous octoate) and open ring polymerization between ε-caprolactone and an alcohol initiator. This reaction occurs at temperatures around 130°C. Examples of alcohol initiators include ethylene glycol or 1,3-propanediol. To enhance its slower degradation as well as improve mechanical properties for use in bone tissue

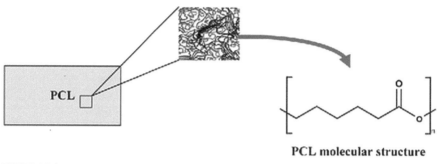

PCL molecular structure

FIGURE 12.3

Polycaprolactone *(PCL)* microstructure and molecular structure.

Reproduced with permission from Hajiali et al.[145]

engineering, PLA can be used in composites with inorganic materials such as calcium silicate or tricalcium phosphate (TCP).[95,96] Other composites that have been put forward include ternary composites of two polymers and one inorganic filler. An example is a ternary composite of poly(3-hydroxylbutyrate-co-3-hydroxylvalerate) (PHBV) with calcium silicate and PHBV-PLGA polymer blend.

Kim et al.[10] reported a ternary composite with PLGA/PCL and TCP, which demonstrated better mechanical strength than in the case of PLGA alone. This study, however, did not report on the degradation rate of the composite, which poses questions related to its application as a biodegradable scaffold. This composite also was proved to have good osteogenic properties, presumably due to the presence of TCP in the structure, combined with high surface roughness. Higher surface roughness is usually desirable for cell adhesion. However, if the degradation rate was slower, this could also explain the superior biocompatibility in the absence of fast release of degradation products, and which could negatively affect cell proliferation. Overall, ternary composites offer some opportunities for tailoring of mechanical properties, degradation rate, and bioactivity of PCL containing bone tissue engineering scaffolds.

PCL has also been reported to be tested in electrospun nanofibrous biodegradable constructs, together with PDO and silk fibroin, and tested for the fabrication of vascular grafts. Fig. 12.4 shows scanning electron microscopy (SEM) images of PCL electrospun fibers. The longer degradation time of PCL was characterized as desirable for vascular application, especially for arterial tissue engineering. PDO also displays a lower degradation rate than PLGA, and also has low inflammatory potential and good mechanical properties for vascular tissue engineering. Silk has also been reported as a candidate for vascular tissue engineering scaffolds with good cellular compatibility and ability to withstand arterial pressure. The as-fabricated tri-component polymer scaffolds were seeded with bone marrow—derived murine mast cells (BMMCs) to fabricate the vascular constructs. Mast cells are involved in angiogenesis, tissue remodeling, collagen production, as well as playing a significant role in various diseases such as cancer and those related to the cardiovascular system.

FIGURE 12.4

Scanning electron microscopy images of polycaprolactone (PCL) (A), polydioxanone (PDO) (B), and silk (C) electrospun fibers.

Reproduced with permission from Garg et al.[146]

The electrospun scaffolds promoted mast cell attachment and proliferation after coating with fibronectin. It is possible that mast cells could have a significant role in processes critical for vascular scaffold success such as angiogenesis and integration with the host tissue. Polydioxanone (PDO) is also used for bone defect repair and it has been reported as a candidate for the repair of orbital wall fracture.[97–100]

PDO, however, has been shown to induce a foreign body response when used in biomedical implants, with the formation of a fibrous capsule at the implant site, which suggests an inflammatory process being activated. PDO was used in a copolymer with PLA and fabricated into meshes (Fig. 12.5), and then tested in orbital wall

FIGURE 12.5

Images of a poly(D,L-lactide) (PDLA) copolymer shaped into an implantable mesh.

Reproduced with permission from Kontio et al.[47]

fracture repair in a rat model with positive results in terms of cellular compatibility, resorption time (slower than that of PDO alone), and mechanical properties.

Biodegradable polymers are viable candidates for applications as transient biomedical implants in both orthopedic and cardiovascular applications. Moreover, they can be used as drug delivery carriers, as well as for diagnostics and therapeutics. Their advantage stays in versatile fabrication processes, availability of chemical groups for copolymerization, and ability to be included in composites with tailored properties. Lower mechanical strength than metals poses a disadvantage, especially for orthopedic implants in load-bearing areas. However, since polymers are chemically modifiable, their degradation rates can be customized much more easily than those of metals, which increases their usage versatility tremendously.

Example 12.2
What are the key considerations when designing a degradable polymeric scaffold?

Solution

- Select a nontoxic material: nontoxic byproducts from degradation.
- Suitable ratio between the concentration of ions released by degradation and the physiological capabilities to metabolize them.
- Appropriate balance between mechanical properties (structural support) and effective surface area (porosity).
- Maintain a balance between the corrosion rate and new tissue ingrowth.

Biodegradable metals

Overview

Trace metallic elements, including chromium, manganese, iron, cobalt, nickel, copper, zinc, and molybdenum, are present in the human body.[21] Among these, magnesium and calcium are major metallic elements.[21] Research on biodegradable metals has predominantly focused on base metals of magnesium or iron alloyed with various other elements. There have also been some interesting investigations reported on zinc alloys for biodegradable implants.[101,102]

There have been no major reports of these particular elements causing cytotoxicity to the body, which heightens their viability as potential biodegradable materials. However, excess concentrations of degradation products above the daily elemental allowance of any of these metals have been found to cause other long-term symptoms. Hypermagnesemia can develop from excess magnesium, which can be linked to diseases such as arrhythmia and asystole; hydrogen gas bubbles formed from degrading magnesium have the possibility of blocking blood vessels or causing death. In the last decade, Seitz et al. illuminated that products from some Mg implants such as hydroxides, oxides, chlorides, and Mg apatites can cause a "burst release" of corrosion products and changes occur in local cell activity due to alkalization.[103] Excessive iron has been shown to generate lesions in the gastro-intestinal tract and cause abdominal pain, fatigue, liver damage, and neurotoxicity, while manganese can cause manganism, which causes negative psychiatric and mo-tor effects.[21,104] Zinc overload can also cause neurotoxicity.[104] Consequently, con-trolling the degradation rate of any of these materials is vital for their practical use in temporary implants.

Magnesium alloys

Magnesium alloys have been primarily investigated as a prospective candidate for use as clinical degradable materials. Their capacity as an orthopedic material has been heightened by the fact that these alloys are biocompatible and have elastic moduli, toughness, and compressive yield strength values comparable to cancellous bone.[105] In general, bone modulus varies with a person's age, the type of bone, and the direction of measurement, but previous studies appear to establish that for cancellous bone the Young's modulus is between 0.01 and 2 GPa and its compressive strength is between 0.2 and 80 MPa, which is comparable to several current synthe-sized Mg alloys.[106] However, even though a low modulus aids against stress shield-ing, this can increase the chance of fracture in high-load situations such as large compressive loads in the spine.[107] Thus a balance must be sought when designing for these particular applications.

The most commonly reported downside of these metallic materials is that they corrode too rapidly to allow full completion of tissue reconstruction as they lack the necessary resistance to the high content of chloride elements in the bodily envi-ronment, which has a pH of around 7.4 to 7.6.[108] Moreover, various experiments have been conducted both in vitro and in vivo that demonstrated that the degradation of most Mg alloys caused the formation of large amounts of gas, resulting in wound interface cavitation and tissue necrosis.[109,110] This is due to the aqueous environ-ment forming products of magnesium hydroxide and hydrogen gas.[111] Currently, additional procedures employing syringes are necessary soon after implantation in order to diffuse out the gas that is generated and alleviate the discomfort of the pa-tient. The reported tolerable rate of hydrogen gas release is 10 μL/cm^2/day.[107] In tailoring the degradation rate for an Mg implant, Yuen and Ip suggest that the daily exposure limit for an average 60-kg adult is about 350 to 400 mg of magnesium.[21]

Li et al. demonstrated how open pores and large surface areas can aid fluid trans-port in the body, accelerating tissue reconstruction along with increasing ingrowth of bone tissues.[112] This study showed how porous metallic materials can enhance long-term fixation between bone tissue and the implants. However, the authors used the shape memory alloy NiTi. Nasab et al. explain that while nickel is necessary for stimulating the immune system, it can be toxic when there is a high dissolution of nickel ions or wear particles in the body.[113] Thus researchers have focused more on the potential of Mg alloys instead, though interestingly enough, past alloys experimented with typically consist of various concentrations of aluminum, which has major concerns related to its toxicology.[21] AZ31 (Al: 3%, Zn: 1%, Mg: balance), AZ91 (Al: 9%, Zn: 1%, Mg: balance), WE43 (Nd: 71%, Ce: 8%, Dy: 8%, Pr: 8%, Mg: balance), and LAE442 (Ce: 51%, La: 22%, Nd: 16%, Pr: 8%, Mg: balance) were tested by Witte et al., who determined that AZ31, AZ91, and WE43 degraded faster in vivo than the alloy LAE442.[114] They also found that alloying magnesium with aluminum and zinc increased the rate of oxidation, while alloying with rare-earth elements decreased the oxidation rate.[114] However, there was still a release of subcutaneous hydrogen gas bubbles that had to be punctured with a syringe, seen in Fig. 12.6.

Hampp et al. also proposed the LAE442 alloy and another type designated as LANd442, which is similar to LAE442 except that the rare-earth mixture was replaced by the individual element neodymium.[115] They also tested their alloys on live rabbits and saw an increase in bone volume and bone porosity, but a decrease in bone density. On the other hand, the control group of rabbits that underwent surgery and received no implant showed active bone remodeling as well, indicating that the surgery method itself leads to some cell activation and also initiates the remodeling processes.[115]

One alloy with biocompatibility potential that was later dismissed by Huehnerschulte et al. was ZEK100.[116] Huehnerschulte et al. describe how this Mg alloy

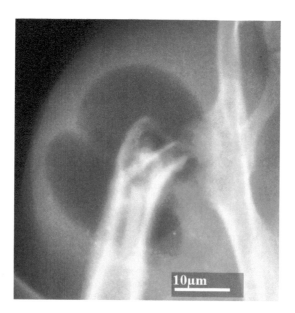

FIGURE 12.6

Postoperative radiograph of subcutaneous gas bubble after 4 weeks of degradation.

Reproduced with permission from Witte et al.[114]

induced an unfavorable osteoclastogenic resorption of old bone and a rushed reactive formation of new bone.[116] Dziuba et al. further support this negative outcome by showing how ZEK100 specifically induced pathological effects on the host tissue following complete degradation and needs to be removed from future biomedical testing.[117]

A Mg-Nd-Zn-Zr (NZK) alloy was tested by Wang et al., who concluded through hemolysis and cytotoxicity tests that the alloy exhibited good hemocompatibility and cytocompatibility in vitro.[118] They also demonstrated that the addition of alloying elements such as Nd, Zn, and Zr increases the high corrosion resistance of magnesium. Li et al. found using Mg-Ca alloys promising, which demonstrated an increase in osteoblasts and osteocytes around the Mg-1Ca alloy pin.[13] The metallic device also degraded within 90 days in vivo and showed no signs of toxicity, elevating it to a significantly high potential for clinical viability.[13]

Xia et al. experimented with an Mg-4.0Zn-0.2Ca alloy that did not exhibit any cytotoxicity on osteoblast cells and furthermore 35% to 38% of the implant had degraded after 90 days in vitro.[119] No inflammation reaction was detected and new bone was observed around the remaining implant.[119] Celarek et al. conducted another study with ZX50 (MgZnCa) and found that the degradation rate was still too fast to maintain its mechanical stability for the remodeling process.[110] Additional experiments with MgZnCa in bulk metallic glass (BMG) form, which had inherent high strength, good elasticity, and a glassy structure showed that BMGs allow alloying of higher fractions of elements compared with crystalline Mg.[110] These qualities of BMGs help make it easier to tailor the degradation rate and hydrogen evolution, but the mechanical stability of this material was lacking. Celarek et al. finally showed through their experiments that WZ21 (MgYZnCa) has a high potential for clinical bioresorbability because it possesses a high shear strength, but the material degrades at a rate slower than that desired for clinical applications.[110]

Methods for decelerating the corrosion rate of Mg alloys have been investigated for quite some time, but no tests have conclusively proved a method to combat the high amount of blood hemolysis caused by the dissolution of Mg ions. In 2007, Witte et al. highlighted how surface treatments have the best chance at slowing down the high corrosion rate of Mg and reducing hemolysis.[120] Witte and colleagues indicated that potential techniques for this include alkaline heat treating, microarc oxidation, phosphate treating, electrodeposition, and polymer coating.[120] Tests conducted by Zhang et al. have shown that alloying Mg with a high percentage of Zn helps create a passivation film, providing protection for the material against chloride ions and body fluid, thereby decreasing the rate of corrosion.[121] Properties such as residual stresses or microstructure of the material have also been found to play a significant role in the degradation kinetics of the implant. For Mg alloys in particular, a finer microstructure has been modeled to help decrease the rate of material degradation.[2] Denkena and Lucas experimentally found that deep rolling an MgCa3.0 alloy decreased its rate of corrosion by a factor of 100 due to an increase in subsurface compressive stresses.[2] Due to the preceding research, an alloy containing Mg and a small amount of Ca appears to have the highest biodegradable potential, but surface and subsurface modifications are still necessary to enhance the mechanical properties of Mg to reduce the chance of brittle fracture while further decreasing hydrogen evolution in order to bring this material up to the clinical standard for hard tissues.

Magnesium has also shown considerable promise for use as a biodegradable metal vascular scaffold material. Preliminary human trials for the treatment of critical limb ischemia with magnesium (>90%, alloyed with rare-earth elements) vascular scaffolds showed success in all 20 patients, with a 3-month clinical patency rate of 89.5% and a 12-month patency rate of 72.4%.[3,122] No toxicity was observed in any of the patients, and the limb salvage rate was 94.7% after 1 year.[3] Schranz et al. implanted a magnesium alloy vascular scaffold (AMS, Biotronik) into the aorta of a 3-week-old male patient.[123] Follow-up angiography revealed that the vessel had begun to return to its original damaged state upon degradation of the magnesium vascular scaffold, requiring implantation of a second scaffold.[123] This suggests that the absorption rate of the magnesium alloy is still too rapid to support the vessel for the duration of the healing process. In spite of the insertion of the second stent, the levels of magnesium in the patient's blood were not elevated.[123] The Biotronik magnesium alloy vascular scaffold was also tested in the Clinical Performance and Angiographic Results of Coronary Stenting with Absorbable Metal Stents trial.[124] The vascular scaffold was completely absorbed within 2 months, with radial support lost possibly within days of implantation.[124] No deaths, thrombosis, or heart attacks were reported as a result of the stenting, but these vascular scaffolds were associated with a high restenosis rate.[124] Based on this information, it is clear that magnesium vascular scaffolds are biocompatible, but for many alloys their high strength is lost too quickly due to their resorption rate being too rapid, causing vessel recoil during the healing process. As a note, strut design is especially important for these highly degrading alloys because pitting or nonuniform breaking down of the struts can significantly decrease the efficacy of the material's mechanical support, especially if the strut size is too small.[104] Thus it is ideal to produce a homogeneous microstructure in the scaffold.

Heublein et al. implanted six magnesium alloy discs, 200 μm thick and 3 mm in diameter, subcutaneously into rats to assess the corrosion effects and inflammation due to these materials in vivo.[125] The material that produced the least detrimental effects, composed of 2% aluminum and 1% rare-earth elements (Ce, Pr, Nd), was then implanted as a coronary vascular scaffold into pigs.[125] No in-stent thrombosis was detected in any of the pigs, while one pig died shortly after implantation due to unknown causes.[125] By contrast, a more recent study by Schumacher et al. showed that extruded pure magnesium stimulated an inflammatory response with nasal epithelial cells.[126] In the same year, Lock et al. looked into another avenue using MgY, AZ31, and pure Mg for ureteral stent applications and found the alloys to degrade quite differently in urine environments compared with blood arteries.[127]

Post-implantation   **4-month follow up**

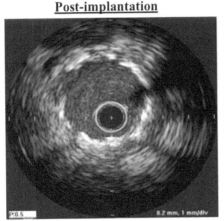 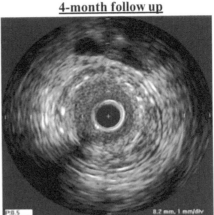

FIGURE 12.7

Bioresorbable metal scaffold after implantation at 4 months' follow-up. Intravascular ultrasound imaging after implantation and at 4 months' follow-up, demonstrating adequate expansion and apposition of the scaffolds to the vessel wall and degradation of the thick struts seen immediately after implantation over the 4 months of follow-up.

Reproduced with permission from Waksman et al.[128]

Waksman et al. reported on the implantation of the PROGRESS-AMS (Clinical Performance and Angiographic Results of Coronary Stenting with Absorbable Metal Stents) magnesium vascular scaffold from Biotronik into 63 patients for the treatment of coronary lesions.[128] Ultrasound imaging showed that the scaffolds degraded in about 4 months, with no adverse effects to the vessel wall, nor was calcification observed.[128] Fig. 12.7 shows ultrasound images of the scaffolds at implantation and after 4 months in vivo.[128] Early recoil causes restenosis. Therefore continuing to improve degradation profiles is necessary in order to make magnesium vascular scaffolds viable candidates for clinical intervention.[128]

In 2005 Zartner et al. implanted a degradable magnesium vascular scaffold, consisting of less than 10% rare-earth elements, into a 6-week-old preterm baby, born at 26 weeks' gestation.[129] In an attempt to ligate the arterial duct, the left pulmonary artery was unintentionally ligated, resulting in respiratory failure and occlusion of the left pulmonary artery.[129] The left pulmonary artery was sharply bent, preventing angiography past the stenosis, so a degradable magnesium vascular scaffold was introduced to re-establish perfusion to the lung 15 days after the initial ligation.[129] Reperfusion was established in the lung, and no toxicity due to the magnesium was observed.[129] In a follow-up paper, Zartner reported that the baby had contracted pneumonia and died 5 months later.[130] However, autopsy showed that the scaffold had completely resorbed with no traces and no necrosis had occurred.[130] The inner lumen diameter of the left pulmonary artery was 3.7 mm, indicating slight growth after implantation (from 3 mm), while the healthy right pulmonary artery had a diameter of 7 mm.[130] Additional in vivo studies are necessary in order to determine whether this intervention is viable for clinical use.

Hydrogen bubble formation appears to be even more rapid in fast-flowing vascular environments, demonstrating the material's problem of rapid degradation and very short lifetime in vivo.[109] Because of this fast corrosion rate, the radial support may be lost too early, which can result in recoil and restenosis.[128] For these reasons, more work is still needed on magnesium vascular scaffolds to determine the ideal alloy composition and scaffold geometry to ensure the degradation time is sufficient to support the vessel during healing, but rapid enough to prevent late stent thrombosis and restenosis. Drug-eluting magnesium-based absorbable scaffolds appear to show the greatest promise for this type of application. However, clinical effectiveness is still limited for bare metallic scaffolds.[131]

Iron alloys

In an almost ironic circumstance, iron, the other degradable metallic material of choice, has a degradation rate in osteogenic environments found to be excessively slow to be practical for biodegradable applications.[132] Therefore an alloying element such as manganese is usually added in an effort to increase iron's corrosion rate through creation of microgalvanic corrosion sites, while also reducing its magnetic susceptibility. Contrary to tests with magnesium alloys, finer microstructures and thereby larger volumes of high-energy grain boundaries in iron alloys appear to increase the corrosion rate.[133]

Yuen and Ip reported that the daily exposure limits of Fe and Mn for an average 60-kg adult are 2.55 mg and 0.42 mg, respectively, which is much lower than that of Mg.[21] Again, one of the main considerations for future investigations into any metal alloy is to measure the concentrations of degradation products and make sure they are within the safe physiological ranges of concentrations. Also, the metals need to degrade without eliciting the release of large flakes of metal oxide from their surface. When any type of particles or flakes are released, their size needs to be under the size that is considered the renal threshold for being able to be eliminated from the body. Typically that refers to particles that are under 6 nm along the long axis, which are able to be filtered, while particles that are larger than that will not be able to be eliminated. Experiments have also shown Fe-Mn alloys displaying antiferromagnetic behavior, which makes it more compatible with magnetic resonance imaging (MRI) scans during in vivo experiments compared with pure iron.[134] Hermawan et al. created a sintered compact of Fe-35Mn and found it to have an ultimate tensile and yield strength of 550 ± 8 MPa and 235 ± 8 MPa, respectively, which is comparable to existing mechanical properties of clinically used SS316L.[134] Moravej and Mantovani obtained similar yield strength results for their Fe35Mn alloys and explained that the generated porosity and MnO inclusions within the material helped accelerate the degradation of the alloy.[133] Moravej and Mantovani further suggested that modifying the degradation rate could be achieved by altering the microstructure and the concentration of Mn.[133] Interestingly, Liu and Zheng found that slightly decreased corrosion rates are associated with alloying small amounts of Mn, Al, or B, illuminating a nonlinear response for composition of alloying elements and their effect on corrosion rate.[135]

Schaffer et al. conducted a more recent investigation into the biocompatibility and mechanical properties of vacuum induction–melted, extruded, and cold-drawn ferrous wires.[136] More specifically, they experimented with different amounts of cold-worked 99.95% Fe, 316L SS for control, Fe-35Mn, a composite Fe-35Mn-drawn-filled-tube (DFT) 25% ZM21, Fe-DFT-25%Mg, and Fe-DFT-57%Mg.[136] Schaffer et al. concluded that the Fe35Mn alloy had the best potential for future biodegradable studies in both vascular and orthopedic applications due to its comparable toughness of 30.5 mJ/mm^3 to that of 316L SS, and fatigue strengths capable of enduring normal structural functions.[136] Furthermore, Schaffer et al. explain that in order for Fe-Mn alloys to become better suited for soft and hard tissue applications, research needs to be conducted into discovering how variations in the grain size, microstructure, and material composition affect the corrosion susceptibility of Fe-Mn alloys.[8]

Besides tests on biocompatibility and corrosion rate, there appears to be a need for further research into better tailoring the surface properties (microstructure, morphology, roughness, and surface patterning) to destabilize iron's ability to quickly form a thick oxide layer that hampers and slows chloride and media attack. Instead of focusing solely on the corrosion interaction of body media on the implant, another alternative to this issue may be to look at somehow facilitating an increased rate of cellular attack on these oxide layers, potentially increasing the pH slightly around the material to increase the degradation rate. Additionally, cell adhesion plays a major role in the degradation rate of these alloys, as attached cells can slow the attack of the surrounding media on the absorbable metal and reduce the degradation rate in vivo considerably. However, a strong physical connection between hard tissues and a biodegradable implant is vital toward promoting osteoblast proliferation and thus encouraging bone growth to replace the implant in the necessary reconstructive timeframe.[134] Future studies on ferrous biodegradable materials would benefit from further developing high surface area and porous substructures in order to increase the kinetics of corrosion in slow-moving environments such as hard tissues. At the present time, Fe-Mn alloys appear to have mechanical properties similar to those of current permanent orthopedics and thus have a high clinical potential for beneficial skeletal reconstruction, but still degrade too slowly to be used in transient applications.

Iron is an intriguing candidate for a degradable metal vascular scaffold as it is necessary in trace concentrations in the blood for proper oxygen transport. Peuster first reported on the in vitro and in vivo degradation of pure iron scaffolds in 2001.[132] Liu and Zheng also discovered that there were higher cell viabilities for endothelial cells cultured on various Fe alloys than smooth muscle cells, indicating good potential as coronary vascular scaffolds.[135] The NOR-I vascular scaffold (pure iron) was implanted into 16 rabbit aortas.[132] Despite the slow degradation rate (struts were still detected 18 months postimplantation), there were no cases of thrombosis or death.[132,137] However, the scaffolds caused considerable damage to the tunica media.[132,137]

As mentioned previously, iron-manganese alloys have the potential to bridge the gap between pure iron's slow degradation rates and pure magnesium's rapid degradation, allowing for tailoring to a more ideal degradation rate.[138] Iron-manganese alloys containing more than 29 wt% Mn are completely austenitic and antiferromagnetic, which makes them more MRI compatible than 316L SS.[134,138] An alloy composed of Fe-35Mn has been shown to have good ductility and a yield strength of up to 200 MPa.[134] Compared with pure iron, the Fe-35Mn alloy has a lower corrosion potential and a corrosion rate of almost three times that of pure iron.[134] Compared with 316L SS, Fe-35Mn alloys have shown similar toughness and greater tensile strength, suggesting that these alloys would provide adequate radial support to the vessel.[136] Preliminary in vitro studies on the endothelial attachment on Fe-35Mn showed a 200% increase in attachment compared with that on 316L.[8] More tests are currently underway to assess the blood compatibility of iron-manganese alloys, along with cytotoxicity and cell adhesion. The degradation rate, biocompatibility, and mechanical properties of iron-manganese alloys make them an intriguing candidate for degradable metal scaffolds, but more research is still necessary to determine the utility of these materials for vascular interventions.

Zinc alloys

Zinc has come to the attention of researchers in the last few years as a possible vascular scaffold material due to its antiinflammatory and antiproliferative properties.[101] Zinc may be effective in reducing the risk of atherosclerosis, as it influences apoptosis of vascular endothelial cells.[139] Ren et al. found that the administration of zinc supplements to New Zealand white rabbits on a high cholesterol diet significantly reduced the size of atherosclerotic lesions compared with rabbits on a high cholesterol diet without zinc, and reduced the levels of Fe detected in the lesions.[140] This could potentially reduce the major problem of in-stent restenosis, which is one of the most common causes of implant failure.[141] It is believed that zinc stabilizes the membrane of endothelial cells, preventing apoptosis.[101] In a more recent study, Bowen et al. found that the cross-sectional area of pure zinc scaffold struts was reduced by >35% after 6 months of implantation in rat aortas.[101] Mechanical integrity of the vascular scaffolds must be maintained for approximately 4 months in order to facilitate vessel healing, and the zinc scaffold retained about 70% of its cross-section at 4 months postimplantation.[101] While zinc has advantages in terms of antiatherogenic properties and degradation rate, it suffers from very low radial strength compared with other alloys.[101] The tensile strength of pure zinc is approximately 120 MPa, whereas a minimum of 300 MPa is desired to provide adequate support in a blood vessel.[101] Fig. 12.8 reproduced from Bowen et al. shows zinc samples after they have been degraded for up to 6 months in vivo.[101]

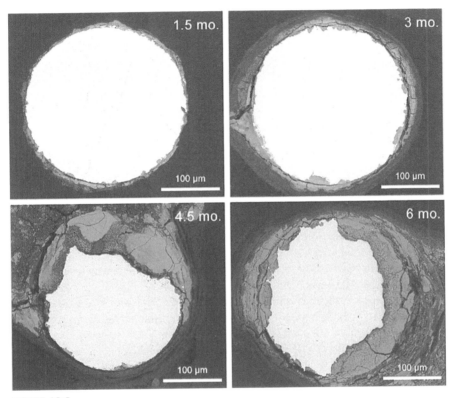

FIGURE 12.8

Backscattered electron images of zinc materials after 1.5, 3, 4.5, and 6 months of implantation, respectively.

Reproduced with permission from Bowen et al.[101]

Pure Zn and Zn-Mg appear to be the only alloys investigated for bone fixation to date, where the latter employs Mg in an attempt to increase the overall corrosion rate, while also increasing the ultimate and yield strengths to make them stronger than bone.[102,142] However, the percent elongation (1%–2%) of these materials is still too low to be able to withstand the load forces in orthopedic applications.[142] In general, pure zinc corrodes much more slowly than Mg alloys and Vojtech et al. found that by adding small concentrations of Mg (1%, 1.5%, and 3% Mg), the corrosion rates increased slightly while immersed in simulated body fluid.[142] Interestingly enough, Prosek et al. found the corrosion rate to decrease with increasing Mg concentrations in humid air, until reaching Zn-32Mn where the corrosion rate was even higher than that of pure Mg.[102]

Overall, zinc alloys are still relatively new to being added as a class of biodegradable metallic materials, so significantly more research is needed to determine their vascular scaffold and orthopedic potential.

Methods for assessing biomaterial degradation

Regardless of the type of application of a certain biodegradable material (implant, tissue engineering scaffold, or drug delivery carrier), once in the physiological environment its properties will be affected by degradation and some kind of tissue or immune system response will also be inevitable. The question that we have to ask is how should we evaluate these effects and understand what are the changes induced by biomaterial degradation.

In the initial biodegradation stages, water and other chemicals and enzymes will diffuse into the surface layers of the biomaterial and initiate either hydrolysis for polymers or corrosion for metals. As discussed previously, proteins will also immediately adhere to the implant surface, which typically results in the slowing down of the degradation. However, we will discuss here the effects of the water and other chemical attack via various mechanisms of corrosion or hydrolysis. The initial stage of the degradation will occur at the biomaterial surface and typically will lead to higher surface roughness and surface energy, resulting from chemical changes or corrosion products. At this point there is no significant molecular weight loss in the biomaterial. The second state of the biodegradation is the stage when weight loss is registered, sometimes along with an increase of porosity of the biomaterial. This is the result of chain scission due to hydrolysis of the polymers, as well as formation and dissolution of corrosion products in metals. Later stages of biodegradation result in the collapse of the structure of the biomaterial and a dramatic increase in the weight loss levels. For polymers, crystallinity is completely disrupted, and thus the mechanical properties also decrease sharply.

To observe these degradation stages, one can use different characterization techniques. Surface techniques such as contact angle measurements, atomic force microscopy (AFM), Fourier transform infrared spectroscopy (FTIR), or X-ray photoelectron spectroscopy (XPS) can be used in the first stage of biodegradation, when the surface of the biomaterial is affected. Later, bulk analysis methods can be used, with a goal of registering weight loss induced by degradation, mechanical property changes, or temperature transitions.

It is possible to perform water uptake measurements to understand the level of hydrophobicity or hydrophilicity of the material, especially polymers. Hydrophilic polymers would degrade faster than hydrophobic ones, and thus this technique can provide a clue in that respect. A very common method to monitor biodegradation is the mass loss measurement. The mass of the biomaterial is measured before and after a degradation period. Residual moisture needs to be eliminated before measurements for accuracy. The drying temperature needs to be low enough to not influence the biomaterial properties, but high enough to eliminate the water. The data presented from these experiments is the percentage of weight loss.

For polymers, one can use gel permeation chromatography (GPC) or viscosity measurements to determine the polymer's molecular weight (M_w).

Since hydrolysis of polymers first occurs in the amorphous regions, measurements of crystallinity can give a clue of how degradation proceeds. In the early stages of degradation, we can observe an increase in crystallinity, due to the hydrolysis of the amorphous regions of the polymers. This can be monitored by differential scanning calorimetry (DSC) or by wide-angle X-ray diffraction (WAXD). DSC is used by measuring the changes in the glass transition temperature (T_g), melting temperature (T_m), and crystallization temperature (T_c).

SEM and AFM can both give clues to the changes in the surface roughness and morphology, the presence of surface oxides in metals, cracks, or pores, and allow for roughness measurements. It is also possible to directly measure dimensional changes in the biomaterial during degradation. Surface chemistry information given by FTIR or XPS can give clues to what type of degradation mechanism occurs for a specific material.

Mechanical properties can also be measured to understand whether the biomaterial could be useful to support tissue healing for applications where mechanical loading is expected, such as in orthopedics. Tensile, bending, and compressive strength of such biomaterials can be measured by using ASTM standards.

In addition to the testing of biomaterials, it may be useful to characterize the degradation products. Depending on the material, degradation stage, and specific application, the degradation solutions may contain enzymes, hydrolysis or corrosion products, proteins, salts, and debris. Chemical and physical analysis methods that can be used include nuclear magnetic resonance (NMR), mass spectroscopy (MS), UV-visible spectroscopy (UV-Vis), or high-performance liquid chromatography (HPLC). Since these are complex solutions, separation of some of the components may be necessary for accurate results, and ultrafiltration membranes can be useful for this purpose.

Summary

Metals possess much higher strengths compared with their polymer counterparts, which leads to better initial mechanical support during tissue healing. They also display higher toughness and do not experience creep and stress relaxation, which are phenomena affecting polymers. Polymers, on the other hand, enjoy higher flexibility and easier pathways to customization of their degradation rate by changes in their functional groups, molecular weight, or molecular backbone structure. Specific challenges must be overcome both for polymeric and metallic alloys to become clinically viable in the future as degradable biomedical implants. For metals, stress shielding, for example, continues to be a challenge for higher modulus materials. However, due to the transient nature of these implants and their use more in non–load-bearing situations, this may not be a major concern. A more significant challenge for metals is that the release rate and concentration of metallic degradation products must be better controlled to eliminate toxic-level metal concentrations. Conversely, a larger array of biodegradable polymers than biodegradable metals and alloys have been found to be tissue compatible and not pose significant concerns regarding degradation products. Currently the two major metallic materials that appear to show the greatest potential for orthopedic applications with a base alloy of magnesium or iron include WZ21 (MgYZnCa)[110] and Fe-35Mn,[134,136] respectively. In the case of Mg alloys, the two major obstacles that still need to be overcome include the high degradation rate and the material's tendency to generate destructive hydrogen bubbles. For biodegradable polymers, PLA, PGA, and their PLGA copolymers are the most used for implantation as fixation and repair devices, tissue engineering scaffolds, or for drug delivery applications. Crystallinity, molecular weight, pH, temperature, and the amenability to hydrolysis will play important roles in modulating the degradation rate of polymers.

In the case of iron alloys, which have been considered for biodegradable implants and degrade very slowly (years), the rate of degradation must be substantially increased, either by altering the composition of biocompatible alloying elements or by modifying the microstructure and surface substructures to promote more rapid corrosion and inherent instability within the iron oxide layer. Other potential alternatives for increasing the degradation rate exist, including manufacturing these materials with higher porosity to increase the surface area, while also making sure the porosity does not detrimentally affect the mechanical properties of the material, thereby increasing surface area and corrosiveness. Increases in porosity and surface area are also strategies used to increase the degradation rate of polymers by offering greater access for the aqueous environment to the polymer and initiating hydrolysis.

Zn alloys also need further investigation into methods of increasing their corrosion rates, while also increasing their ultimate tensile, compressive, and yield strengths to support the surrounding tissue. Research into these alloys is still relatively new. Despite several existing challenges, the future of metallic materials being used for biodegradable implantation remains promising.

A more complex problem that remains to be identified and agreed upon in the literature is the specification of *ideal* degradation rates for different biodegradable materials. A degradation limit of anything less than a few years is the unofficial guideline that researchers aim for, as anything that degrades for too long is considered permanent. However, rates of biodegradability will differ for vascular and orthopedic applications, in addition to the tissue healing time variances in different locations within the body. It would be beneficial for scientists in the field to determine target degradation rates for regions of the body so that more accurate comparisons can be better drawn to all past and future experimental corrosion data. These agreed-upon goal rates would additionally help in tailoring future biodegradable materials to the variety of clinical applications currently available.

Problems

1. There are three metallic scaffolds delivered to you for preliminary testing as biodegradable materials to be applied to transient cardiovascular devices, such as stents. The table below shows the general performance of each material. Based on the data provided, which would be the best candidate? Explain your rationale.

	Sample A	Sample B	Sample C
Porosity (%)	20	75	90
Pore size (nm)	100	100	150
Corrosion time (days)[a]	350	120	250
Released ions	Fe	Mg	Al

[a] *Time required to corrode at least 90% of the implanted material.*

2. You need to determine which degradable material to use as a tissue engineering scaffold for repair of a critical-sized bone defect. You are planning to seed a polymeric scaffold with cells and then implant it into the defect.
   a) What materials would you propose? Explain.
   b) How would you evaluate biomaterial degradation in vitro?

3. Magnesium is a biomaterial candidate that is a serious contender in the design and clinical application of biodegradable vascular stents. Explain the advantages and disadvantages for the selection of this material for this type of application.

4. You are asked to choose between porous hydroxyapatite and beta-TCP as candidate materials for bone defect repair. Which one would you recommend and why?

5. You are asked to design a polymeric scaffold that will maintain its mechanical strength for at least 3 months and degrade within a year. The candidates for this application are PGA and PLA. Which polymer would you choose and why?

6. Will PGA, a polymer with a crystallinity level of about 50%, degrade faster or slower than PLA, a polymer with a crystallinity level of approximately 40%? Explain.

7. Explain the origin of the bone-like interaction that takes place on the interface of Ti-alloy implants and ions dissolved in surrounding body fluids.

8. When conducting testing on a new biodegradable copolymer, you found that the degradation rate you measured via in vitro mass loss experiments was much higher than the one observed by performing X-ray imaging after subcutaneous implantation in rats. What could be the reason behind this finding?

9. If a degradable material is considered to be used in vivo, what characteristics must the material have besides serving as a mechanical support? Explain.

10. Explain the main differences between biodegradable and bioresorbable materials.

11. What type of bioresorbable materials degrade by bulk vs. surface erosion? Explain the difference in the degradation rates between the two mechanisms.

References

1. Pietrzak WS, Sarver D, Verstynen M. Bioresorbable implants—practical considerations. *Bone.* 1996;19(1):S109–S119.
2. Denkena B, Lucas A. Biocompatible magnesium alloys as absorbable implant materials—adjusted surface and subsurface properties by machining processes. *CIRP Annals.* 2007;56(1):113–116.
3. Waksman R. Update on bioabsorbable stents: from bench to clinical. *J Interv Cardiol.* 2006;19(5):414–421.
4. Bonan RA, Asgar AW. Biodegradable stents—where are we in 2009? *Intervent Cardiol.* 2009;6:81–84.
5. Feng Q, Zhang D, Xin C, et al. Characterization and in vivo evaluation of a bio-corrodible nitrided iron stent. *J Mater Sci Mater Med.* 2013;24(3):713–724.
6. Savage P, O'Donnell BP, McHugh PE, Murphy BP, Quinn DF. Coronary stent strut size dependent stress-strain response investigated using micromechanical finite element models. *Ann Biomed Eng.* 2004;32(2):202–211.
7. Duraiswamy NC, Cesar JM, Schoephoerster RT, Moore Jr JE. Effects of stent geometry on local flow dynamics and resulting platelet deposition in an in vitro model. *Biorheology.* 2008;45(5):547–561.
8. Schaffer JE, Nauman EA, Stanciu LA. Cold drawn bioabsorbable ferrous and ferrous composite wires: an evaluation of in vitro vascular cytocompatibility. *Acta Biomater.* 2013;9(10):8574–8584.
9. Alexy RD, Levi DS. Materials and manufacturing technologies available for production of a pediatric bioabsorbable stent. *BioMed Res Int.* 2013;2013:1–11.
10. Kim K, Yu M, Zong X, et al. Control of degradation rate and hydrophilicity in electro-spun non-woven poly(D,L-lactide) nanofiber scaffolds for biomedical applications. *Biomaterials.* 2003;24(27):4977–4985.
11. Hermawan H, Dube D, Mantovani D. Developments in metallic biodegradable stents. *Acta Biomater.* 2010;6(5):1693–1697.
12. Lee S, Porter M, Wasko S, et al. Potential bone replacement materials prepared by two methods. *MRS Proceedings.* 2012;1418:mrsf11-1418-mm06-02.
13. Li Z, Gu X, Lou S, Zheng Y. The development of binary Mg-Ca alloys for use as biodegradable materials within bone. *Biomaterials.* 2008;29(10):1329–1344.
14. Drogset JO, Grontvedt T, Myhr G. Magnetic resonance imaging analysis of bioabsorbable interference screws used for fixation of bone-patellar tendon-bone autografts in endoscopic reconstruction of the anterior cruciate ligament. *Am J Sports Med.* 2006;34(7):1164–1169.
15. Drogset JO, Grontvedt T, Tegnander A. Endoscopic reconstruction of the anterior cruciate ligament using bone-patellar tendon-bone grafts fixed with bioabsorbable or metal interference screws: a prospective randomized study of the clinical outcome. *Am J Sports Med.* 2005;33(8):1160–1165.
16. Vaccaro AR, Singh K, Haid R, et al. The use of bioabsorbable implants in the spine. *Spine J.* 2003;3(3):227–237.
17. Cox S, Mukherjee DP, Ogden AL, et al. Distal tibiofibular syndesmosis fixation: a cadaveric, simulated fracture stabilization study comparing bioabsorbable and metallic single screw fixation. *J Foot Ankle Surg.* 2005;44(2):144–151.
18. Maurus PB, Kaeding CC. Bioabsorbable implant material review. *Operat Tech Sports Med.* 2004;12(3):158–160.

19. An YH, Woolf SK, Friedman RJ. Pre-clinical in vivo evaluation of orthopaedic bio-absorbable devices. *Biomaterials*. 2000;21(24):2635−2652.

20. Hofmann GO, Wagner FD. New implant designs for bioresorbable devices in orthopaedic surgery. *Clin Mater.* 1993;14(3):207−215.

21. Yuen CK, Ip WY. Theoretical risk assessment of magnesium alloys as degradable biomedical implants. *Acta Biomater*. 2010;6(5):1808−1812.

22. Zhu S, Huang N, Xu L, et al. Biocompatibility of pure iron: in vitro assessment of degradation kinetics and cytotoxicity on endothelial cells. *Mater Sci Eng C*. 2009;29(5):1589−1592.

23. Santos ICT, Rodrigues A, Figueiredo L, Rocha LA, Tavares JMRS. Mechanical properties of stent−graft materials. *Proc IME J Mater Des Appl*. 2012;226(4):330−341.

24. Niculescu M, Antoniac A, Vasile E, et al. Evaluation of biodegradability of surgical synthetic absorbable suture materials: an in vitro study. *Mater Plast*. 2016;53(4):642−645.

25. Viju S, Thilagavathi G, Gupta B. Preparation and properties of PLLA/PLCL fibres for potential use as a monofilament suture. *J Textil Inst*. 2010;101(9):835−841.

26. Hassan K, Kim SH, Park I, et al. Small diameter double layer tubular scaffolds using highly elastic PLCL copolymer for vascular tissue engineering. *Macromol Res*. 2011;19(2):122−129.

27. Kroeze RJ, Helder MN, Govaert LE, Smit TH. Biodegradable polymers in bone tissue engineering. *Materials*. 2009;2(3):833−856.

28. Tran RT, Thevenot P, Zhang Y, Gyawali D, Tang LP, Yang J. Scaffold sheet design strategy for soft tissue engineering. *Materials*. 2010;3(2):1375−1389.

29. Huh BK, Kim BH, Kim SN, et al. Surgical suture braided with a diclofenac-loaded strand of poly(lactic-co-glycolic acid) for local, sustained pain mitigation. *Mater Sci Eng C Mater Biol Appl*. 2017;79:209−215.

30. Soares JS. Bioabsorbable polymeric drug-eluting endovascular stents. A clinical review. *Minerva Biotecnol*. 2009;21(4):217−230.

31. Rajendran M, Selvaraj AS. Recent advances in the application of biopolymer scaffolds for 3D culture of cells. In: Kalarikkal N, Augustine R, Oluwafemi OS, Joshy KS, Thomas S, eds. *Nanomedicine and Tissue Engineering: State of the Art and Recent Trends*. Cambridge, MA: Apple Academic Press; 2016:307−364.

32. Ono K, Williams GR, Clem M, et al. Repair of soft tissue to bone using a biodegradable suture anchor. *Orthopedics*. 1997;20(11):1051−1055.

33. Tynan J, Ward P, Byrne G, Dowling DP. Deposition of biodegradable polycaprolactone coatings using an in-line atmospheric pressure plasma system. *Plasma Process Polym*. 2009;6:S51−S56.

34. Im JN, Kim JK, Kim HK, In CH, Lee KY, Park WH. In vitro and in vivo degradation behaviors of synthetic absorbable bicomponent monofilament suture prepared with poly(p-dioxanone) and its copolymer. *Polym Degrad Stabil*. 2007;92(4):667−674.

35. Xu XY, Liu T, Zhang K, et al. Biodegradation of poly(L-lactide-co-glycolide) tube stents in bile. *Polym Degrad Stabil*. 2008;93(4):811−817.

36. Velema J, Kaplan D. Biopolymer-based biomaterials as scaffolds for tissue engineering. *Adv Biochem Eng Biotechnol*. 2006;102:187−238.

37. Nair LS, Laurencin CT. Polymers as biomaterials for tissue engineering and controlled drug delivery. *Adv Biochem Eng Biotechnol*. 2006;102:47−90.

38. Gupta B, Revagade N, Hilborn J. Poly(lactic acid) fiber: an overview. *Prog Polym Sci*. 2007;32(4):455−482.

39. Qi ZP, Guo WL, Zheng S, et al. Enhancement of neural stem cell survival, proliferation and differentiation by IGF-1 delivery in graphene oxide-incorporated PLGA electrospun nanofibrous mats. *RSC Adv.* 2019;9(15):8315−8325.

40. Zhang J, Li JN, Jia GL, et al. Improving osteogenesis of PLGA/HA porous scaffolds based on dual delivery of BMP-2 and IGF-1 via a polydopamine coating. *RSC Adv.* 2017;7(89):56732−56742.

41. Shomali AA, Guillory RJ, Seguin D, Goldman J, Drelich JW. Effect of PLLA coating on corrosion and biocompatibility of zinc in vascular environment. *Surf Innov.* 2017;5(4): 211−220.

42. Zwingmann J, Mehlhorn AT, Sudkamp N, Stark B, Dauner M, Schmal H. Chondrogenic differentiation of human articular chondrocytes differs in biodegradable PGA/PLA scaffolds. *Tissue Eng.* 2007;13(9):2335−2343.

43. Lee JW, Hyun H, Cho JS, Kim MS, Khang G, Lee HB. The effect of crystallinity on release behavior of albumin from MPEG-polyesters vehicles. *Tissue Eng Regen Med.* 2007;4(3):399−405.

44. Bodde EWH, Habraken WJEM, Mikos AG, Spauwen PHM, Jansen JA. Effect of polymer molecular weight on the bone biological activity of biodegradable polymer/calcium phosphate cement composites. *Tissue Eng.* 2009;15(10):3183−3191.

45. Cuddihy MJ, Kotov NA. Poly(lactic-co-glycolic acid) bone scaffolds with inverted colloidal crystal geometry. *Tissue Eng.* 2008;14(10):1639−1649.

46. Liu L, Yan YN, Xiong Z, Zhang RJ, Wang XH. A novel poly (lactic-co-glycolic acid)-collagen hybrid scaffold fabricated via multi-nozzle low-temperature deposition. In: *Virtual and Rapid Manufacturing.* London: Taylor & Francis; 2008:57.

47. Kontio R, Ruuttila P, Lindroos L, et al. Biodegradable polydioxanone and poly(L/D)lactide implants: an experimental study on peri-implant tissue response. *Int J Oral Maxillofac Surg.* 2005;34(7):766−776.

48. Makela EA, Vainionpaa S, Vihtonen K, et al. The effect of a penetrating biodegradable implant on the growth plate—an experimental-study on growing-rabbits with special reference to polydioxanone. *Clin Orthop Relat Res.* 1989;(241):300−308.

49. Salgado CL, Sanchez EMS, Zavaglia CAC, Almeida AB, Granja PL. Injectable biodegradable polycaprolactone-sebacic acid gels for bone tissue engineering. *Tissue Eng.* 2012;18(1−2):137−146.

50. Li LH, Narayanan TSNS, Kim YK, et al. Deposition of microarc oxidation-polycaprolactone duplex coating to improve the corrosion resistance of magnesium for biodegradable implants. *Thin Solid Films.* 2014;562:561−567.

51. Wise SG, Byrom MJ, Waterhouse A, Bannon PG, Ng MKC, Weiss AS. A multilayered synthetic human elastin/polycaprolactone hybrid vascular graft with tailored mechanical properties. *Acta Biomater.* 2011;7(1):295−303.

52. Yang WJ, Fu J, Wang DX, et al. Study on chitosan/polycaprolactone blending vascular scaffolds by electrospinning. *J Biomed Nanotechnol.* 2010;6(3):254−259.

53. McClure MJ, Sell SA, Ayres CE, Simpson DG, Bowlin GL. Electrospinning-aligned and random polydioxanone-polycaprolactone-silk fibroin-blended scaffolds: geometry for a vascular matrix. *Biomed Mater.* 2009;4(5):055010.

54. Tillman BW, Yazdani SK, Lee SJ, Geary RL, Atala A, Yoo JJ. The in vivo stability of electrospun polycaprolactone-collagen scaffolds in vascular reconstruction. *Biomaterials.* 2009;30(4):583−588.

55. McClure MJ, Sell SA, Bowlin GL. Multi-layered polycaprolactone-elastin-collagen small diameter conduits for vascular tissue engineering. *Proceedings of the Asme Summer Bioengineering Conference*. 2008;Pts a and B, 2009:77−78.

56. Williamson MR, Woollard KJ, Griffiths HR, Coombes AGA. Gravity spun polycaprolactone fibers for applications in vascular tissue engineering: proliferation and function of human vascular endothelial cells. *Tissue Eng*. 2006;12(1):45−51.

57. Venugopal J, Zhang YZ, Ramakrishna S. Fabrication of modified and functionalized polycaprolactone nanofibre scaffolds for vascular tissue engineering. *Nanotechnology*. 2005;16(10):2138−2142.

58. Azevedo HS, Gama FM, Reis RL. In vitro assessment of the enzymatic degradation of several starch based biomaterials. *Biomacromolecules*. 2003;4(6):1703−1712.

59. Xiong Z, Yan YN, Zhang RJ, Sun L. Fabrication of porous poly(L-lactic acid) scaffolds for bone tissue engineering via precise extrusion. *Scripta Mater*. 2001;45(7):773−779.

60. Sui G, Yang XP, Mei F, et al. Poly-L-lactic acid/hydroxyapatite hybrid membrane for bone tissue regeneration. *J Biomed Mater Res*. 2007;82a(2):445−454.

61. Lee JB, Park HN, Ko WK, et al. Poly(L-lactic acid)/hydroxyapatite nanocylinders as nanofibrous structure for bone tissue engineering scaffolds. *J Biomed Nanotechnol*. 2013;9(3):424−429.

62. Weng WZ, Song SJ, Cao LH, et al. A comparative study of bioartificial bone tissue poly-L-lactic acid/polycaprolactone and PLLA scaffolds applied in bone regeneration. *J Nanomater*. 2014;2014:1−7.

63. Lee JB, Kim JE, Balikov DA, et al. Poly(L-lactic acid)/gelatin fibrous scaffold loaded with simvastatin/beta-cyclodextrin-modified hydroxyapatite inclusion complex for bone tissue regeneration. *Macromol Biosci*. 2016;16(7):1027−1038.

64. Nakano Y, Hori Y, Sato A, et al. Evaluation of a poly(L-lactic acid) stent for sutureless vascular anastomosis. *Ann Vasc Surg*. 2009;23(2):231−238.

65. Petas A, Talja M, Tammela TLJ, Taari K, Valimaa T, Tormala P. The biodegradable self-reinforced poly-DL-lactic acid spiral stent compared with a suprapubic catheter in the treatment of post-operative urinary retention after visual laser ablation of the prostate. *Br J Urol*. 1997;80(3):439−443.

66. Lan ZY, Lyu YN, Xiao JM, et al. Novel biodegradable drug-eluting stent composed of poly-L-lactic acid and amorphous calcium phosphate nanoparticles demonstrates improved structural and functional performance for coronary artery disease. *J Biomed Nanotechnol*. 2014;10(7):1194−1204.

67. Wu YZ, Shen L, Wang QB, et al. Comparison of acute recoil between bioabsorbable poly-L-lactic acid XINSORB stent and metallic stent in porcine model. *J Biomed Biotechnol*. 2012;2012:413956.

68. Isotalo T, Talja M, Valimaa T, Tormala P, Tammela TLJ. A bioabsorbable self-expandable, self-reinforced poly-L-lactic acid urethral stent for recurrent urethral strictures: long-term results. *J Endourol*. 2002;16(10):759−762.

69. Tschon M, Finai M, Giavaresi G, et al. In vitro and in vivo behaviour of biodegradable and injectable PLA/PGA copolymers related to different matrices. *Int J Artif Organs*. 2007;30(4):352−362.

70. Kwun IS, Cho YE, Kim HJ, et al. Beneficial effect of biodegradable polyglycolide (PGA) or polylactide (PLA) polymers on extracellular matrix mineralization on osteoblastic MC3T3-E1 cells. *Faseb J*. 2006;20(5):A1062.

71. Rimondini L, Nicoli-Aldini N, Fini M, Guzzardella G, Tschon M, Giardino R. In vivo experimental study on bone regeneration in critical bone defects using an injectable

biodegradable PLA/PGA copolymer. *Oral Surg Oral Med Oral Pathol Oral Radiol Endod.* 2005;99(2):148–154.

72. Athanasiou KA, Agrawal CM, Barber FA, Burkhart SS. Orthopaedic applications for PLA-PGA biodegradable polymers. *Arthroscopy.* 1998;14(7):726–737.
73. Agrawal CM, Athanasiou KA, Heckman JD. Biodegradable PLA-PGA polymers for tissue engineering in orthopaedics. *Porous Mater Tissue Eng.* 1997;250:115–128.
74. Hollinger JO. Preliminary report on the osteogenic potential of a biodegradable copolymer of polyactide (PLA) and polyglycolide (PGA). *J Biomed Mater Res.* 1983; 17(1):71–82.
75. Oshima Y, Harwood FL, Coutts RD, Kubo T, Amiel D. Variation of mesenchymal cells in polylactic acid scaffold in an osteochondral repair model. *Tissue Eng C Methods.* 2009;15(4):595–604.
76. Yan H, Yu CL. Repair of full-thickness cartilage defects with cells of different origin in a rabbit model. *Arthroscopy.* 2007;23(2):178–187.
77. Chu CR, Coutts RD, Yoshioka M, Harwood FL, Monosov AZ, Amiel D. Articular-cartilage repair using allogeneic perichondrocyte-seeded biodegradable porous polylactic acid (PLA)—a tissue-engineering study. *J Biomed Mater Res.* 1995;29(9):1147–1154.
78. Shen X, Li TT, Xie XX, et al. PLGA-based drug delivery systems for remotely triggered cancer therapeutic and diagnostic applications. *Front Bioeng Biotechnol.* 2020;8:1–19.
79. Dhar S, Kolishetti N, Lippard SJ, Farokhzad OC. Targeted delivery of a cisplatin pro-drug for safer and more effective prostate cancer therapy in vivo. *Proc Natl Acad Sci U S A.* 2011;108(5):1850–1855.
80. Qian F, Stowe N, Liu EH, Saidel GM, Gao JM. Quantification of in vivo doxorubicin transport from PLGA millirods in thermoablated rat livers. *J Contr Release.* 2003; 91(1–2):157–166.
81. Betancourt T, Brown B, Brannon-Peppas L. Doxorubicin-loaded PLGA nanoparticles by nanoprecipitation: preparation, characterization and in vitro evaluation. *Nanomedicine.* 2007;2(2):219–232.
82. Fonseca C, Simoes S, Gaspar R. Paclitaxel-loaded PLGA nanoparticles: preparation, physicochemical characterization and in vitro anti-tumoral activity. *J Contr Release.* 2002;83(2):273–286.
83. Wang J, Ng CW, Win KY, et al. Release of paclitaxel from polylactide-co-glycolide (PLGA) microparticles and discs under irradiation. *J Microencapsul.* 2003;20(3): 317–327.
84. Mezu-Ndubuisi OJ, Wang YY, Schoephoerster J, et al. Intravitreal delivery of VEGF-A(165)-loaded PLGA microparticles reduces retinal vaso-obliteration in an in vivo mouse model of retinopathy of prematurity. *Curr Eye Res.* 2019;44(3):275–286.
85. Mezu-Ndubuisi OJ, Wang YY, Schoephoerster J, Gong SQ. Intravitreal delivery of VEGF-A165-loaded PLGA microparticles reduces retinal vaso-obliteration in an in vivo mouse model of retinopathy of prematurity. *Invest Ophthalmol Vis Sci.* 2018;59(9):275–286.
86. Zhang Q, Hubenak J, Iyyanki T, et al. Engineering vascularized soft tissue flaps in an animal model using human adipose derived stem cells and VEGF plus PLGA/PEG microspheres on a collagen-chitosan scaffold with a flow-through vascular pedicle. *Biomaterials.* 2015;73:198–213.
87. Wang JH, Xu YM, Fu Q, et al. Continued sustained release of VEGF by PLGA nanospheres modified BAMG stent for the anterior urethral reconstruction of rabbit. *Asian Pac J Trop Med.* 2013;6(6):481–484.

88. Nejaddehbashi F, Hashemitabar M, Bayati V, Abbaspour M, Moghimipour E, Orazizadeh M. Application of polycaprolactone, chitosan, and collagen composite as a nanofibrous mat loaded with silver sulfadiazine and growth factors for wound dressing. *Artif Organs*. 2019;43(4):413−423.

89. Thomas R, Soumya KR, Mathew J, Radhakrishnan EK. Electrospun polycaprolactone membrane incorporated with biosynthesized silver nanoparticles as effective wound dressing material. *Appl Biochem Biotechnol*. 2015;176(8):2213−2224.

90. Boonkong W, Petsom A, Thongchul N. Rapidly stopping hemorrhage by enhancing blood clotting at an opened wound using chitosan/polylactic acid/polycaprolactone wound dressing device. *J Mater Sci Mater Med*. 2013;24(6):1581−1593.

91. Williams JM, Adewunmi A, Schek RM, et al. Tissue engineered bone using polycaprolactone scaffolds made by selective laser sintering. *Nanoscale Mater Sci Bio Med*. 2005; 845:119−128.

92. Song HH, Yoo MK, Moon HS, Choi YJ, Lee HC, Cho CS. A novel polycaprolactone/hydroxyapatite scaffold for bone tissue engineering. *Advanced Biomaterials VII*. 2007: 342−343: p. 265−268.

93. Rentsch C, Rentsch B, Breier A, et al. Long-bone critical-size defects treated with tissue-engineered polycaprolactone-co-lactide scaffolds: a pilot study on rats. *J Biomed Mater Res*. 2010;95a(3):964−972.

94. Ahn SH, Lee HJ, Kim GH. Polycaprolactone scaffolds fabricated with an advanced electrohydrodynamic direct-printing method for bone tissue regeneration. *Biomacromolecules*. 2011;12(12):4256−4263.

95. Bao CLM, Chong MSK, Qin L, et al. Effects of tricalcium phosphate in polycaprolactone scaffold for mesenchymal stem cell-based bone tissue engineering. *Mater Technol*. 2019;34(6):361−367.

96. Kumar A, Zhang YR, Terracciano A, et al. Load-bearing biodegradable polycaprolactone-poly (lactic-co-glycolic acid)-beta tri-calcium phosphate scaffolds for bone tissue regeneration. *Polym Adv Technol*. 2019;30(5):1189−1197.

97. Otsuka NY, Mah JY, Orr FW, Martin RF. Biodegradation of polydioxanone in bone tissue—effect on the epiphyseal plate in immature rabbits. *J Pediatr Orthop*. 1992; 12(2):177−180.

98. Lee JH, Kim JH, Oh SH, et al. Tissue-engineered bone formation using periosteal-derived cells and polydioxanone/pluronic F127 scaffold with pre-seeded adipose tissue-derived CD146 positive endothelial-like cells. *Biomaterials*. 2011;32(22): 5033−5045.

99. Pihlajamaki HK, Salminen ST, Tynninen O, Bostman OM, Laitinen O. Tissue restoration after implantation of polyglycolide, polydioxanone, polylevolactide, and metallic pins in cortical bone: an experimental study in rabbits. *Calcif Tissue Int*. 2010;87(1): 90−98.

100. Pihlajamaki H, Salminen S, Laitinen O, Tynninen O, Bostman O. Tissue response to polyglycolide, polydioxanone, polylevolactide, and metallic pins in cancellous bone: an experimental study on rabbits. *J Orthop Res*. 2006;24(8):1597−1606.

101. Bowen PK, Drelich J, Goldman J. Zinc exhibits ideal physiological corrosion behavior for bioabsorbable stents. *Adv Mater*. 2013;25(18):2577−2582.

102. Prosek T, Nazarov A, Bexell U, Thierry D, Serak J. Corrosion mechanism of model zinc—magnesium alloys in atmospheric conditions. *Corrosion Sci*. 2008;50(8):2216−2231.

103. Seitz JM, Eifler R, Bach FW, Maier HJ. Magnesium degradation products: effects on tissue and human metabolism. *J Biomed Mater Res*. 2014;102(10):3744−3753.

104. Zheng YF, Gu XN, Witte F. Biodegradable metals. *Mater Sci Eng R Rep*. 2014;77: 1—34.

105. Gibson LJ. The mechanical behaviour of cancellous bone. *J Biomech*. 1985;18(5): 317—328.

106. Wen CE, Yamada Y, Shimojima K, Chino Y, Hosokawa H, Mabuchi M. Compressibility of porous magnesium foam: dependency on porosity and pore size. *Mater Lett*. 2004; 58(3—4):357—360.

107. Kirkland NT. Magnesium biomaterials: past, present and future. *Corrosion Eng Sci Technol*. 2013;47(5):322—328.

108. Shaw BA, Sikora E, Virtanen S. Fix, heal, and disappear: A new approach to using metals in the human body. *Electrochem Soci Interface*. 2008;17(2):45—49.

109. Zeng R, Dietzel W, Witte F, Hort N, Blawert C. Progress and challenge for magnesium alloys as biomaterials. *Adv Eng Mater*. 2008;10(8):B3—B14.

110. Celarek A, Kraus T, Tschegg EK, et al. PHB, crystalline and amorphous magnesium alloys: promising candidates for bioresorbable osteosynthesis implants? *Mater Sci Eng C Mater Biol Appl*. 2012;32(6):1503—1510.

111. Gill P, Munroe N, Dua R, Ramaswamy S. Corrosion and biocompatibility assessment of magnesium alloys. *J Biomateri Nanobiotechnol*. 2012;03(01):10—13.

112. Li Y-H, Rao G-B, Rong L-J, Li Y-Y. The influence of porosity on corrosion characteristics of porous NiTi alloy in simulated body fluid. *Mater Lett*. 2002;57(2):448—451.

113. Nasab MB, Hassan MR, Sahari BB. Metallic biomaterials of knee and hip—a review. *Trends Biomater Artif Organs*. 2010;24(1):69—82.

114. Witte F, Kaese V, Haferkamp H, et al. In vivo corrosion of four magnesium alloys and the associated bone response. *Biomaterials*. 2005;26(17):3557—3563.

115. Hampp C, Angrisani N, Reifenrath J, Bormann D, Seitz JM, Meyer-Lindenberg A. Evaluation of the biocompatibility of two magnesium alloys as degradable implant materials in comparison to titanium as non-resorbable material in the rabbit. *Mater Sci Eng C Mater Biol Appl*. 2013;33(1):317—326.

116. Huehnerschulte TA, Reifenrath J, von Rechenberg B, et al. In vivo assessment of the host reactions to the biodegradation of the two novel magnesium alloys ZEK100 and AX30 in an animal model. *Biomed Eng Online*. 2012;11:14—20.

117. Dziuba D, Meyer-Lindenberg A, Seitz JM, Waizy H, Angrisani N, Reifenrath J. Long-term in vivo degradation behaviour and biocompatibility of the magnesium alloy ZEK100 for use as a biodegradable bone implant. *Acta Biomater*. 2013;9(10): 8548—8560.

118. Wang Y, He Y, Zhu Z, et al. In vitro degradation and biocompatibility of Mg-Nd-Zn-Zr alloy. *Chin Sci Bull*. 2012;57(17):2163—2170.

119. Xia Y, Zhang B, Wang Y, Qian M, Geng L. In-vitro cytotoxicity and in-vivo biocompatibility of as-extruded Mg—4.0Zn—0.2Ca alloy. *Mater Sci Eng C*. 2012;32(4):665—669.

120. Witte F, Feyerabend F, Maier P, et al. Biodegradable magnesium-hydroxyapatite metal matrix composites. *Biomaterials*. 2007;28(13):2163—2174.

121. Zhang E, Yin D, Xu L, Yang L, Yang K. Microstructure, mechanical and corrosion properties and biocompatibility of Mg—Zn—Mn alloys for biomedical application. *Mater Sci Eng C*. 2009;29(3):987—993.

122. Peeters P, Bosiers M, Verbist J, Deloose K, Heublein B. Preliminary results after application of absorbable metal stents in patients with critical limb ischemia. *J Endovasc Ther*. 2005;12(1):1—5.

123. Schranz D, Zartner P, Michel-Behnke I, Akinturk H. Bioabsorbable metal stents for percutaneous treatment of critical recoarctation of the aorta in a newborn. *Catheter Cardiovasc Interv.* 2006;67(5):671−673.

124. Ormiston JA, Serruys PW. Bioabsorbable coronary stents. *Circ Cardiovasc Interv.* 2009;2(3):255−260.

125. Heublein B, Rohde R, Kaese V, Niemeyer M, Hartung W, Haverich A. Biocorrosion of magnesium alloys: a new principle in cardiovascular implant technology? *Heart.* 2003; 89(6):651−656.

126. Schumacher S, Roth I, Stahl J, Baumer W, Kietzmann M. Biodegradation of metallic magnesium elicits an inflammatory response in primary nasal epithelial cells. *Acta Biomater.* 2014;10(2):996−1004.

127. Lock JY, Wyatt E, Upadhyayula S, et al. Degradation and antibacterial properties of magnesium alloys in artificial urine for potential resorbable ureteral stent applications. *J Biomed Mater Res.* 2014;102(3):781−792.

128. Waksman R, Erbel R, Di Mario C, et al. Early- and long-term intravascular ultrasound and angiographic findings after bioabsorbable magnesium stent implantation in human coronary arteries. *JACC Cardiovasc Interv.* 2009;2(4):312−320.

129. Zartner P, Cesnjevar R, Singer H, Weyand M. First successful implantation of a biodegradable metal stent into the left pulmonary artery of a preterm baby. *Catheter Cardiovasc Interv.* 2005;66(4):590−594.

130. Zartner P, Buettner M, Singer H, Sigler M. First biodegradable metal stent in a child with congenital heart disease: evaluation of macro and histopathology. *Catheter Cardiovasc Interv.* 2007;69(3):443−446.

131. Kitabata H, Waksman R, Warnack B. Bioresorbable metal scaffold for cardiovascular application: current knowledge and future perspectives. *Cardiovasc Revasc Med.* 2014;15(2):109−116.

132. Peuster M. A novel approach to temporary stenting: degradable cardiovascular stents produced from corrodible metal—results 6-18 months after implantation into New Zealand white rabbits. *Heart.* 2001;86(5):563−569.

133. Moravej M, Mantovani D. Biodegradable metals for cardiovascular stent application: interests and new opportunities. *Int J Mol Sci.* 2011;12(7):4250−4270.

134. Hermawan H, Alamdari H, Mantovani D, Dubé D. Iron−manganese: new class of metallic degradable biomaterials prepared by powder metallurgy. *Powder Metall.* 2013;51(1):38−45.

135. Liu B, Zheng YF. Effects of alloying elements (Mn, Co, Al, W, Sn, B, C and S) on biodegradability and in vitro biocompatibility of pure iron. *Acta Biomater.* 2011;7(3): 1407−1420.

136. Schaffer JE, Nauman EA, Stanciu LA. Cold-drawn bioabsorbable ferrous and ferrous composite wires: an evaluation of mechanical strength and fatigue durability. *Metall Mater Trans B.* 2012;43(4):984−994.

137. Griffiths H, Peeters p, Verbist J, et al. Future devices: bioabsorbable stents. *Br J Cardiol.* 2004;11:80−84.

138. Hermawan H, Purnama A, Dube D, Couet J, Mantovani D. Fe-Mn alloys for metallic biodegradable stents: degradation and cell viability studies. *Acta Biomater.* 2010; 6(5):1852−1860.

139. Hennig B, Toborek M, McClain CJ. Antiatherogenic properties of zinc: implications in endothelial cell metabolism. *Nutrition.* 1996;12(10):711−717.

140. Ren M, Rajendran R, Ning P, et al. Zinc supplementation decreases the development of atherosclerosis in rabbits. *Free Radic Biol Med.* 2006;41(2):222−225.

141. Berger M, Rubinraut E, Barshack I, Roth A, Keren G, George J. Zinc reduces intimal hyperplasia in the rat carotid injury model. *Atherosclerosis.* 2004;175(2):229−234.

142. Vojtech D, Kubasek J, Serak J, Novak P. Mechanical and corrosion properties of newly developed biodegradable Zn-based alloys for bone fixation. *Acta Biomater.* 2011;7(9): 3515−3522.

143. Kenny C. PLA, PGA, and PLGA as biomaterials. https://openwetware.org; 2016.

144. Chemical Retrieval on the Web (CROW), CROW Polymer Property Database. www.polymerdatabase.com, 2016.

145. Hajiali F, Tajbakhsh S, Shojaei A. Fabrication and properties of polycaprolactone composites containing calcium phosphate-based ceramics and bioactive glasses in bone tissue engineering: a review. *Polym Rev.* 2018;58(1):164−207.

146. Garg K, Ryan JJ, Bowlin GL. Modulation of mast cell adhesion, proliferation, and cytokine secretion on electrospun bioresorbable vascular grafts. *J Biomed Mater Res.* 2011; 97 A:405−413.

147. Erne P, Schier M, Resink TJ. The road to bioabsorbable stents: reaching clinical reality? *Cardiovasc Intervent Radiol.* 2006;29(1):11−16.

148. Anselme K, Ponche A, Bigerelle M. Relative influence of surface topography and surface chemistry on cell response to bone implant materials. Part 2: Biological aspects. *Proc Inst Mech Eng H.* 2010;224(12):1487−1507.

149. Schinhammer M, Hanzi AC, Loffler JF, Uggowitzer PJ. Design strategy for biodegradable Fe-based alloys for medical applications. *Acta Biomater.* 2010;6(5):1705−1713.

150. Alicea LaA, J.I., López I a, Mulero LE, Sánchez LA. *Mechanics Biomaterials: Stents. Applications of Engineering Mechanics in Medicine.* 2004:F1−F21.

151. Hermawan HR, Ramdan D, Djuansjah JRP. *Metals for Biomedical Applications.* London: IntechOpen; 2009.

152. Levesque JD, Dubé' D, Fiset M, Mantovani D. Materials and propertires for coronary stents. *Adv Mater Process* 162(9):45−48.

153. Schaffer JE. *Development and Characterization of Vascular Prosthetics for Controlled Bioabsorption [Ph.D. Thesis].* West Lafayette, Indiana: Purdue University; 2012: 1−236.

Index

Note: Page numbers followed by "b" indicate boxes, those followed by "f" indicate figures, and those followed by "t" indicate tables.